D1228251

Industrial Safety and Health for People-Oriented Services

Handbook of Safety and Health for the Service Industry

Industrial Safety and Health for Goods and Materials Services

Industrial Safety and Health for Infrastructure Services

Industrial Safety and Health for Administrative Services

Industrial Safety and Health for People-Oriented Services

Industrial Safety and Health for People-Oriented Services

Charles D. Reese

CRC Press
Taylor & Francis Group
Boca Raton   London   New York

CRC Press is an imprint of the
Taylor & Francis Group, an **informa** business

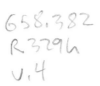

CRC Press
Taylor & Francis Group
6000 Broken Sound Parkway NW, Suite 300
Boca Raton, FL 33487-2742

© 2009 by Taylor & Francis Group, LLC
CRC Press is an imprint of Taylor & Francis Group, an Informa business

No claim to original U.S. Government works
Printed in the United States of America on acid-free paper
10 9 8 7 6 5 4 3 2 1

International Standard Book Number-13: 978-1-4200-5384-5 (Hardcover)

Library of Congress Cataloging-in-Publication Data

Reese, Charles D.
     Industrial safety and health for people-oriented services / Charles D. Reese.
          p. cm.
     Includes bibliographical references and index.
     ISBN 978-1-4200-5384-5 (alk. paper)

     1. Service industries--United States--Safety measures. 2. Service industries--Employees--Health and hygiene--United States. 3. Human services--United States--Employees. I. Title.
     HD7269.S452U6744 2009

                                                                      658.3'82--dc22

Visit the Taylor & Francis Web site at
http://www.taylorandfrancis.com

and the CRC Press Web site at
http://www.crcpress.com

Contents

Preface

Industrial Safety and Health for People-Oriented Services deals with education; health care and social assistance; arts, entertainment, and recreation; accommodation and food services; and other services sectors. These sectors provide a host of services to the general public including all types of educational services such as schooling from the elementary to the university level. The health care and social assistance areas include physicians; dentists; ambulances; other health professionals; hospitals; and mental health; and residential care for substance abuser, the mentally ill, and the elderly. The arts, entertainment, and recreation consist of performing arts, sports events, theaters, museums, historic sites, amusement parks, gambling establishments, and fitness centers. Accommodation and food services are comprised of hotels, motels, and restaurants while other services encompass automotive repair, most other repair services, laundry, death care, religious organizations, labor unions, and civic organizations.

Since these sectors are very people oriented, they face hazards that are inherent in the specific services that they provide. The workers in these sectors face similar occupationally related safety and health hazards such as biological hazards, chemical hazards, electrical hazards, lifting, ergonomic issues, equipment, fire, hot processes, ionizing radiation, nonionizing radiation, violent individuals, and slip/trips/falls.

Although the aforementioned hazards are not the only hazards faced by workers in these sectors, they are the most common ones. Because of the diversity in people-oriented services, other job-specific hazards may result.

This workforce constantly interacts with the public and must therefore work carefully so as to guarantee the safety of their patrons and clients. At times the training needed may be very extensive to perform in a safe and efficient manner. However, these hazards can be managed by applying the principles identification, intervention, and prevention, all of which are proven techniques of occupational safety and health.

In today's work environment, workers must be trained for emergencies and security must be provided for both workers and patrons. By adhering to acceptable safe work practices and occupational safety and health regulations, the safety and health of the workforce can be ensured while running a productive business. This book serves as a guide in achieving this objective.

Dr. Charles D. Reese

Author

For 30 years, **Charles D. Reese, PhD,** has been involved with occupational safety and health as an educator, manager, and consultant. In his early career, Dr. Reese was an industrial hygienist at the National Mine Health and Safety Academy. He later became manager for the nation's occupational trauma research initiative at the National Institute for Occupational Safety and Health's Division of Safety Research. Dr. Reese has played an integral role in trying to ensure workplace safety and health. As the managing director for the Laborers' Health and Safety Fund of North America, he was responsible for the welfare of the 650,000 members of the laborers' union in the United States and Canada.

Dr. Reese has developed many occupational safety and health training programs, which range from radioactive waste remediation to confined space entry. He has also written numerous articles, pamphlets, and books on related issues.

Dr. Reese, professor emeritus, was a member of the graduate and undergraduate faculty at the University of Connecticut, where he taught courses on Occupational Safety and Health Administration regulations, safety and health management, accident-prevention techniques, industrial hygiene, and ergonomics. As professor of environmental/occupational safety and health, he was instrumental in coordinating the safety and health efforts at the University of Connecticut. He is often invited to consult with industry on safety and health issues and is asked for expert consultation in legal cases.

Dr. Reese is also the principal author of the *Handbook of OSHA Construction Safety and Health (Second Edition)*; *Material Handling Systems: Designing for Safety and Health*; *Annotated Dictionary of Construction Safety and Health*; *Occupational Health and Safety Management: A Practical Approach*; and *Office Building Safety and Health and Accident/Incident Prevention Techniques*.

1 Introduction to the Service Industry

Serving people is a major function of the service industry. Veterinary services are a part of the service industry.

The service industry consists of many different places of work, called establishments. Establishments are physical locations in which people work, such as a branch office of a bank, a gasoline station, a school, a department store, or an electricity generation facility. Establishments can range from large retail operations with corporate office complexes employing thousands of workers to small community stores, restaurants, professional offices, and service businesses employing only a few workers. Establishments should not be confused with companies or corporations that are legal entities. Thus, a company or corporation may have a single establishment or more than one establishment.

Establishments that use and provide the same services are organized together into industries. Industries are in turn organized together into industry sectors. These are further organized into subsectors. Each of the industry groups requires workers with varying skills and employs unique service techniques. An example of this is found in utilities, which employs workers in establishments that provides electricity, natural gas, and water. The service industry is broken down into the following supersectors:

Trade
    Retail trade (44 and 45)
    Wholesale trade (42)

Transportation and utilities
    Transportation (48)
    Warehousing (49)
    Utilities (22)
Information services (51)
Financial activities
    Financial and insurance sector (52)
    Real estate and rental and leasing sector (53)
Professional and business services
    Professional, scientific, and technical services (54)
    Management of companies and enterprises (55)
    Administrative and support and waste management and remediation (56)
Education and health services
    Educational services sector (61)
    Health care and social assistance sector (62)
Leisure and hospitality
    Arts, entertainment, and recreation sector (71)
    Accommodation and food sector (72)
Other services (81)

The service industry is the fastest growing industrial sector in the United States and has seen growth in the international arena. The service industry accounts for approximately 70% of the total economic activity in the United States according to the U.S. Bureau of Census. This non-goods-producing industry, which includes retail trade, wholesale trade, and other service-related industries as previously mentioned, has a very diverse grouping.

1.1 NAICS

With the passage of the North American Free Trade Agreement (NAFTA), it became apparent that the long employed standard industrial classification (SIC) was no longer very useful when dealing with industries found in Canada and Mexico. Consequently, the Bureau of Labor Statistics has developed a new system entitled the North American Industrial Classification System (NAICS).

NAICS uses a six-digit hierarchical coding system to classify all economic activity into 20 industry sectors. Five sectors are mainly goods-producing sectors and 15 are entirely services-producing sectors.

1.2 EMPLOYMENT IN THE SERVICE INDUSTRY

To have some idea of the numbers of employees addressed when speaking of the service industry, the worker population in each of the service industry sector is provided. The fast growing service industry as well as the number and variety of

TABLE 1.1
Employment in the Service Industry (2004)

Service Industry Sectors	Employment	Percentage of Service Industry (84,896,300)	Percentage of All Private Industries (107,551,800)
Wholesale trade	5,642,500	6.6	5.2
Retail trade	15,060,700	17.7	14.0
Warehousing	555,800	0.65	0.52
Transportation	3,450,400	4.1	3.2
Utilities	583,900	0.69	0.54
Administrative and support and waste management and remediation	7,829,400	9.2	7.3
Information	3,099,600	3.7	2.9
Finance and insurance	5,813,300	6.8	5.4
Real estate	2,077,500	2.4	1.9
Management of companies and enterprises	1,696,500	2.0	1.6
Professional, scientific, and technical services	6,768,900	8.0	6.3
Education services	2,079,200	2.4	1.9
Health and social services	14,005,700	16.5	13.0
Art, entertainment, and recreation	1,852,900	2.2	1.7
Accommodation and food services	10,614,700	12.5	9.9
Other services	3,785,200	4.5	3.5

Source: From Bureau of Labor Statistics. United States Department of Labor. Available at http://www. bls.gov, Washington, 2007.

occupations within each sector provides a window into the safety and health hazards that need to be addressed within each sector of the service industry's workforce (Table 1.1).

1.3 SAFETY

One of the most telling indicators of working condition is an industry's injury and illness rates. Overexertion, being struck by an object, and falls on the same level are among the most common incidents causing work-related injuries.

The service industry is a large umbrella that encompasses many types of businesses, each of which has its own safety and health issues. Some of the service industries' businesses are more hazardous than others. This book does not address each sector independently, but provides the tools and information needed to address the hazards and safety and health issues within each sector of the service industry.

The service industry is made up of a large number of widely dissimilar industry sectors. Each sector has its own unique functions that result in each sector having its own set of unique hazards that the individual workforces must face and that their employers must address.

The functioning of each sector results in different types of energy being released, and therefore the differences in the types of accidents, incidents, injuries, and illnesses that occur. Thus, the hazards and energy sources dictate the specific Occupational Safety and Health Administration (OSHA) regulations that each sector is compelled to be in compliance with.

For these reasons, it is not possible to describe or address the service industry in the same manner as construction, shipyard, or office building industries, where workforces perform similar tasks and thus face similar hazards.

Each hazard is discussed based upon the type of energy released, and its ability to cause specific accidents or incidents. In each section that describes hazards, the best practices for intervention and prevention of the release of the specific energy are emphasized. This approach will allow for the identification and prevention of hazards, and for framing regulations by any service industry sector. It will allow for a similar approach to address areas where the service industry sectors are alike and can be addressed as a collective industry rather than as individual sectors, for example, the same sectors will need to address areas such as compliance with OSHA regulations, conducting training, and effective management of safety and health.

The intent of this book is to provide a source for the identification and prevention of most of the injuries and illnesses occurring in the service industry. Also, it summarizes applicable safety standards that impact the service industry as well as address how to work with and around OSHA to comply with its regulations. The book covers safety hazards involved with confined spaces, electrical equipment, falls, fork-lifts, highway vehicles, preventive maintenance activities, handling chemicals, radiation, welding, etc. The content describes the safety hazard as applied to the type of energy released or to the unique event that occurs from exposure to the hazard.

The question that we should be asking, "is the kind of safety being practiced prevent the destruction that we see in the American workforce?" Maybe we ought to ask how we define safety for a start. Here are some definitions of safety that may be useful:

1. Doing things in a manner so that no one will get hurt and so the equipment and product will not get damaged
2. Implementation of good engineering design, personnel training, and the common sense to avoid bodily harm or material damage
3. Systematic planning and execution of all tasks so as to produce safe products and services with relative safety to people and property
4. Protection of persons and equipment from hazards that exceed normal risk
5. Application of techniques and controls to minimize the hazards involved in a particular event or operation, considering both potential personal injury and property damage
6. Employing processes to prevent accidents both by conditioning the environment as well as conditioning the person toward safe behavior
7. Function with minimum risk to personal well-being and to property
8. Controlling exposure to hazards that could cause personal injury and property damage
9. Controlling people, machines, and the environment that could cause personal injury or property damage

10. Performing your daily tasks in the manner that they should be done, or when you do not know, seek the necessary knowledge
11. Elimination of foreseen hazards and the necessary training to prevent accidents or to provide limited acceptable risk to personnel and facilities

Do any of these definitions match the safety guidelines practiced in workplaces where you have responsibility? If so, have you had any deaths or injuries to any of your workforce? When injuries and illnesses are not occurring anymore then the ultimate goals have been reached. Now comes the task of maintaining what has been gained.

As one can see, safety can be defined in many different ways. Nearly all of these definitions include property damage as well as personal injury. It shows that the thinking is in the right direction and that safety consists of a total loss-control activity. The book's content contains only one facet of a comprehensive safety effort, which is a never ending journey.

1.4 WHY TRAUMA PREVENTION?

There are very real advantages when addressing trauma prevention, which do not exist when addressing illness prevention. The advantages are

- Trauma occurs in real time with no latency period (an immediate sequence of events).
- Accident/incident outcomes are readily observable (only have to reconstruct a few minutes to a few hours).
- Root or basic causes are more clearly identified.
- It is easy to detect cause and effect relationships.
- Traumas are not difficult to diagnose.
- Trauma is highly preventable.

1.5 ACCIDENTS OR INCIDENTS

The debate over the use of the term "accidents" versus "incidents" has been long and continual. Although these terms are used virtually interchangeably in the context of this book, you should be aware of the distinction between the two. Accidents are usually defined as an unexpected, unplanned, or uncontrollable event or mishap. This undesired event results in personal injury and property damage or both and may also lead to equipment failure. An incident is all of the above as well as the adverse effects on production.

This definition for an accident underlies the basic foundation of this book. The philosophy behind this book is that we can control these types of events or mishaps by addressing the existence of hazards and taking steps to remove or mitigate them as part of the safety effort. This is why we spend time identifying hazards and determining risk. Thus, the striving for a safe workplace, where the associated risks are judged to be acceptable, is the goal of safety. This will result in freedom from those circumstances that can cause injury or death to workers, and damage to or loss of equipment or property.

The essence of this book's approach is that we can control those factors, which are the causing agents of accidents. Hazard prevention described in this book is addressed both from practical and regulatory approaches.

1.6 COMPREHENSIVE ACCIDENT PREVENTION

Accident prevention is very complex because of interactions that transpire within the workplace. These interactions are between

- Workers
- Management
- Equipment/machines
- Environment

The interaction between workers, management, equipment/machinery, and the workplace environment have enough complexity themselves as they try to blend together in the physical workplace environment. However, this physical environment is not the only environment that has an impact upon the accident prevention effort in companies. The social environment is also an interactive factor that encompasses our lives at work and beyond. Government entities that establish rules and regulations leave their mark upon the workplace. But others in the social arena such as unions, family, peer pressure, friends, and associates also exert pressure on the workplace environment The extent of the interactions that must be attended to for having a successful accident prevention effort is paramount.

Many workplaces have high accident incidence rates because they are hazardous. Hazards are dangerous situations or conditions that can lead to accidents. The more hazards present, the greater the chance of accidents. Unless safety procedures are followed, there will be a direct relationship between the number of hazards in the workplace and the number of accidents that will occur there.

In most industries, people work together with machines in an environment that causes employees to face hazards that can lead to injury, disability, or even death. To prevent industrial accidents, the people, machines, and other factors, which can cause accidents, including the energies associated with them, must be controlled. This can be done through education and training, good safety engineering, and enforcement.

Many accidents can be prevented. One study showed that 88% were caused by human failure (unsafe acts), 10% by mechanical failure (unsafe conditions), and only 2% were beyond human control (acts of God).

If workers are aware of what hazards are, and what can be done to eradicate them, many accidents can be prevented. For a situation to be called an accident, it must have certain characteristics. The personal injury may be considered minor when it requires no treatment or only first aid. Personal injury is considered serious if it results in a fatality or in a permanent, partial, or temporary total disability (lost-time injuries). Property damage may also be minor or serious.

1.7   FATALITY AND INJURY PROFILE FOR THE SERVICE INDUSTRY

In 2005, there were 5702 occupationally related deaths in all of private industry, while the service industry had 2736 (48%) of these fatalities the goods-producing industry had 42% fatalities. In Table 1.2, the major contributors to these fatalities are depicted.

Injuries are examined somewhat differently and the statistical data are presented usually in four different ways. These are as follows:

1. Nature of injury or illness names the principal physical characteristic of a disabling condition, such as sprain/strain, cut/laceration, or carpal tunnel syndrome.
2. Part of body affected is directly linked to the nature of injury or illness cited, for example, back sprain, finger cut, or wrist and carpal tunnel syndrome.
3. Source of injury or illness is the object, substance, exposure, or bodily motion that directly produced or inflicted the disabling condition cited. Examples are a heavy box, a toxic substance, fire/flame, and bodily motion of injured/ill worker.
4. Event or exposure (type of accident) signifies the manner in which the injury or illness was produced or inflicted, for example, overexertion while lifting or fall from a ladder (see Appendix A).

Tables 1.3 through 1.6 allow us to start identifying the most common facets of an injury profile. The total employment for the service industry in 2004 was 84,896,300 and the total number of injuries was 850,930. The data in the tables denote the most frequently occurring factor resulting in the injury/incident or resulting from the injury/incident.

It would appear from a rough observation of Tables 1.3 through 1.6 that a service industry employee would suffer a sprain or strain to the trunk and in most cases the back or possibly the lower or upper extremities because of one of the three causes: worker motion/position; floors, walkways, or ground surfaces; or containers that resulted in an overexertion/lifting or fall on the same level. As it can be seen, these

TABLE 1.2
Occupational Death Cause in Percent

Cause	Service Industry (%)	All Private Industries (%)
Highway	34	25
Homicides	16	10
Falls	9	13
Struck-by	7	11

Source: From Bureau of Labor Statistics, United States Department of Labor. *National Census of Fatal Occupational Injuries in 2005.* Available at http://bls.gov.

TABLE 1.3
Nature of Injury by Number and Percent
for the Service Industry

Nature of Injury	Number	Percent
Sprains/strains[a]	377,760	44
Fractures[a]	55,450	6.5
Cuts/punctures[a]	63,220	7
Bruises[a]	82,610	10
Heat burns	12,780	1.5
Chemical burns	4,330	0.5
Amputations	2,710	0.3
Carpal tunnel syndrome	10,810	1.3
Tendonitis	3,950	0.4
Multiple trauma[a]	34,450	4
Back pain (only)	28,600	3

Source: From Bureau of Labor Statistics, United States Department of Labor. *Workplace Injuries and Illnesses in 2004.* Available at http://bls.gov.

[a] Five most frequently occurring conditions.

TABLE 1.4
Body Part Injured by Number and Percent
for the Service Industry

Body Part Injured	Number	Percent
Head	51,500	6
Eyes	19,070	2
Neck	15,960	1.8
Trunk[a]	314,190	37
Back[a]	204,240	24
Shoulder	56,350	7
Upper extremities[a]	173,260	20
Finger	58,080	6.8
Hand	30,810	3.6
Wrist	38,000	4.5
Lower extremities[a]	183,780	22
Knee	69,250	8
Foot and toe	39,050	4.6
Body systems	10,940	1.3
Multiple body parts[a]	95,490	11

Source: From Bureau of Labor Statistics, United States Department of Labor. *Workplace Injuries and Illnesses in 2004.* Available at http://bls.gov.

[a] Five most frequently injured body parts.

TABLE 1.5

**Source of Injury by Number and Percent
for the Service Industry**

Sources of Injuries	Number	Percent
Parts and materials	51,680	6
Worker motion/position[a]	119,340	14
Floor, walkways, or ground surfaces[a]	168,620	20
Hand tools	29,420	3.5
Vehicles[a]	88,830	10
Health care patient[a]	57,220	6.7
Chemicals and chemical products	11,070	1.3
Containers[a]	124,700	15
Furniture and fixtures	36,700	4
Machinery	40,940	4.8

Source: From Bureau of Labor Statistics, United States Department
of Labor. *Workplace Injuries and Illnesses in 2004.*
Available at http://bls.gov.

[a] Five most frequent sources of injury.

TABLE 1.6

**Exposure/Accident Type by Number and Percent
for the Service Industry**

Type of Accidents	Number	Percent
Struck by an object[a]	101,390	12
Struck against an object	51,670	6
Caught in or compressed or crushed	25,290	3
Fall to lower level	46,820	5.5
Fall on same level[a]	130,260	15
Slips or trips without a fall	27,400	3
Overexertion[a]	227,350	27
Lifting[a]	126,380	15
Repetitive motion	27,180	3.2
Exposure to harmful substance or environment	36,070	4
Transportation accidents[a]	51,070	6
Fires and explosions	1,100	0.1
Assaults/violent acts	22,790	2.7

Source: From Bureau of Labor Statistics, United States Department
of Labor. *Workplace Injuries and Illnesses in 2004.*
Available at http://bls.gov.

[a] Five most frequent exposures or type of accidents that led to an
injury.

data give us some information to start our search for the hazards that contributed to these injuries.

1.8   OCCUPATIONAL ILLNESSES IN THE SERVICE INDUSTRY

Occupational illnesses have always been underreported. For this reason, they do not seem to get the same attention as injuries since their numbers or causes are not of epidemic proportions. The reasons why illnesses are not reported include the following:

- Not occurring in real time and usually having a latency period before signs and symptoms occur.
- Not readily observable and have been linked to personal habits and exposure from hobbies. There is the question of multiple exposures and synergistic effects on-the-job and off-the-job.
- Not always easy to detect cause and effect relationships.
- Often difficult to diagnose since many exhibit flu or cold symptoms.

These are not excuses for not pursuing preventive strategies, but an explanation of why occupational illnesses are more difficult to accurately describe and identify their root cause. The 84,896,300 employees experienced 131,500 (53%) cases of illness during 2004 according to the Bureau of Labor Statistics. This compares to the total number of illnesses reported by all of industry that equaled 249,000 of which 53% was attributed to the service industry. The most common reported types of occupational illnesses for this period are found in Table 1.7.

The remainder of this book is directed toward managing, preventing, and controlling hazards that occur within the goods and material service sector of the service industry. This includes the wholesale trade, retail trade, and warehousing sectors.

It is important to keep in mind that because of the complexity and diversity within the industry sectors of the service industry, no cookie cutter approach could be used nor is a one-size-fits-all approach possible. There has to be a mixing of

TABLE 1.7
Occupational Illnesses by Number of Cases and Percent
for the Service Industry

Illness Type	Number	Percent
Skin diseases and disorders	24,900	19
Respiratory conditions	13,000	10
Poisoning	2,000	1.5
Hearing loss	4,000	3
All others	87,400	66.5

Source: From Bureau of Labor Statistics, United States Department of Labor. *Workplace Injuries and Illnesses in 2004.* Available at http://bls.gov.

information and data from diverse sources such as the NAICS and the outdated SIC, since all agencies have not changed to the new system. Also, even within the supersectors and the sectors themselves there is not a common approach to the management of safety and health, identification of hazards compatible, or the same approach for each varied sector, nor should we expect these to be. This is the reason that by consulting the table of contents of this book and *Industrial Safety and Health for Goods and Materials Services*, *Industrial Safety and Health for Infrastructure Services*, and *Industrial Safety and Health for Administrative Services*, decisions can be made regarding which book would be most useful to your particular business. In some cases, one book will fulfill a company's safety and health needs while in other cases all the four books will be most beneficial.

REFERENCES

Bureau of Labor Statistics, United States Department of Labor. Available at http://www.bls.gov, Washington, 2007.

Bureau of Labor Statistics, United States Department of Labor. *National Census of Fatal Occupational Injuries in 2005*. Available at http://bls.gov.

Bureau of Labor Statistics, United States Department of Labor. *Workplace Injuries and Illnesses in 2004*. Available at http://bls.gov.

2 Educational Services

A typical university building with students coming and going.

The education and health services supersector comprises two divisions: the educational services sector (sector 61), and the health care and social assistance sector (sector 62). Only privately owned establishments are included in this discussion; publicly owned establishments that provide education or health services are included in government.

The educational services sector (61) comprises establishments that provide instruction and training in a wide variety of subjects. Instruction and training are provided by specialized establishments, such as schools, colleges, universities, and training centers.

The North American Industrial Classification System (NAICS) categorizes the health care and social assistance sector in the following manner:

Educational services (61)
  Educational services (611000)
    Elementary and secondary schools (611100)
    Junior colleges (6112000)
    College, universities, and professional Schools (611300)
    Business school and computer and management training (611400)
    Technical and trade schools (611500)
    Other schools and instruction (611600)
    Educational support services (611700)

2.1 EDUCATIONAL SERVICES

Educational Services have some factors that make them unique. About one in four Americans are enrolled in educational institutions. Educational services is the second largest industry, accounting for about 13 million jobs. Most teaching positions—which constitute almost half of all educational services jobs—require at least a bachelor's degree, and some require a master's or doctoral degree. Retirements in a number of education professions will create many job openings.

Education is an important part of life. The amount and type of education that individuals receive are a major influence on both the types of jobs they are able to hold and their earnings. Lifelong learning is important in acquiring new knowledge and upgrading one's skills, particularly in this age of rapid technological and economic changes. The educational services industry includes a variety of institutions that offer academic education, vocational or career and technical instruction, and other education and training to millions of students each year.

Because school attendance is compulsory until at least age 16 in all 50 states and the District of Columbia, elementary, middle, and secondary schools are the most numerous of all educational establishments. They provide academic instruction to students in kindergarten through grade 12 in a variety of settings, including public schools, parochial schools, boarding and other private schools, and military academies. Some secondary schools offer a combination of academic and career and technical instruction.

Postsecondary institutions—universities, colleges, professional schools, community or junior colleges, and career and technical institutes—provide education and training in both academic and technical subjects for high-school graduates and other adults. Universities offer bachelors, masters, and doctoral degrees, while colleges generally offer only the bachelor's degree. Professional schools offer graduate degrees in fields such as law, medicine, business administration, and engineering. The undergraduate bachelor's degree typically requires 4 years of study, while graduate degrees require additional years of study. Community and junior colleges and technical institutes offer associate degrees, certificates, or other diplomas, typically involving 2 years of study or less. Career and technical schools provide specialized training and services primarily related to a specific job. They include computer and cosmetology training institutions, business and secretarial schools, correspondence schools, and establishments that offer certificates in commercial art and practical nursing.

This industry also includes institutions that provide training and services to schools and students, such as curriculum development, student exchanges, and tutoring. Also included are schools or programs that offer nonacademic or self-enrichment classes, such as automobile driving and cooking instruction, among other things.

School conditions often vary from town to town. Some schools in poorer neighborhoods may be rundown, have few supplies and equipment, and lack air conditioning. Other schools may be new, well equipped, and well maintained. Conditions at postsecondary institutions are generally very good. Regardless of the type of conditions in elementary and secondary schools, seeing students develop and enjoy learning can be rewarding for teachers and other education workers. However, dealing with unmotivated students or those with social or behavioral problems can be stressful and require patience and understanding.

Most educational institutions operate 10 months a year, but summer sessions for special education or remedial students are not uncommon; institutions that cater to adult students, and those that offer educational support services, such as tutoring, generally operate year-round as well. Education administrators, office and administrative support workers, and janitors and cleaners often work the entire year. Night and weekend work is common for teachers of adult literacy and remedial and self-enrichment education, for postsecondary teachers, and for library workers in postsecondary institutions. Part-time work is common for this same group of teachers, as well as for teacher assistants and school bus drivers. The latter often work a split shift, driving one or two routes in the morning and afternoon. Drivers are assigned to drive students on field trips, to athletic and other extracurricular activities, or to midday kindergarten programs work additional hours during or after school (see Figure 2.1). Many teachers spend significant time outside of school preparing for class, doing administrative tasks, conducting research, writing articles and books, and pursuing advanced degrees.

Despite occurrences of violence in some schools, educational services is a relatively safe industry. There were 2.7 cases of occupational injury and illness per 100 full-time workers in private educational establishments in 2003, compared with 5.0 in all industries combined.

FIGURE 2.1   Qualified drivers utilized buses like these to safely transport students to all types of events.

2.2   PROFILE OF EDUCATIONAL SERVICE WORKERS' DEATHS, INJURIES, AND ILLNESSES

2.2.1   DEATHS

There were 45 occupationally related deaths to educational services workers in 2005. Educational services accounted for 1.6% of the service industry deaths (2736). Table 2.1 shows the percent values from each major category of those deaths.

2.2.2   INJURIES

There were 10,070 reported injuries among educational services workers in 2004. This was approximately 12% of the total injuries for service industries (850,930). In Tables 2.2 through 2.5 the distributions of the nature, body part, source, and exposure (accident type) of the 10,070 injuries are presented.

2.2.3   ILLNESSES

In the educational services sector there were 1700 cases of occupationally related illnesses; this is 1.3% of the total for the service industry (131,500) (see Table 2.6).

2.3   HAZARDS FACED BY EDUCATIONAL SERVICES WORKERS

Working in a very open and public forum exposes the educational services workers to a number of hazards that one would not usually connect with this type of work. The hazards covered in this book are the primary ones that affect educational services workers found in public and private educational institutions. In most cases the most frequent hazards faced by educational services employees are

- Walking and working surfaces
- Slips, trips, and falls
- Strains/sprains
- Fires

TABLE 2.1
Occupational Death Cause by Percent for Educational Services

Cause	Educational Services %
Highway	13
Homicides	7
Falls	11
Struck by	0

Source: From Bureau of Labor Statistics, U.S. Department of Labor. *National Census of Fatal Occupational Injuries in 2005.* Available at http://bls.gov.

TABLE 2.2
Nature of Injury by Number and Percent
for Educational Services

Nature of Injury	Number	Percent
Sprains/strains[a]	4290	43
Fractures[a]	890	8.8
Cuts/punctures	360	3.6
Bruises[a]	830	8
Heat burns	120	1.2
Chemical burns	30	0.3
Amputations	0	0
Carpal tunnel syndrome	150	1.5
Tendonitis	180	1.8
Multiple trauma[a]	550	5.5
Back pain	280	2.8

Source: From Bureau of Labor Statistics, U.S. Department of Labor. *Workplace Injuries and Illnesses in 2004.* Available at http://bls.gov.

[a] Five most frequently occurring conditions.

TABLE 2.3
Body Part Injured by Number and Percent
for Educational Services

Body Part Injured	Number	Percent
Head	840	3.4
Eyes	200	2
Neck	240	2.4
Trunk[a]	3120	31
Back[a]	1930	19
Shoulder	630	6.3
Upper extremities[a]	1770	17.6
Finger	470	4.7
Hand	280	2.8
Wrist	380	3.8
Lower extremities[a]	2840	28
Knee[a]	1410	14
Foot, toe	350	3.5
Body systems	150	1.5
Multiple body parts	1110	11

Source: From Bureau of Labor Statistics, U.S. Department of Labor. *Workplace Injuries and Illnesses in 2004.* Available at http://bls.gov.

[a] Five most frequently injured body parts.

TABLE 2.4
Source of Injury by Number and Percent for Educational Services

Sources of Injuries	Number	Percent
Parts and materials	250	2.5
Worker motion/position[a]	1760	17
Floor, walkways, or ground surfaces[a]	2690	27
Handtools	270	2.7
Vehicles[a]	700	7
Health care patient	90	0.9
Chemicals and chemical products	120	1.2
Containers[a]	930	9
Furniture and fixtures[a]	680	6.8
Machinery	340	3.4

Source: From Bureau of Labor Statistics, U.S. Department of Labor. *Workplace Injuries and Illnesses in 2004.* Available at http://bls.gov.

[a] Five most frequent sources of injury.

- Office hazards
- Repetitive/cumulative trauma
- Violence and security hazards
- Biological hazards

TABLE 2.5
Exposure/Accident Type by Number and Percent for Educational Services

Type of Accidents	Number	Percent
Struck by object[a]	850	8
Struck against Object	510	5
Caught in or compressed or crushed	160	1.6
Fall to lower level[a]	820	8
Fall on same level[a]	1970	19.7
Slips or trips without a fall	330	3.3
Overexertion[a]	2290	22.7
Lifting	1230	12
Repetitive motion[a]	570	5.7
Exposure to harmful substance or environment	400	4
Transportation accident	480	4.8
Fires and explosions	0	0
Assaults/violent acts	500	5

Source: From Bureau of Labor Statistics, U.S. Department of Labor. *Workplace Injuries and Illnesses in 2004.* Available at http://bls.gov.

[a] Five most frequent exposures or types of accidents that led to an injury.

TABLE 2.6

Occupational Illnesses by Number of Cases and Percent for Educational Services

Illness Type	Number	Percent
Skin diseases and disorders	300	17.6
Respiratory conditions	300	17.6
Poisoning	100	5.9
Hearing loss	0	0
All others	900	53

Source: From Bureau of Labor Statistics, U.S. Department of Labor. *Workplace Injuries and Illnesses in 2004.* Available at http://bls.gov.

2.4 OCCUPATIONS

The educational services industry was the second largest industry in the economy in 2004, providing jobs for about 13 million workers—more than 12.8 million wage and salary workers and 199,000 self-employed and unpaid family workers. Most jobs are found in elementary and secondary schools and the types of occupations are usually bus drivers, counselors, education administrators, instructional coordinators, librarians, teacher assistants, teachers—adult literacy and remedial education, enrichment programs, postsecondary, preschool, kindergarten, elementary, middle, and secondary, and special education.

Public schools also employ more workers than private schools at both levels, because most students attend public educational institutions. According to the latest data from the Department of Education's National Center for Education Statistics, close to 90% of students attend public primary and secondary schools, and about 75% attend public postsecondary institutions.

Workers in the educational services industry take part in all aspects of education, from teaching and counseling students to driving school buses and serving cafeteria lunches. Although two out of three workers in educational services are employed in professional and related occupations, the industry employs many administrative support, managerial, service, and other workers.

Preschool, kindergarten, and elementary school teachers play a critical role in the early development of children. They usually instruct one class in a variety of subjects, introducing the children to mathematics, language, science, and social studies. Often, they use games, artwork, music, computers, and other tools to teach basic skills.

Middle and secondary school teachers help students delve more deeply into subjects introduced in elementary school. Middle and secondary school teachers specialize in a specific academic subject, such as English, mathematics, or history, or a career and technical area, such as automobile mechanics, business education, or computer repair. Some supervise after-school extracurricular activities, and some help students deal with academic problems, such as choosing courses, colleges, and careers.

Special education teachers work with students—from toddlers to those in their early twenties—who have a variety of learning and physical disabilities. While most work in traditional schools and assist students who require extra support, some work in schools specifically designed to serve students with the most severe disabilities. With all but the most severe cases, special education teachers modify the instruction of the general education curriculum and, when necessary, develop alternative assessment methods to accommodate a student's special needs. They also help special education students develop emotionally, feel comfortable in social situations, and be aware of socially acceptable behavior.

Postsecondary teachers, or faculty, as they are usually called, are generally organized into departments or divisions, based on their subject or field. They teach and advise college students and perform a significant part of our nation's research. They prepare lectures, exercises, and laboratory experiments; grade exams and papers; and advise and work with students individually. Postsecondary teachers keep abreast of developments in their field by reading current literature, talking with colleagues and businesses, and participating in professional conferences. They also consult with government, business, nonprofit, and community organizations. In addition, they do their own research to expand knowledge in their field, often publishing their findings in scholarly journals, books, and electronic media.

Adult literacy and remedial education teachers teach English to speakers of other languages (ESOL), prepare sessions for the General Educational Development (GED) exam, and give basic instruction to out-of-school youths and adults. Self-enrichment teachers teach classes that students take for personal enrichment, such as cooking or dancing.

Education administrators provide vision, direction, leadership, and day-to-day management of educational activities in schools, colleges and universities, businesses, correctional institutions, museums, and job training and community service organizations. They set educational standards and goals and aid in establishing the policies and procedures to carry them out. They develop academic programs; monitor students' educational progress; hire, train, motivate, and evaluate teachers and other staff; manage counseling and other student services; administer record-keeping; prepare budgets; and handle relations with staff, parents, current and prospective students, employers, and the community.

Instructional coordinators evaluate school curricula and recommend changes to them. They research the latest teaching methods, textbooks, and other instructional materials and coordinate and provide training to teachers. They also coordinate equipment purchases and assist in the use of new technology in schools.

Educational, vocational, and school counselors work at the elementary, middle, secondary, and postsecondary school levels and help students evaluate their abilities, talents, and interests so that the students can develop realistic academic and career options. Using interviews, counseling sessions, tests, and other methods, secondary school counselors also help students understand and deal with their social, behavioral, and personal problems. They advise on college majors, admission requirements, and entrance exams and on trade, technical school, and apprenticeship programs. Elementary school counselors do more social and personal counseling and less career and academic counseling than do secondary school counselors.

School counselors may work with students individually or in small groups, or they may work with entire classes.

Librarians help people find information and learn how to use it effectively in their scholastic, personal, and professional pursuits. Librarians manage library staff and develop and direct information programs and systems for the public, as well as oversee the selection and organization of library materials; and retrieve information from computer databases. Clerical library technicians help librarians acquire, prepare, and organize material; direct library users to standard references. Assistants check out and receive library materials, collect overdue fines, and shelve materials.

Teacher assistants, also called teacher aides or instructional aides, provide instructional and clerical support for classroom teachers, allowing the teachers more time to plan lessons and to teach. Using the teacher's lesson plans, they provide students with individualized attention, tutoring and assisting children—particularly special education and non-English speaking students—in learning class material. Assistants also aid and supervise students in the cafeteria, in the schoolyard, in hallways, or on field trips. They record grades, set up equipment, and prepare materials for instruction.

School bus drivers transport students to and from schools and related activities. The educational services industry employs many other workers who are found in a wide range of industries. For example, office and administrative support workers such as secretaries, administrative assistants, and general office clerks account for about 12% of jobs in educational services. The educational services industry employs some of the most highly educated workers in the labor force.

2.5 APPLICABLE OSHA REGULATIONS

Another way to gather an understanding of the hazards faced by educational services is to see the types of violations that Occupational Safety and Health Administration (OSHA) have found during their inspections of educational institutions. These violations provide another way of targeting hazards that has the potential to cause injury, illness, and death of workers. As can be seen from the 25 most frequently cited violations, OSHA cites this industry under the general industry standard (29 CFR 1910) and the recordkeeping standard (29 CFR 1904) (see Table 2.7).

Although the previous violations were the 25 most frequently issued violations, OSHA has cited other hazards with less frequency. Some of these are as follows:

- Hazardous locations
- Automatic sprinklers
- Ladders
- Fixed ladders
- Fire prevention
- Confined spaces
- Eye and face protection

TABLE 2.7

Fifty Most Frequent OSHA Violations for Education Services (82)

CFR Standard	Number Cited	Description
1910.1200	7	Hazard communication
1910.134	4	Respiratory protection
1910.305	4	Electrical, wiring methods, components and equipment
1910.36	3	Design and construction requirements for exit routes
1910.132	3	Personal protective equipment, general requirements
1910.213	3	Woodworking machinery
1910.333	3	Electrical, selection and use of work practices
1910.23	2	Guarding floor and wall openings and holes
1910.37	2	Maintenance, safeguards, and operational features of exit routes
1910.147	2	The control of hazardous energy, lockout/tagout
1910.157	2	Portable fire extinguishers
1910.219	2	Mechanical power-transmission apparatus
1910.303	2	Electrical systems design, general requirements
1910.1025	2	Lead
1910.1030	2	Bloodborne pathogens
1910.1450	2	Occupational exposure to hazardous chemical in laboratories
1904.29	1	Forms
1910.24	1	Fixed industrial stairs
1910.138	1	Hand protection
1910.141	1	Sanitation
1910.151	1	Medical services and first aid
1910.304	1	Electrical, wiring design and protection
1910.332	1	Electrical, training
1926.62	1	Lead
1926.1101	1	Asbestos

Note: Standards cited by Federal OSHA for the retail service sector from October 2005 to September 2006 are included here.

With the hazards faced by this sector it is imperative that safety and health be an integral part of the educational process, with the specific purpose of protecting its employees.

REFERENCES

Bureau of Labor Statistics, U.S. Department of Labor. *National Census of Fatal Occupational Injuries in 2005.* Available at http://bls.gov.

Bureau of Labor Statistics, U.S. Department of Labor. *Workplace Injuries and Illnesses in 2004.* Available at http://bls.gov.

3 Health Care and Social Assistance

Health care is an important people-oriented service as this hospital building shows.

The education and health services supersector comprises two sectors: the educational services sector (sector 61), and the health care and social assistance sector (sector 62). Only privately owned establishments are included in this discussion; publicly owned establishments that provide education or health services are included in government.

The health care and social assistance sector (62) comprises establishments providing health care and social assistance to individuals. The industries in this sector are arranged on a continuum starting with establishments providing medical care exclusively, continuing with those providing health care and social assistance, and finally finishing with those providing only social assistance. NAICS's breakdown for the health care and social assistance sector is in the following manner:

Health care and social assistance (62)
    Ambulatory health care services (62100)
        Offices of physicians (621100)
        Offices of dentists (621200)
    Offices of other health practitioners (621300)
        Offices of chiropractors (6213310)
        Offices of optometrists (621320)

      Offices of mental health practitioners (except physicians) (621330)
      Offices of physical, occupational, speech therapists and audiologists
         (621340)
      Offices of other health practitioners (621390)
    Outpatient care centers (621400)
    Medical and diagnostic laboratories (621500)
    Home health care services (621600)
    Other ambulatory health care services (621900)
      Ambulance services (521910)
      All other ambulatory health care services (621990)
Hospitals (622000)
    General medical and surgical hospitals (622100)
    Psychiatric and substance abuse hospitals (622200)
    Specialty (except psychiatric and substance abuse hospitals (622300)
Nursing and residential care facilities (623000)
    Nursing care facilities (623100)
    Residential mental retardation, mental health, and substance Abuse
      facilities (623200)
      Residential mental retardation facilities (623210)
      Residential mental health and substance abuse facilities (623220)
    Community care facilities for the elderly (623300)
    Other residential care facilities (623900)
Social assistance (624000)
    Individual and family services (624100)
      Services for the elderly and persons with disabilities (624120)
    Community food and housing, and emergency and other relief services
      (624200)
    Vocational rehabilitation services (624300)
    Child day care services (624400)

3.1 HEALTH CARE

As the largest industry in 2004, health care provided 13.5 million jobs—13.1 million jobs for wage and salary workers and about 411,000 jobs for the self-employed. It is interesting to note that 8 out of 20 occupations projected to grow the fastest are in health care. Most workers have jobs that require less than 4 years of college education, but health diagnosing and treating practitioners are among the most educated workers.

Combining medical technology and the human touch, the health care industry administers care around the clock, responding to the needs of millions of people— from newborns to the critically ill.

About 545,000 establishments make up the health care industry; they vary greatly in terms of size, staffing patterns, and organizational structures. About 76% of health care establishments are offices of physicians, dentists, or other health practitioners. Although hospitals constitute only 2% of all health care establishments, they employ 40% of all workers.

The health care industry includes establishments ranging from small-town private practices of physicians who employ only one medical assistant to busy inner-city hospitals that provide thousands of diverse jobs. In 2004, about half of non-hospital health care establishments employed fewer than five workers (chart 1). By contrast, 7 out of 10 hospital employees were in establishments with more than 1000 workers.

The health care industry consists of the following nine segments: hospitals; nursing and residential care facilities; offices of physicians; offices of dentists; home health care services; offices of other health practitioners; outpatient care centers; other ambulatory health care services; and medical and diagnostic laboratories (Figure 3.1).

Average weekly hours of nonsupervisory workers in private health care varies among the different segments of the industry. Workers in offices of dentists averaged only 26.9 h per week in 2004, while those in psychiatric and substance abuse hospitals averaged 36.4 h, compared with 33.7 h for all private industry.

Health care services employed about 20% of the workforce as a whole in 2004, but accounted for 39% of workers in offices of dentists and 33% of those in offices of other health practitioners. Students, parents with young children, dual jobholders, and older workers make up much of the part-time workforce.

Many health care establishments operate around the clock and need staff at all hours. Shift work is common in some occupations, such as registered nurses. Numerous health care workers hold more than one job.

In 2004, the incidence of occupational injury and illness in hospitals was 8.7 cases per 100 full-time workers, compared with an average of 5.0 for private industry overall. Nursing care facilities had a much higher rate of 10.1.

FIGURE 3.1   Some medical and dental offices are small operations that have taken over private homes.

Health care workers involved in direct patient care must take precautions to prevent back strain from lifting patients and equipment; to minimize exposure to radiation and caustic chemicals; and to guard against infectious diseases, such as AIDS, tuberculosis, and hepatitis. Home care personnel who make house calls are exposed to the possibility of being injured in highway accidents, all types of overexertion when assisting patients, and falls inside and outside homes.

3.2 SOCIAL ASSISTANCE

Social assistance makes up about one out of three jobs in this professional and service occupation. Average earnings are low because of the large number of part-time and low-paying service jobs.

Careers in social assistance appeal to persons with a strong desire to make life better and easier for others. Workers in this industry are usually good communicators and enjoy interacting with people (see Figure 3.2). Social assistance establishments provide a wide array of services, which include helping the homeless, counseling troubled and emotionally disturbed individuals, training the unemployed or under-employed, and helping the needy to obtain financial assistance. About 61,000 establishments in the private sector provided social assistance in 2004. Thousands of other establishments, mainly in state and local governments, provided additional social assistance.

Social assistance consists of four segments—individual and family services; community food and housing, and emergency and other relief services; vocational rehabilitation services; and child day care services. The child day care services segment includes day care and preschool care centers. The services provided are individual and family services, community food and housing, and emergency and other relief services, and vocational rehabilitation services.

FIGURE 3.2   Social services require good communications skills.

Some social assistance establishments operate around the clock. Thus, evening, weekend, and holiday work is common. Some establishments may be understaffed, resulting in large caseloads for each worker. Jobs in voluntary, nonprofit agencies often are part-time.

Some workers spend a substantial amount of time traveling within the local area. For example, home health and personal care aides routinely visit clients in their homes; social workers and social and human service assistants also may make home visits. Social assistance workers were somewhat older than workers in other industries. About 40% are 45 years old or older, compared with 39% of all workers.

3.3 PROFILE OF HEALTH CARE AND SOCIAL ASSISTANCE WORKERS' DEATHS, INJURIES, AND ILLNESSES

3.3.1 DEATHS

There were 104 occupationally related deaths among health and social care services workers in 2005. The health care and social assistance sector accounted for 3.8 % of the service industry deaths (2736). Table 3.1 shows the percent values from each major category of those deaths.

3.3.2 INJURIES

There were 179,910 reported injuries to health and social care services workers in 2004. This is 21% of the total injuries for the service industry (850,930). In Tables 3.2 through 3.5 the distributions of the nature, body part, source, and exposure (accident type) of the 179,910 injuries are presented.

3.3.3 ILLNESSES

In the health and social care services sector there were 45,900 cases of occupationally related illnesses, this is 35% of the total for the service industry (see Table 3.6).

TABLE 3.1

Occupational Death Cause by Percent for Health and Social Care Services

Causes	Health and Social Care Services (%)
Highway	31
Homicides	11
Falls	11
Struck by	—

Source: From Bureau of Labor Statistics, U.S. Department of Labor. *National Census of Fatal Occupational Injuries in 2005.* Available at http://bls.gov.

TABLE 3.2

Nature of Injury by Number and Percent for Health and Social Care Services

Nature of Injury	Number	Percent
Sprains/strains[a]	95,500	53
Fractures[a]	9,320	5.2
Cuts/punctures	4,920	2.7
Bruises[a]	16,370	9
Heat burns	1,880	1
Chemical burns	1,120	0.6
Amputations	40	0.1
Carpal tunnel syndrome	1,990	1
Tendonitis	660	0.4
Multiple trauma[a]	6,390	3.5
Back pain[a]	7,690	4

Source: From Bureau of Labor Statistics, U.S. Department of Labor, *Workplace Injuries and Illnesses in 2004.* Available at http://bls.gov.

[a]   Five most frequently occurring conditions.

TABLE 3.3

Body Part Injured by Number and Percent for Health and Social Care Services

Body Part Injured	Number	Percent
Head	9,020	5
Eyes	3,150	1.7
Neck	4,070	2.3
Trunk[a]	78,100	43
Back[a]	55,330	31
Shoulder	12,700	7
Upper extremities[a]	28,370	16
Finger	7,510	4
Hand	3,790	2
Wrist	8,380	4.8
Lower extremities[a]	32,770	18
Knee	14,160	7.9
Foot, toe	5,250	2.9
Body systems	2,960	1.6
Multiple body parts[a]	23,350	13

Source: From Bureau of Labor Statistics, U.S. Department of Labor, *Workplace Injuries and Illnesses in 2004.* Available at http://bls.gov.

[a]   Five most frequently injured body parts.

TABLE 3.4
Source of Injury by Number and Percent for Health and Social Care Services

Sources of Injury	Number	Percent
Parts and materials	2,040	1
Worker motion/position[a]	23,770	13
Floor, walkways, or ground surfaces[a]	35,720	20
Handtools	1,920	1
Vehicles[a]	8,810	5
Health care patients[a]	55,710	31
Chemicals and chemical products	3,130	1.7
Containers[a]	9,660	5.4
Furniture and fixtures	8,710	4.8
Machinery	3,870	2.2

Source: From Bureau of Labor Statistics, U.S. Department of Labor, *Workplace Injuries and Illnesses in 2004.* Available at http://bls.gov.

[a] Five most frequent sources of injury.

TABLE 3.5
Exposure/Accident Type by Number and Percent for Health and Social Care Services

Type of Accidents	Number	Percent
Struck by object[a]	11,330	6.3
Struck against object	7,980	4.4
Caught in or compressed or crushed	2,730	1.5
Fall to lower level	5,000	2.8
Fall on same level[a]	31,700	17.6
Slips or trips without a fall	5,640	3.1
Overexertion[a]	64,500	36
Lifting[a]	30,890	17
Repetitive motion	5,160	2.9
Exposure to harmful substances or environment[a]	8,400	4.7
Transportation accidents	5,380	3
Fires and explosions	50	0.2
Assaults/violent acts[a]	12,320	7

Source: From Bureau of Labor Statistics, U.S. Department of Labor, *Workplace Injuries and Illnesses in 2004.* Available at http://bls.gov.

[a] Five most frequent exposures or types of accidents that lead to an injury.

TABLE 3.6
Occupational Illnesses by Number of Cases
and Percent for Health and Social Care Services

Illness Type	Number	Percent
Skin diseases and disorders	7,100	15
Respiratory conditions	5,700	12
Poisoning	400	0.8
Hearing loss	100	0.2
All others	32,600	71

Source: From Bureau of Labor Statistics, U.S. Department of Labor, *Workplace Injuries and Illnesses in 2004.* Available at http://bls.gov.

3.4   HAZARDS FACED BY HEALTH AND SOCIAL CARE SERVICE WORKERS

Working with patients and clients as well as interacting with the general public contributes to the hazards faced by health and social care service workers.

The hazards covered in this book are the primary ones that affect retail workers found in facility and store operations. In most cases the most frequent hazards faced by retail workers are

- Walking and working surfaces
- Equipment dangers
- Material handling/lifting of containers
- Slips, trips, and falls
- Strains/sprains
- Trauma injuries
- Vehicle accidents
- Fires
- Office hazards
- Repetitive/cumulative trauma
- Violence and security
- Radiation
- Biological hazards

3.5   OCCUPATIONS

3.5.1   HEALTH CARE SERVICES

As the largest industry in 2004, health care provided 13.5 million jobs—13.1 million jobs for wage and salary workers and about 411,000 jobs for self-employed and unpaid family workers. Of the 13.1 million wage and salary jobs, 41% were in hospitals; 22% were in nursing and residential care facilities; and 16% were in offices of physicians. About 92% of wage and salary jobs were in private industry; the rest were in state and local government hospitals. The majority of jobs for self-employed and unpaid family workers in health care were in offices of physicians, dentists, and other health practitioners—about 282,000 out of the 411,000 total self-employed.

Workers in health care tend to be older than workers in other industries. Health care workers are also more likely to remain employed in the same occupation, due, in part, to the high level of education and training required for many health occupations.

Health care firms employ large numbers of workers in professional and service occupations. Together, these two occupational groups account for three out of four jobs in the industry. The next largest share of jobs, 18%, is in office and administrative support. Management, business, and financial operations occupations account for only 4% of employment. Other occupations in health care make up only 3% of the total. Some of the occupations found in health care services are audiologists, cardiovascular technologists and technicians, chiropractors, clinical laboratory technologists and technicians, dental assistants, dental hygienists, dentists, diagnostic medical sonographers, dietitians and nutritionists, emergency medical technicians and paramedics, licensed practical and licensed vocational nurses, medical and health services managers, medical assistants, medical records and health information technicians, medical secretaries, medical transcriptionists, nuclear medicine technologists, nursing, psychiatric, and home health aides, occupational therapist assistants and aides, occupational therapists, medical, dental and ophthalmic laboratory technicians, opticians, dispensing, optometrists, personal and home care aides, pharmacists, pharmacy assistants, pharmacy technicians, physical therapist assistants and aides, physical therapists, physician assistants, physicians, surgeons, podiatrists, psychologists, radiologic technologists and technicians, receptionists and information clerks, recreational therapists, registered nurses, respiratory therapists, social and human service assistants, social workers, speech-language pathologists, and surgical technologists.

Professional occupations, such as physicians and surgeons, dentists, registered nurses, social workers, and physical therapists, usually require at least a bachelor's degree in a specialized field or higher education in a specific health field, although registered nurses also enter through associate degree or diploma programs.

Other health professionals and technicians work in many fast growing occupations, such as medical records and health information technicians and dental hygienists. These workers may operate technical equipment and assist in health diagnosing and treating practitioners. Graduates of 1 year or 2 year training programs often fill such positions; the jobs usually require specific formal training beyond high school, but less than 4 years of college.

Service occupations attract many workers with little or no specialized education or training. For instance, some of these workers are nursing aides, home health aides, building cleaning workers, dental assistants, medical assistants, and personal and home care aides. Nursing or home health aides provide health-related services for ill, injured, disabled, elderly, or infirm individuals either in institutions or in their homes. By providing routine personal care services, personal and home care aides help the elderly, disabled, and ill to live in their own homes instead of in an institution.

Most workers in health care jobs provide clinical services, but many also are employed in occupations with other functions. Numerous workers in management and administrative support jobs keep organizations running smoothly. Although many medical and health service managers have a background in a clinical specialty or training in health care administration, some enter these jobs with a general business education.

Each segment of the health care industry provides a different mix of wage and salary health-related jobs.

3.5.2 Social Assistance

More than one third of nongovernmental social assistance jobs are in professional and related occupations. Social workers counsel and assess the needs of clients, refer them to the appropriate sources of help, and monitor their progress. They may specialize in child welfare and family services, mental health, medical social work, school social work, community organization activities, or clinical social work. Social and human service assistants work in a variety of social and human service delivery settings. Job titles and duties of these workers vary, but they include human service worker, case management aide, social work assistant, mental health aide, child abuse worker, community outreach worker, and gerontology aide. Counselors help people evaluate their interests and abilities and advise and assist them with personal and social problems.

Many occupations are found in social assistance services such as childcare workers, counselors, education administrators, nursing, psychiatric, and home health aides, personal and home care aides, social and human service assistants, social workers, teachers—adult literacy and remedial education, teachers—self-enrichment education, and teacher assistants.

Almost one third of employment in the social assistance industry is in many of the service occupations. Personal and home care aides help the elderly, disabled, and ill live in their own homes, instead of in an institution, by providing routine personal care services. Although some are employed by public or private agencies, many are self-employed. Persons in food preparation and serving-related occupations serve residents at social assistance institutions. Home health aides provide health-related services for ill, injured, disabled, or elderly individuals in their homes.

Obtaining affordable, quality child day care, especially for children under age 5, is a major concern for many parents. Child day care needs are met in different ways. Care in a child's home, care in an organized child care center, and care in a provider's home—known as family child care—are all common arrangements for preschool-aged children. Older children also may receive child day care services when they are not in school, generally through before- and after-school programs or private summer school programs. With the increasing number of households in which both parents work full time, this industry has been one of the fastest growing in the U.S. economy.

As in most industries, office and administrative support workers—secretaries and bookkeepers, for example—as well as managers account for many jobs. However, the social assistance sector employs a much smaller percentage of production, installation, maintenance, repair, and sales jobs than does the economy as a whole.

Certain occupations are more heavily concentrated in some segments of the industry than in others. Individual and family services employ the greatest numbers of social workers, social and human service assistants, and personal and home care aides. Vocational rehabilitation services provide the most jobs for adult literacy and remedial and self-enrichment education teachers.

3.6   APPLICABLE OSHA REGULATIONS

Another way to garner an understanding of the hazards faced by health and social care services workers is to see the types of violations that Occupational Safety and Health Administration (OSHA) have found during their inspections of health and social care services establishments. These violations provide another way of targeting hazards that have the potential to cause injury, illness, and death of workers. As can be seen from the 50 most frequently cited violations, OSHA cites this industry under the general industry standard (29 CFR 1910) and the recordkeeping standard (29 CFR 1904) (see Tables 3.7 and 3.8).

TABLE 3.7
Twenty-Five Most Frequent OSHA Violations for Health Care Services

CFR Standard	Number Cited	Description
1910.1030	885	Bloodborne pathogens
1910.1200	169	Hazard communication
1904.29	104	Forms
1910.132	82	Personal protective equipment, general requirements
1910.147	73	The control of hazardous energy, lockout/tagout
1910.305	67	Electrical, wiring methods, components and equipment
1910.37	65	Maintenance, safeguards, and operation features for exit routes
1910.151	59	Medical services and first aid
1910.303	55	Electrical systems design, general requirements
1904.32	45	Annual summary
1910.134	43	Respiratory protection
1904.41	35	Annual OSHA injury and illness survey of ten or more employees
1910.133	33	Eye and face protection
1910.146	32	Permit-required confined spaces
1910.22	31	Working/walking surfaces, general requirements
1910.212	27	Machines, general requirements
1904.8	22	Recording criteria for needlestick and sharps injuries
1910.1001	22	Asbestos
1910.157	20	Portable fire extinguishers
1910.304	20	Electrical, wiring design and protection
1910.215	18	Abrasive wheel machinery
1910.219	17	Mechanical power-transmission apparatus
1910.23	16	Guarding floor and wall openings and holes
1910.1048	14	Formaldehyde
1904.7	11	General recording criteria

Note: Standards cited by federal OSHA for the health care services sector from October 2005 to September 2006 are included here.

TABLE 3.8

Twenty-Five Most Frequent OSHA Violations for Social Care Services

CFR Standard	Number Cited	Description
1910.1030	16	Bloodborne pathogens
1910.1200	12	Hazard communication
1910.305	5	Electrical, wiring methods, components and equipment
1910.134	4	Respiratory protection
1910.22	3	Working/walking surfaces, general requirements
1910.23	3	Guarding floor and wall openings and holes
1910.132	3	Personal protective equipment, general requirements
1910.151	3	Medical services and first aid
1910.178	3	Powered industrial trucks (forklifts)
1910.36	2	Design and construction requirements for exit routes
1910.147	2	The control of hazardous energy, lockout/tagout
1910.212	2	Machines, general requirements
1910.303	2	Electrical systems design, general requirements
1910.1001	1	Asbestos
5A1	1	General duty clause (section of OSHA act)
1904.2	1	Partial exemption for establishments in certain industries
1904.39	1	Reporting fatalities and multiple hospitalization incidents to OSHA
1904.40	1	Providing documents to government representatives
1904.41	1	Annual OSHA injury and illness survey of ten or more employees
1910.38	1	Emergency action plans
1910.133	1	Eye and face protection
1910.332	1	Electrical, training
1910.334	1	Electrical, use of equipment
1926.21	1	Safety training and education
1926.50	1	Medical services and first aid

Note: Standards cited by federal OSHA for the social care services sector from October 2005 to September 2006 are included here.

3.6.1 Health and Social Care Services

Although the previous were the 25 most frequently issued violations, OSHA has cited other hazards with less frequency. Some of these are as follows:

- Hand protection
- Ethylene oxide
- Compressed gases
- Hazardous locations
- Lead
- Fixed industrial stairs
- Sanitation
- Asbestos

- Automatic sprinklers
- Ladders
- Fixed ladders
- Air receivers
- Safeguard for personnel protection
- Bakery equipment

With the hazards faced by this sector it is imperative that safety and health be an integral part of doing business, with the specific purpose of protecting its employees.

REFERENCES

Bureau of Labor Statistics, U.S. Department of Labor. *National Census of Fatal Occupational Injuries in 2005*. Available at http://bls.gov.
Bureau of Labor Statistics, U.S. Department of Labor, *Workplace Injuries and Illnesses in 2004*. Available at http://bls.gov.

4 Leisure and Hospitality Sectors

Museums are places of learning as well as entertainment.

The leisure and hospitality supersector comprises two sectors: the arts, entertainment, and recreation sector (sector 71), and the accommodation and food services sector (sector 72).

The arts, entertainment, and recreation sector includes a wide range of establishments that operate facilities or provide services to meet varied cultural, entertainment, and recreational interests of their patrons. This sector comprises (1) establishments that are involved in producing, promoting, or participating in live performances, events, or exhibits intended for public viewing; (2) establishments that preserve and exhibit objects and sites of historical, cultural, or educational interest; and (3) establishments that operate facilities or provide services that enable patrons to participate in recreational activities or pursue amusement, hobby, and leisure-time interests. The North American Industry Classification System (NAICS) for arts, entertainment, and recreation sector divides it into the following categories:

Arts, entertainment, and recreation (71)
    Performing arts, spectator sports and related industries (711000)
        Performing arts companies (711100)
            Theater companies and diner theaters (711110)
        Spectator sports (711200)
        Promoters of performing arts, sports and similar events (711300)

Agents and managers for artists, athletes, entertainers, and other public
    figures (711400)
  Independent artists, writers, and performers (711500)
Museums, historical sites and similar institutions (712000)
  Museums, historical sites and similar institutions (712100)
Amusement, gambling, and recreation industries (713000)
  Amusement parks and arcades (713100)
  Gambling industries (713200)
  Other amusement and recreation industries (713900)
    Fitness and recreational sports centers (713940)

4.1 ARTS, ENTERTAINMENT, AND RECREATION

The industry is characterized by a large number of seasonal and part-time jobs and
relatively young workers. About 40% of all workers have no formal education
beyond high school. Earnings are relatively low.

As leisure time and personal incomes have grown across the nation, so has the
arts, entertainment, and recreation industry. This industry includes about 115,000
establishments, ranging from art museums to fitness centers. Practically any activity
that occupies a person's leisure time, excluding the viewing of motion pictures and
video rentals, is part of the arts, entertainment, and recreation industry. The diverse
range of activities offered by this industry can be categorized into three broad
groups: live performances or events; historical, cultural, or educational exhibits;
and recreation or leisure-time activities.

4.1.1 LIVE PERFORMANCES OR EVENTS

This segment of the industry includes professional sports and establishments providing sports facilities and services to amateurs. Commercial sports clubs operate
professional and amateur athletic clubs and promote athletic events. All kinds of
popular sports can be found in these establishments, including baseball, basketball,
boxing, football, ice hockey, soccer, wrestling, and even auto racing. Professional
and amateur companies involved in sports promotion are also part of this industry
segment, as are sports establishments in which gambling is allowed, such as dog and
horse racetracks and jai alai courts.

A variety of businesses and groups involved in live theatrical and musical
performances are included in this segment. Theatrical production companies, for
example, coordinate all aspects of producing a play or theater event, including
employing actors and actresses and costume designers and contracting with lighting
and stage crews who handle the technical aspects of productions. Agents and
managers, who represent actors and entertainers and assist them in finding jobs or
engagements, are also included. Booking agencies line up performance engagements
for theatrical groups and entertainers.

Performers of live musical entertainment include popular music artists, dance
bands, disc jockeys, orchestras, jazz musicians, and rock-and-roll bands. Orchestras
range from major professional orchestras with million dollar budgets to community

orchestras, which often have part-time schedules. The performing arts segment also includes dance companies, which produce all types of live theatrical dances. The majority of these dance troupes perform ballet, folk dance, or modern dance.

4.1.2 Historical, Cultural, or Educational Exhibits

Privately owned museums, zoos, botanical gardens, nature parks, and historical sites make up this segment of the industry; publicly owned facilities are included in sections on federal, state, or local government elsewhere in the Career Guide. Each institution in this segment preserves and exhibits objects, sites, and natural wonders with historical, cultural, or educational value.

4.1.3 Recreation or Leisure Activities

A variety of establishments provide amusement for a growing number of customers. Some of these businesses provide video game, pinball, and gaming machines for the public at amusement parks, arcades, and casinos. Casinos and other gaming establishments offering off-track betting are a rapidly growing part of this industry segment (see Figure 4.1). This segment also includes amusement and theme parks, which range in size from local carnivals to multi-acre parks (see Figure 4.2). These establishments may have mechanical rides, shows, and refreshment stands. Other recreation and leisure-time services include golf courses, skating rinks, ski lifts, marinas, day camps, go-cart tracks, riding stables, waterslides, and establishments offering rental sporting goods.

FIGURE 4.1 In recent years, legal gambling has become more prevalent and in some locations a large workforce is employed.

FIGURE 4.2   Theme parks present hazards to the patrons and workers who ride, operate, and maintain these large and somewhat dangerous attractions.

This segment of the industry also includes physical fitness facilities that feature exercise and weight loss programs, gyms, health clubs, and day spas. These establishments also frequently offer aerobics, dance, yoga, and other exercise classes. Other recreation and leisure-time businesses include bowling centers that rent lanes and equipment for tenpin, duckpin, or candlepin bowling.

These facilities may be open to the public or available on a membership basis. Sports and recreation clubs that are open only to members and their guests include some golf courses, country clubs, and yacht, tennis, racquetball, hunting and fishing, and gun clubs. Public golf courses and marinas, unlike private clubs, provide facilities to the general public on a fee-per-use basis.

Technology is a major part of producing arts, entertainment, and recreation activities; for example, lighting and sound are vital for concerts and themed events and elaborate sets are often required for plays. However, most of this work is contracted to firms outside of the arts, entertainment, and recreation industry. (For more information about entertainment technology jobs, see the sources of additional information at the end of this statement.)

4.1.4 WORK CONDITIONS

Jobs in arts, entertainment, and recreation are more likely to be part-time than those in other industries. In fact, the average nonsupervisory worker in the arts, entertainment, and recreation industry worked 25.7 h a week in 2004, as compared to an average of 33.7 h for all private industry. Musical groups and artists were likely to work the fewest hours due to the large number of performers competing for a limited number of engagements, which may require a great amount of travel. The majority of performers are unable to support themselves in this profession alone and are forced to supplement their income through other jobs.

Many types of arts, entertainment, and recreation establishments dramatically increase employment during the summer and either scale back employment during the winter or close down completely. Workers may be required to work nights, weekends, and holidays because that is when most establishments are busiest. Some jobs require extensive travel. Music and dance troupes, for example, frequently tour or travel to major metropolitan areas across the country, in hopes of attracting large audiences.

Many people in this industry work outdoors, whereas others may work in hot, crowded, or noisy conditions. Some jobs, such as those at fitness facilities or in amusement parks, involve some manual labor and, thus, require physical strength and stamina. Also, athletes, dancers, and many other performers must be in particularly good physical condition. Many jobs include customer service responsibilities, so employees must be able to work well with the public.

In 2003, cases of work-related illness and injury averaged 5.9 for every 100 full-time workers, higher than the average of 5.0 for the entire private sector. Risks of injury are high in some jobs, especially those of athletes. Although most injuries are minor, including sprains and muscle pulls, they may prevent an employee from working for a period.

The arts, entertainment, and recreation industry provided about 1.8 million wage-and-salary jobs in 2004. About 58% of these jobs were in the industry segment other amusement and recreation industries—which includes golf courses, membership sports and recreation clubs, and physical fitness facilities.

Although most establishments in the arts, entertainment, and recreation industry are small, 42% of all jobs were in establishments that employ more than 100 workers.

The arts, entertainment, and recreation industry is characterized by a large number of seasonal and part-time jobs and by workers who are younger than the average for all industries. About 46% of all workers are under 35. Many businesses in the industry increase hiring during the summer, often employing high-school-age and college-age workers. Most establishments in the arts, entertainment, and recreation industry contract out lighting, sound, set-building, and exhibit-building work to firms not included in this industry.

4.2 PROFILE OF ARTS, ENTERTAINMENT, AND RECREATION WORKERS' DEATHS, INJURIES, AND ILLNESSES

4.2.1 Deaths

There were 76 occupationally related deaths among arts, entertainment, and recreation workers in 2005. The arts, entertainment, and recreation sector accounted for 2.8% of the service industry deaths (2736). Table 4.1 shows the percent values from each major category of those deaths.

TABLE 4.1
Occupational Death Cause by Percent for Arts, Entertainment, and Recreation

Cause	Arts, Entertainment, and Recreation (%)
Highways	11
Homicides	7
Falls	9
Struck by	13

Source: From Bureau of Labor Statistics, United States Department of Labor. *Workplace Injuries and Illnesses in 2004.* Available at http://bls.gov.

4.2.2 INJURIES

There were 95,380 reported injuries to arts, entertainment, and recreation workers in 2004. This was 11% of the total injuries for the service industries (850,930). In Tables 4.2 through 4.5, the distributions of the nature, body part, source, and exposure (accident type) of the 95,380 injuries are presented.

4.2.3 ILLNESSES

In the arts, entertainment, and recreation sector, there were 4000 cases of occupationally related illnesses; this is 3% of the total for the service industry (see Table 4.6).

TABLE 4.2
Nature of Injury by Number and Percent for Arts, Entertainment, and Recreation

Nature of Injury	Number	Percent
Sprains/strains[a]	6930	39
Fractures[a]	1580	9
Cuts/punctures[a]	1460	8
Bruises[a]	1960	11
Heat burns	360	2
Chemical burns	100	0.5
Amputations	70	0.4
Carpal tunnel syndrome	230	1.3
Tendonitis	90	0.5
Multiple trauma[a]	520	2.9
Back pain	370	2

Source: From Bureau of Labor Statistics, United States Department of Labor. *Workplace Injuries and Illnesses in 2004.* Available at http://bls.gov.

[a] Five most frequently occurring conditions.

TABLE 4.3
Body Part Injured by Number and Percent for Arts, Entertainment, and Recreation

Body Part Injured	Number	Percent
Head	1220	6.9
Eyes	430	2.4
Neck	220	1.2
Trunk[a]	5120	29
Back[a]	2930	17
Shoulder	1080	6
Upper extremities[a]	3970	22
Finger	1530	8.6
Hand	450	2.5
Wrist	1010	5.7
Lower extremities[a]	4970	28
Knee	1720	10
Foot, toe	1120	6
Body systems	330	1.9
Multiple body parts[a]	1850	10

Source: From Bureau of Labor Statistics, United States Department of Labor. *Workplace Injuries and Illnesses in 2004.* Available at http://bls.gov.

[a] Five most frequently occurring conditions.

TABLE 4.4
Source of Injury by Number and Percent for Arts, Entertainment, and Recreation

Sources of Injuries	Number	Percent
Parts and materials[a]	890	5
Worker motion/position[a]	3560	20
Floor, walkways, or ground surfaces[a]	3960	22
Handtools	680	3.8
Vehicles	1460	8
Health care patients	0	0
Chemicals and chemical products	270	1.5
Containers[a]	1510	8.5
Furniture and fixtures	790	4.5
Machinery[a]	900	5

Source: From Bureau of Labor Statistics, United States Department of Labor. *Workplace Injuries and Illnesses in 2004.* Available at http://bls.gov.

[a] Five most frequent sources of injury.

TABLE 4.5
Exposure/Accident Type by Number and Percent for Arts,
Entertainment, and Recreation

Type of Accidents	Number	Percent
Struck by an object[a]	2160	12
Struck against an object[a]	1230	7
Caught in or compressed or crushed	480	2.7
Fall to lower level	1150	6.5
Fall on same level[a]	2810	16
Slips or trips without a fall	770	4.3
Overexertion[a]	3080	17
Lifting[a]	1720	10
Repetitive motion	680	3.8
Exposure to harmful substance or environment	1000	5.6
Transportation accident	1050	5.9
Fires and explosions	110	0.6
Assaults/violent acts	450	2.5

Source: From Bureau of Labor Statistics, United States Department of Labor. *Workplace Injuries and Illnesses in 2004.* Available at http://bls.gov.

[a] Five most frequent exposures or types of accidents that led to an injury.

4.3 HAZARDS FACED BY ARTS, ENTERTAINMENT, AND RECREATION WORKERS

Working by performing for and functioning with the general public in what may be a high-activity forum contributes to the hazards facing the workers in arts, entertainment, and recreation sector. The hazards covered in this book are the primary ones that affect workers in the arts, entertainment, and recreation sector. In most cases, the most frequent hazards faced by arts, entertainment, and recreation workers are

TABLE 4.6
Occupational Illnesses by Number of Cases and
Percent for Arts, Entertainment, and Recreation

Illness Type	Number	Percent
Skin diseases and disorders	1100	27.5
Respiratory conditions	300	7.5
Poisoning	100	2.5
Hearing loss	0	0
Others	2400	60

Source: From Bureau of Labor Statistics, United States Department of Labor. *Workplace Injuries and Illnesses in 2004.* Available at http://bls.gov.

- Walking and working surfaces
- Electrocutions
- Material handling/lifting
- Slips, trips, and falls
- Strains/sprains
- Trauma injuries
- Fires
- Power tools
- Repetitive/cumulative trauma
- Violence and security

4.4   OCCUPATIONS

About 59% of wage-and-salary workers in the industry are employed in service occupations. Amusement and recreation attendants—the largest occupation in the arts, entertainment, and recreation industry—perform a variety of duties depending on where they are employed. Common duties include setting up games, handing out sports equipment, providing caddy services for golfers, collecting money, and operating amusement park rides. The most common occupations in this sector are actors, producers, directors; archivists, curators, and museum technicians; athletes, coaches, umpires, and related workers; broadcast and sound engineering technicians and radio operators; dancers and choreographers; fitness workers; gaming cage workers; gaming services occupations; grounds maintenance workers; musicians, singers, and related workers; recreation workers; and security guards and gaming surveillance officers (see Figure 4.3).

FIGURE 4.3   Security has escalated as an issue and there is a need to ensure the safety of guests and workers. Thus, the need for security personnel is needed more often.

Fitness trainers and aerobics instructors lead or coach groups or individuals in exercise activities and in the fundamentals of sports. Recreation workers organize and promote activities such as arts and crafts, sports, games, music, drama, social recreation, camping, and hobbies. They are generally employed by schools, theme parks and other tourist attractions, or health, sports, and other recreational clubs. Recreation workers schedule organized events to structure leisure time.

Gaming services workers assist in the operation of games, such as keno, bingo, and gaming table games. They may calculate and pay off the amount of winnings or collect players' money or chips.

Tour and travel guides escort individuals or groups on sightseeing tours or through places of interest, such as industrial establishments, public buildings, and art galleries. They may also plan, organize, and conduct long-distance cruises, tours, and expeditions for individuals or groups.

Animal care and service workers feed, water, bathe, exercise, or otherwise care for animals in zoos, circuses, aquariums, or other settings. They may train animals for riding or performance.

Other service workers include waiters and waitresses, who serve food in entertainment establishments; fast-food and counter workers and cooks and food preparation workers, who may serve or prepare food for patrons; and bartenders, who mix and serve drinks in arts, entertainment, and recreation establishments.

Building grounds, cleaning, and maintenance occupations include building cleaning workers, who clean up after shows or sporting events and are responsible for the daily cleaning and upkeep of facilities. Landscaping and groundskeeping workers care for athletic fields and golf courses. These workers maintain artificial and natural turf fields, mark boundaries, and paint team logos. They also mow, water, and fertilize natural athletic fields and vacuum and disinfect synthetic fields. Establishments in this industry also employ workers in protective service occupations. Security guards patrol the property and guard against theft, vandalism, and illegal entry. At sporting events, guards maintain order and direct patrons to various facilities. Gaming surveillance officers and gaming investigators observe casino operations to detect cheating, theft, or other irregular activities by patrons or employees.

Professional and related occupations account for 11% of all jobs in this industry. Some of the most well-known members of these occupations, athletes and sports competitors, perform in any of a variety of sports. Professional athletes compete in events for compensation, either through salaries or prize money. Organizations such as the Women's National Basketball Association (WNBA) and the National Football League (NFL) sanction events for professionals. Few athletes are able to make it to the professional level, where high salaries are common. In some professional sports, minor leagues offer lower salaries with a chance to develop skills through competition before advancing to major league play.

Coaches and scouts train athletes to perform at their highest level. Often, they are experienced athletes who have retired and are able to provide insight from their own experiences to players. Although some umpires, referees, and other sports officials work full time, the majority usually works part time and often has other full-time jobs. For example, many professional sport referees and umpires also officiate at amateur games.

Musicians and singers may play musical instruments, sing, compose, arrange music, or conduct groups in instrumental or vocal performances. The specific skills and responsibilities of musicians vary widely by type of instrument, size of ensemble, and style of music. For example, musicians can play jazz, classical, or popular music, either alone or in groups ranging from small rock bands to large symphony orchestras.

Actors entertain and communicate with people through their interpretation of dramatic and other roles. They can belong to a variety of performing groups, ranging from those appearing in community and local dinner theaters to those playing in full-scale Broadway productions. Dancers express ideas, stories, rhythm, and sound with their bodies through different types of dance, including ballet, modern dance, tap, folk, and jazz. Dancers usually perform in a troupe, although some perform solo. Many become teachers when their performing careers end. Choreographers create and teach dance, and they may be called upon to direct and stage presentations. Producers and directors select and interpret plays or scripts and give directions to actors and dancers. They conduct rehearsals, audition cast members, and approve choreography. They also arrange financing, hire production staff members, and negotiate contracts with personnel.

Archivists, curators, and museum technicians play an important role in preparing museums for display. Archivists appraise, edit, and direct safekeeping of permanent records and historically valuable documents. They may also participate in research activities based on archival materials. Curators administer a museum's affairs and conduct research programs. Museum technicians and conservators prepare specimens, such as fossils, skeletal parts, lace, and textiles, for museum collection and exhibits. They may also take part in restoring documents or installing and arranging materials for exhibits.

Audio and video equipment technicians set up and operate audio and video equipment, including microphones, sound speakers, video screens, projectors, video monitors, recording equipment, connecting wires and cables, sound and mixing boards, and related electronic equipment for theme parks, concerts, and sports events. They may also set up and operate associated spotlights and other custom lighting systems.

About 8% of all jobs in this industry are in sales and related occupations. The largest of these, cashiers, often use a cash register to receive money and give change to customers. In casinos, gaming change persons and booth cashiers exchange coins and tokens for patrons' money. Counter and rental clerks check out rental equipment to customers, receive orders for service, and handle cash transactions.

Another 9% of jobs in this industry are in office and administrative support occupations. Receptionists and information clerks, one of the larger occupations in this category, answer questions and provide general information to patrons. Other large occupations in this group include general office clerks and secretaries and administrative assistants. Gaming cage workers conduct financial transactions for patrons in gaming establishments. For example, they may accept a patron's credit application and verify credit references to provide check-cashing authorizations or to establish house credit accounts. In addition, they may reconcile daily summaries of transactions to balance books or sell gambling chips, tokens, or tickets to patrons. At a patron's request, gaming cage workers may convert gaming chips, tokens, or tickets to currency.

Management, business, and financial occupations make up 6% of employment in this industry. Managerial duties in the performing arts include marketing, business management, event booking, fund-raising, and public outreach. Agents and business managers of artists, performers, and athletes represent their clients to prospective employers and may handle contract negotiations and other business matters. Recreation supervisors and park superintendents oversee personnel, budgets, grounds and facility maintenance, and land and wildlife resources. Some common administrative jobs in sports are tournament director, health club manager, and sports program director.

Installation, maintenance, and repair occupations make up 4% of this industry's employment. General maintenance and repair workers are the largest occupation in this group.

About 40% of all workers in the arts, entertainment, and recreation industry have no formal education beyond high school. In the case of performing artists or athletes, talent and years of training are more important than education. However, upper-level management jobs usually require a college degree.

Most service jobs require little or no previous training or education beyond high school. Many companies hire young, unskilled workers, such as students, to perform low-paying seasonal jobs. Employers look for people with the interpersonal skills necessary to work with the public.

In physical fitness facilities, fitness trainer and aerobic instructor positions are usually filled by persons who develop an avid interest in fitness and then become certified to teach. Certification from a professional organization may require knowledge of cardiopulmonary resuscitation (CPR); an associate degree or experience as an instructor at a health club; and successful completion of written and oral exams covering a variety of areas, including anatomy, nutrition, and fitness testing. Sometimes, fitness workers become health club managers or owners. To advance to a management position, a degree in physical education, sports medicine, or exercise physiology is useful.

In the arts, employment in professional and related occupations usually requires a great deal of talent. There are many highly talented performers, creating intense competition for every opening. Performers such as musicians, dancers, and actors often study their professions most of their lives, taking private lessons and spending hours practicing. Usually, performers have completed some college or related study. Musicians, dancers, and actors often go on to become teachers after completing the necessary requirements for at least a bachelor's degree. Musicians who complete a graduate degree in music sometimes move on to a career as a conductor. Dancers sometimes become choreographers, and actors can advance into producer and director jobs.

Almost all arts administrators have completed 4 years of college, and the majority possess a master's or a doctoral degree. Experience in marketing and business is helpful because promoting events is a large part of the job.

Entry-level supervisory or professional jobs in recreation sometimes require completion of a 2 year associate degree in parks and recreation at a junior college.

Completing a 4 year bachelor's degree in this field is necessary for high-level supervisory positions. Students can specialize in such areas as aquatics, therapeutic recreation, aging and leisure, and environmental studies. Those who obtain graduate degrees in the field and have years of experience may obtain administrative or university teaching positions. The National Recreation and Parks Association (NRPA) certifies individuals who meet eligibility requirements for professional and technical jobs. Certified park and recreation professionals must pass an exam; earn a bachelor's degree with a major in recreation, park resources, or leisure services from a program accredited by the NRPA or by the American Association for Leisure and Recreation; or earn a bachelor's degree and have at least 5 years of relevant full-time work experience, depending on the major field of study.

4.5 APPLICABLE OSHA REGULATIONS

Another way to gather an understanding of the hazards faced by retail workers is to see the types of violations that Occupational Safety and Health Administration (OSHA) have found during their inspections of art, entertainment, and recreation operations.

These violations provide another way of targeting hazards that have the potential to cause injury, illness, and death of workers. As can be seen from the 35 most frequently cited violations, OSHA cites this industry under the general industry standard (29 CFR 1910) and the recordkeeping standard (29 CFR 1904) (see Table 4.7).

TABLE 4.7
Thirty-Five Most Frequent OSHA Violations for Arts, Entertainment, and Recreation

CFR Standard	Number Cited	Description
1910.1200	22	Hazard communication
1910.305	19	Electrical, wiring methods, components and equipment
1910.1025	18	Lead
1910.178	16	Powered industrial trucks (forklifts)
1910.95	15	Occupational noise exposure
1910.23	13	Guarding floor and wall openings and holes
1910.157	11	Portable fire extinguishers
1910.219	10	Mechanical power-transmission apparatus
1910.303	10	Electrical systems design, general requirements
1910.1001	10	Asbestos
1928.1101	10	Asbestos
1910.22	8	Working/walking surfaces, general requirements
1910.147	8	The control of hazardous energy, lockout/tagout
1910.132	7	Personal protective equipment, general requirements
1910.134	7	Respiratory protection
1910.213	7	Woodworking machinery requirements
1910.37	6	Maintenance, safeguards, and operational features of exit routes

(continued)

TABLE 4.7 (continued)
Thirty-Five Most Frequent OSHA Violations for Arts, Entertainment, and Recreation

CFR Standard	Number Cited	Description
1910.151	5	Medical services and first aid
1904.29	4	Forms
1910.67	4	Vehicle-mounted elevating/rotating work platforms
1910.215	4	Abrasive wheel machinery
1910.243		Guarding of portable powered tools
1910.253	4	Oxygen-fuel gas welding and cutting
1910.1030	4	Bloodborne pathogens
5A1	4	General duty clause (section of OSHA Act)
1904.2	3	Partial exemption for establishments in certain industries
1904.39	3	Reporting fatalities and multiple hospitalization incidents to OSHA
1910.28	3	Safety requirements for scaffolding
1910.176	3	Handling materials, general requirements
1910.180	3	Crawler locomotive and truck cranes
1910.212	3	Machines, general requirements
1910.334	3	Electrical, use of equipment
1910.24	2	Fixed industrial stairs
1910.38	2	Emergency action plans
1910.141	2	Sanitation
1910.145	2	Specifications for accident prevention signs and tags

Note: Standards cited by federal OSHA for the retail service sector from October 2005 to September 2006 are included here.

Although the previous were the 35 most frequently issued violations, OSHA has cited other hazards with less frequency. Some of these are as follows:

- Arc welding and cutting
- Fixed ladders
- Flammable and combustible liquids
- Permit-required confined spaces
- Overhead and gantry cranes
- Hand protection
- Occupational foot protection
- Abrasive wheels

With the hazards faced by this sector it is imperative that safety and health be an integral part of doing business, with the specific purpose of protecting its employees.

REFERENCE

Bureau of Labor Statistics, U.S. Department of Labor, *Workplace Injuries and Illnesses in 2004* at http://bls.gov.

5 Accommodation and Food Services

Hotels come in many forms today from motels like this to small bed-and-breakfast inns.

The leisure and hospitality supersector comprises two sectors: the arts, entertainment, and recreation sector (sector 71), and the accommodation and food services sector (sector 72).

The accommodation and food service sector comprises establishments providing customers with lodging and preparing meals, snacks, and beverages for immediate consumption. The sector includes both accommodation and food services establishments because the two activities are often combined at the same establishment. The North American Industrial Classification System's (NAICS's) breakdown for the accommodation and food service sector is as follows:

Accommodation and food services (72)
    Accommodation (721000)
        Traveler accommodation (721100)
            Casino hotels (721120)
            RV (recreational vehicle) parks and recreational camps (721200)
            Rooming and boarding houses (721300)
        Full service and drinking places (722000)
            Full-service restaurants (722100)

Limited-service eating places (722200)
Special food services (722300)
Drinking places (alcoholic beverages) (722400)

5.1 HOTELS AND OTHER ACCOMMODATIONS

Service occupations, by far the largest occupational group, account for 65% of the industry's employment. Hotels employ many young workers and others in part-time and seasonal jobs. Average earnings are lower than in most other industries.

Hotels and other accommodations are as diverse as the many family and business travelers they accommodate. The industry includes all types of lodging, from upscale hotels to recreational vehicle (RV) parks. Motels, resorts, casino hotels, bed-and-breakfast inns, and boarding houses also are included. In fact, in 2004 nearly 62,000 establishments provided overnight accommodations to suit many different needs and budgets.

Establishments vary greatly in size and in the services they provide. Hotels and motels comprise the majority of establishments and tend to provide more services than other lodging places. There are five basic types of hotels—commercial, resort, residential, extended-stay, and casino. Most hotels and motels are commercial properties that cater mainly to business people, tourists, and other travelers who need accommodations for a brief stay. Commercial hotels and motels are usually located in cities or suburban areas and operate year round. Larger properties offer a variety of services for their guests, including a range of restaurant and beverage service options—from coffee bars and lunch counters to cocktail lounges and formal fine-dining restaurants. Some properties provide a variety of retail shops on the premises, such as gift boutiques, newsstands, drug and cosmetics counters, and barber and beauty shops. An increasing number of full-service hotels now offer guests access to laundry and valet services, swimming pools, and fitness centers or health spas. A small, but growing, number of luxury hotel chains also manage condominium units in combination with their transient rooms, providing both hotel guests and condominium owners with access to the same services and amenities.

Larger hotels and motels often have banquet rooms, exhibit halls, and spacious ballrooms to accommodate conventions, business meetings, wedding receptions, and other social gatherings. Conventions and business meetings are major sources of revenue for these hotels and motels. Some commercial hotels are known as conference hotels—fully self-contained entities specifically designed for meetings. They provide physical fitness and recreational facilities for meeting attendees, in addition to state-of-the-art audiovisual and technical equipment, a business center, and banquet services.

Resort hotels and motels offer luxurious surroundings with a variety of recreational facilities, such as swimming pools, golf courses, tennis courts, game rooms, and health spas, as well as planned social activities and entertainment. Resorts are typically located in vacation destinations or near natural settings, such as mountains, the seashore, theme parks, or other attractions. As a result, the business of many resorts fluctuates with the season. Some resort hotels and motels provide additional convention and conference facilities to encourage customers to combine business with pleasure. During the off-season, many of these establishments solicit conventions, sales meetings, and incentive tours to fill their otherwise empty rooms; some resorts even close for the off-season.

Residential hotels provide living quarters for permanent and semi permanent residents. Extended-stay hotels combine features of a resort and a residential hotel. Typically, guests use these hotels for a minimum of five consecutive nights. Casino hotels provide lodging in hotel facilities with a casino on the premises. The casino provides table-wagering games and may include other gambling activities, such as slot machines and sports betting. In addition to hotels and motels, bed-and-breakfast inns, recreational vehicle (RV) parks, campgrounds, and rooming and boarding houses provide lodging for overnight guests. Other short-term lodging facilities in this industry include guesthouses, or small cottages located on the same property as a main residence, and youth hostels—dormitory-style hotels with few frills, occupied mainly by students traveling on limited budgets.

Increases in competition and in the sophistication of travelers have induced the chains to provide lodging to serve a variety of customer budgets and accommodation preferences. In general, these lodging places may be grouped into properties that offer luxury, all-suite, moderately priced, and economy accommodations (see Figure 5.1).

Work in hotels and other accommodations can be demanding and hectic. Hotel staffs provide a variety of services to guests and must do so efficiently, courteously, and accurately. They must maintain a pleasant demeanor even during times of stress or when dealing with an impatient or irate guest. Alternately, work at slower times, such as the off-season or overnight periods, can seem slow and tiresome without the

FIGURE 5.1 Economic motels dot the country for travelers.

constant presence of hotel guests. Still, hotel workers must be ready to provide guests and visitors with gracious customer service at any hour.

Because hotels are open around the clock, employees frequently work varying shifts or variable schedules. Employees who work the late shift generally receive additional compensation. Many employees enjoy the opportunity to work part-time, nights or evenings, or other schedules that fit their availability for work and the hotel's needs. Hotel managers and many department supervisors may work regularly assigned schedules, but they also routinely work longer hours than scheduled, especially during peak travel times or when multiple events are scheduled. Also, they may be called in to work on short notice in the event of an emergency or to cover a position. Those who are self-employed, often owner-operators, tend to work long hours and often live at the establishment.

Food preparation and food service workers in hotels must withstand the strain of working during busy periods and being on their feet for many hours. Kitchen workers lift heavy pots and kettles and work near hot ovens and grills. Job hazards include slips and falls, cuts, and burns, but injuries are seldom serious. Food service workers often carry heavy trays of food, dishes, and glassware. Many of these workers work part time, including evenings, weekends, and holidays (Figure 5.2).

Office and administrative support workers generally work scheduled hours in an office setting, meeting with guests, clients, and hotel staff. Their work can become hectic processing orders and invoices, dealing with demanding guests, or servicing requests that require a quick turnaround, but job are hazards typically limited to muscle and eye strain common to working with computers and office equipment.

In 2003, work-related injuries and illnesses averaged 6.7 for every 100 full-time workers in hotels and other accommodations, compared with 5.0 for workers

FIGURE 5.2   Food preparation workers are exposed to many hazards.

throughout private industry. Work hazards include burns from hot equipment, sprained muscles and wrenched backs from heavy lifting, and falls on wet floors.

5.2   FOOD SERVICES AND DRINKING PLACES

Food services and drinking places provide many young people with their first jobs— in 2004, more than 21% of workers in these establishments were aged 16 through 19, about five times the proportion for all industries. Cooks, waiters and waitresses, and combined food preparation and serving workers comprised more than half of employment on the industry. About two of five employees worked parttime, more than twice the proportion for all industries.

Food services and drinking places may be the world's most widespread and familiar industry. These establishments include all types of restaurants, from casual fast-food eateries to formal, elegant dining establishments. The food services and drinking places industry comprises about 500,000 places of employment in large cities, small towns, and rural areas across the United States.

About 45% of establishments in this industry are limited-service eating places, such as fast-food restaurants, cafeterias, and snack and nonalcoholic beverage bars, which primarily serve patrons who order or select items and pay before eating. Full-service restaurants account for about 39% of establishments and cater to patrons who order, are served, and consume their food while seated, and then pay after eating. Drinking places (alcoholic beverages)—bars, pubs, nightclubs, and taverns—primarily prepare and serve alcoholic beverages for consumption on the premises. Drinking places comprise about 11% of all establishments in this industry. Special food services, such as food service contractors, caterers, and mobile food service vendors, account for less than 6% of establishments in the industry.

The most common type of a limited-service eating place is a franchised operation of a nationwide restaurant chain that sells fast food. Features that characterize these restaurants include a limited menu, the absence of waiters and waitresses, and emphasis on limited service. Menu selections usually offer limited variety and are prepared by workers with minimal cooking skills. Food is typically served in disposable, take-out containers that retain the food's warmth, allowing restaurants to prepare orders in advance of customers' requests. A growing number of fast-food restaurants provide drive-through and walk-up services.

Cafeterias are another type of limited-service eating place and usually offer a somewhat limited selection that varies daily. Full-service restaurants offer more menu categories, including appetizers, entrées, salads, side dishes, desserts, and beverages, and varied choices within each category. Chefs and cooks prepare items to order, which may run from grilling a simple hamburger to composing a more complex and sophisticated menu item. Waiters and waitresses offer table service in comfortable surroundings. Midscale or family-type restaurants offer a reasonable selection at a reasonable price compared to more elegant dining establishments. National chains are a growing segment of full-service restaurants. Some drinking places also offer patrons limited dining services in addition to providing alcoholic beverages. In some states, they also sell packaged alcoholic beverages for consumption off the premises.

Many food services and drinking places establishments in this industry are open for long hours. Staff are typically needed to work during evenings, weekends, and holiday hours. Full-time employees, often head or executive chefs and food service managers, typically work longer hours—12 h days are common—and also may be on call to work at other times when needed. Part-time employees, usually waiters and waitresses, dining-room attendants, hosts and hostesses, and fast-food employees, typically work shorter days (4–6 h/day) or fewer days per week than most full-time employees (see Figure 5.3).

Food services and drinking places employ more part-time workers than other industries. About two of five workers in food services and drinking places worked part time in 2004, more than twice the proportion for all industries. This allows some employees flexibility in setting their work hours, affording them a greater opportunity to tailor work schedules to personal or family needs. Some employees may rotate work on some shifts to ensure proper coverage at unpopular work times or to fully staff restaurants during peak demand times.

Food services and drinking places must comply with local fire, safety, and sanitation regulations. They must also provide appropriate public accommodations and ensure that employees use safe food handling measures. These practices require establishments to maintain supplies of chemicals, detergents, and other materials that may be harmful if not used properly.

Typical establishments have well-designed kitchens with state-of-the-art cooking and refrigeration equipment and proper electrical, lighting, and ventilation systems to keep everything functioning. However, kitchens are usually noisy and may be very hot near stoves, grills, ovens, or steam tables. Chefs, cooks, food preparation workers, and other kitchen staff, such as dishwashers, may suffer minor cuts or

FIGURE 5.3   Food preparation is often accomplished in very tight working conditions.

burns, be subject to scalding or steaming liquids, and spend most of their time standing in a relatively confined area. Chefs and cooks are under extreme pressure to work quickly to stay on top of orders in a busy restaurant. The fast pace requires employees to be alert and quick-thinking, but may also result in muscle strains from trying to move heavy pots or force pressurized containers open without safely taking the proper precautions.

Dining areas also may be well designed, but can become crowded and noisy when busy. Servers, attendants, and other dining-room staff, such as bartenders and hosts or hostesses, need to protect against falls, spills, or burns while serving diners and keeping service areas stocked. Also, dining-room staff must be aware of stairs, raised platforms, or other obstacles when directing patrons through narrow areas or to distant seating areas.

In most food services and drinking places, workers spend most of their time on their feet—preparing meals, serving diners, or transporting dishes and supplies throughout the establishment. Upper body strength is often needed to lift heavy items, such as trays of dishes, platters of food, or cooking pots. Work during peak dining hours can be very hectic and stressful.

Employees who have direct contact with customers, such as waiters and waitresses or hosts and hostesses, should have a neat appearance and maintain a professional and pleasant manner. Professional hospitality is required from the moment guests enter the restaurant until the time they leave. Sustaining a proper demeanor during busy times or over the course of a long shift may be difficult.

Kitchen staff also needs to be able to work as a team and to communicate with each other. Timing is critical in preparing complex dishes. Coordinating orders to ensure that an entire table's meals are ready at the same time is essential, particularly in a large restaurant during busy dining periods.

In 2003, the rate of work-related injuries and illnesses was 4.6 per 100 full-time workers in eating and drinking places, slightly less than the average of 5.0 for the private sector. Work hazards include the possibility of burns from hot equipment, sprained muscles and wrenched backs from heavy lifting, and falls on slippery floors.

5.3   PROFILE OF ACCOMMODATION AND FOOD SERVICES WORKERS' DEATHS, INJURIES, AND ILLNESSES

5.3.1   Deaths

There were 134 occupationally related deaths among accommodation and food service sector workers in 2005. The retail sector accounted for 4.9% of the service industry deaths (2736). Table 5.1 shows the percent values from each major category of those deaths.

5.3.2   Injuries

There were 77,620 reported injuries to accommodation and food service sector workers in 2004. This was 9% of the total injuries for the service industries (850,930). In Tables 5.2 through 5.5, the distributions of the nature, body part, source, and exposure (accident type) of the 77,620 injuries are presented.

TABLE 5.1

Occupational Death Cause by Percent for the Accommodation and Food Service Sector

Cause	Accommodation and Food Service Sector (%)
Highway	13
Homicides	63
Falls	7
Struck by	0

Source: From Bureau of Labor Statistics, United States Department of Labor. *Workplace Injuries and Illnesses in 2004.* Available at http://bls.gov.

5.3.3 ILLNESSES

In the accommodation and food service sector there were 9200 cases of occupationally related illnesses; this is 6% of the total for the service industry (see Table 5.6).

5.4 OCCUPATIONS

5.4.1 HOTELS AND OTHER ACCOMMODATIONS

The vast majority of workers in this industry—more than 8 out of 10 in 2004—were employed in service and office and administrative support occupations. Workers in

TABLE 5.2

Nature of Injury by Number and Percent for the Accommodation and Food Service Sector

Nature of Injury	Number	Percent
Sprains/strains[a]	25,780	33
Fractures[a]	4,350	5.6
Cuts/punctures[a]	12,740	16
Bruises[a]	7,450	10
Heat burns[a]	6,130	7.9
Chemical burns	330	0.4
Amputations	140	0.2
Carpal tunnel syndrome	540	0.7
Tendonitis	0	0
Multiple trauma	3,150	4
Back pain	2,480	3.2

Source: From Bureau of Labor Statistics, United States Department of Labor. *Workplace Injuries and Illnesses in 2004.* Available at http://bls.gov.

[a] Five most frequently occurring conditions.

TABLE 5.3
Body Part Injured by Number and Percent for the
Accommodation and Food Service Sector

Body Part Injured	Number	Percent
Head	4,440	5.7
Eyes	1,600	2
Neck	710	0.9
Trunk[a]	20,780	27
Back[a]	12,780	16
Shoulder	4,100	5
Upper extremities[a]	24,520	32
Finger[a]	10,350	13
Hand	6,090	7.8
Wrist	3,180	4
Lower extremities[a]	16,310	21
Knee	6,360	8.2
Foot, toe	3,440	4
Body systems	750	0.9
Multiple body parts	480	12

Source: From Bureau of Labor Statistics, United States Department of Labor. *Workplace Injuries and Illnesses in 2004.* Available at http://bls.gov.

[a] Five most frequently injured body parts.

TABLE 5.4
Source of Injury by Number and Percent for the
Accommodation and Food Service Sector

Sources of Injuries	Number	Percent
Parts and materials	1,240	1.6
Worker motion/position[a]	8,870	11
Floor, walkways, or ground surfaces[a]	21,010	27
Handtools[a]	6,430	8
Vehicles	3,360	4
Health care patients	0	0
Chemicals and chemical products	450	0.5
Containers[a]	3,010	3.9
Furniture and fixtures[a]	2,080	2.7
Machinery	1,040	1.3

Source: From Bureau of Labor Statistics, United States Department of Labor. *Workplace Injuries and Illnesses in 2004.* Available at http://bls.gov.

[a] Five most frequent sources of injury.

TABLE 5.5

Exposure/Accident Type by Number and Percent for the Accommodation and Food Service Sector

Type of Accidents	Number	Percent
Struck by an object[a]	12,090	16
Struck against an object	6,000	7.7
Caught in or compressed or crushed	1,950	2.5
Fall to lower level	2,880	9.4
Fall on same level[a]	19,670	25
Slips or trips without a fall	3,000	3.9
Overexertion[a]	12,380	16
Lifting[a]	8,020	10
Repetitive motion	1,400	1.4
Exposure to harmful substances or environment[a]	7,890	10
Transportation accidents	1,890	2.4
Fires and explosions	0	0
Assaults/violent acts	750	0.9

Source: From Bureau of Labor Statistics, United States Department of Labor. *Workplace Injuries and Illnesses in 2004.* Available at http://bls.gov.

[a] Five most frequent exposures or types of accidents that led to an injury.

these occupations usually learn their skills on the job. Postsecondary education is not required for most entry-level positions; however, college training may be helpful for advancement in some of these occupations. For many administrative support and service occupations, personality traits and a customer-service orientation may be more important than formal schooling. Traits most important for success in the hotel

TABLE 5.6

Occupational Illnesses by Number of Cases and Percent for the Accommodation and Food Service Sector

Illness Type	Number	Percent
Skin diseases and disorders	3300	36
Respiratory conditions	800	8.7
Poisoning	100	1
Hearing loss	0	0
Others	5000	54

Source: From Bureau of Labor Statistics, United States Department of Labor. *Workplace Injuries and Illnesses in 2004.* Available at http://bls.gov.

and motel industry are good communication skills; the ability to get along with people in stressful situations; a neat, clean appearance; and a pleasant manner.

Employment is concentrated in densely populated cities and resort areas. Compared with establishments in other industries, hotels, motels, and other lodging places tend to be small. About 91% employ fewer than 50 people and about 56% employ fewer than 10 workers. As a result, lodging establishments offer opportunities for those who are interested in owning and running their own business. Although establishments tend to be small, the majority of jobs are in larger hotels and motels with more than 100 employees.

Hotels and other lodging places often provide first jobs to many new entrants to the labor force. As a result, many of the industry's workers are young. In 2004, about 19% of the workers were younger than age 25, compared with about 14% across all industries.

Service occupations, by far the largest occupational group in the industry, account for 65% of the industry's employment. Most service jobs are in housekeeping occupations—including maids and housekeeping cleaners, janitors and cleaners, and laundry workers—and in food preparation and service jobs—including chefs and cooks, waiters and waitresses, bartenders, fast-food and counter workers, and various other kitchen and dining-room workers. The industry also employs many baggage porters and bellhops, gaming services workers, and grounds maintenance workers. Other occupations found in the accommodation sector are building cleaning workers; chefs, cooks, and other food preparation workers; food and beverage serving and related workers; food service managers; hotel, motel, and resort desk clerks; gaming cage workers; gaming services occupations; lodging managers; recreation and fitness workers; and security guards and gaming surveillance officers.

Workers in cleaning and housekeeping occupations ensure that the lodging facility is clean and in good condition for the comfort and safety of guests. Maids and housekeepers clean lobbies, halls, guestrooms, and bathrooms. Janitors help with the cleaning of the public areas of the facility, empty trash, and perform minor maintenance work. Workers in the various food service occupations deal with customers in the dining room or at a service counter. Waiters and waitresses take customers' orders, serve meals, and prepare checks. Hosts and hostesses welcome guests, show them to their tables, and give them menus. Bartenders fill beverage orders for customers seated at the bar or from waiters and waitresses who serve patrons at tables. Cooks and food preparation workers prepare food in the kitchen. Many full-service hotels employ a uniformed staff to assist arriving and departing guests. Baggage porters and bellhops carry bags and escort guests to their rooms. Concierges arrange special or personal services for guests. Doorkeepers help guests into and out of their cars, summon taxis, and carry baggage into the hotel lobby. Hotels also employ the largest percentage of gaming services workers because much of gaming takes place in casino hotels.

Office and administrative support positions accounted for 18% of the jobs in hotels and other accommodations in 2004. Hotel desk clerks, secretaries, bookkeeping and accounting clerks, and telephone operators ensure that the front office operates smoothly.

Hotels and lodging managers or general and operations managers in large hotels often have several other lodging places employing many different types of managers to direct and coordinate the activities of the front office, kitchen, dining room, and other departments, such as housekeeping, accounting, personnel, purchasing, publicity, sales, security, and maintenance.

Hotels and other lodging places employ a variety of workers found in many other industries. Maintenance workers, such as stationary engineers, plumbers, and painters, fix leaky faucets, do some painting and carpentry, see that heating and air-conditioning equipment works properly, mow lawns, and exterminate pests. The industry also employs cashiers, accountants, personnel workers, entertainers, and recreation workers. Also, many additional workers inside a hotel may work for other companies under contract to the hotel or may provide personal or retail services directly to hotel guests from space rented by the hotel. This group includes guards and security officers, barbers, cosmetologists, fitness trainers and aerobics instructors, valets, gardeners, and parking attendants.

Although the skills and experience needed by workers in this industry depend on the specific occupation, most entry-level jobs require little or no previous training. Basic tasks usually can be learned in a short time. Almost all workers in the hotel and other accommodations industries undergo on-the-job training, which is usually provided under the supervision of an experienced employee or manager. Some large chain operations have formal training sessions for new employees; many also provide video or online training.

5.4.2 FOOD SERVICES AND DRINKING PLACES

The food services and drinking places industry, with about 8.9 million wage and salary jobs in 2004, ranks among the nation's leading employers. Food services and drinking places tend to be small; about 72% of the establishments in the industry employ fewer than 20 workers. As a result, this industry is often considered attractive to individuals who want to own and run their own businesses. An estimated 248,000 self-employed and unpaid family workers are employed in the industry, representing about 3% of total employment.

Establishments in this industry, particularly fast-food establishments, are leading employers of teenagers—aged 16 through 19—providing first jobs for many new entrants to the labor force. In 2004, about 21% of all workers in food services and drinking places were teenagers, about five times the proportion in all industries. About 45% were under age 25, more than three times the proportion in all industries.

Workers in this industry perform a variety of tasks. They prepare food items from a menu or according to a customer's order, keep food preparation and service areas clean, accept payment from customers, and provide the establishment managerial or office services, such as bookkeeping, ordering, and advertising. Cooks, waiters and waitresses, and combined food preparation and serving workers account for more than half of food services jobs.

Employees in the various food services and related occupations deal with customers in a dining area or at a service counter. Waiters and waitresses take customers' orders, serve food and beverages, and prepare itemized checks. In fine-dining restaurants, they may describe chef's specials and take alcoholic beverage orders. In some establishments, they escort customers to their seats, accept payments, and set up and clear tables. In many larger restaurants, however, these tasks may be assigned to, or shared with, other workers.

Other food services occupations include hosts and hostesses, who welcome customers, show them to their tables, and offer them menus. Bartenders fill drink orders for waiters and waitresses and from customers seated at the bar. Dining-room attendants and bartender helpers assist waiters, waitresses, and bartenders by clearing, cleaning, and setting up tables, as well as keeping service areas stocked with supplies. Counter attendants take orders and serve food at counters, cafeteria steam tables, and fast-food counters. Depending on the size and type of establishment, attendants may also operate cash registers.

Combined food preparation and serving workers, including fast food, prepare and serve items in fast-food restaurants. Most take orders from customers at counters or drive-through windows at fast-food restaurants. They assemble orders, hand them to customers, and accept payment. Many of these workers also cook and package food, make coffee, and fill beverage cups using drink-dispensing machines.

Workers in the various food preparation occupations prepare food in the kitchen. Institution and cafeteria cooks work in the kitchens of schools, hospitals, industrial cafeterias, and other institutions where they prepare large quantities of a small variety of menu items. Restaurant cooks usually prepare a wider selection of dishes for each meal, cooking individual servings to order. Short-order cooks prepare grilled items and sandwiches in establishments that emphasize fast service. Fast-food cooks prepare and package a limited selection of food that is either prepared to order or kept warm until sold in fast-food restaurants. Food preparation workers clean and prepare basic food ingredients, such as meats, fish, and vegetables for use in making complex meals, keep work areas clean, and perform simple cooking tasks under the direction of the chef or head cook. Dishwashers clean dishes, glasses, pots, and kitchen accessories by hand or by machine.

Food service managers hire, train, supervise, and discharge workers in food services and drinking places establishments. They also purchase supplies, deal with vendors, keep records, and help whenever an extra hand is needed. Executive chefs oversee the kitchen, select the menu, train cooks and food preparation workers, and direct the preparation of food. In fine-dining establishments, maitre d's may serve as hosts or hostesses while overseeing the dining room. Larger establishments may employ general managers, as well as a number of assistant managers. Many managers and executive chefs are part owners of the establishments they manage.

Food services and drinking places may employ a wide range of other workers, including accountants, advertising and public relations workers, bookkeepers, dietitians, mechanics and other maintenance workers, musicians and other entertainers, human resources workers, and various clerks. However, many establishments may choose to contract this work to outside establishments who also perform these tasks for several food services and drinking places outlets.

The skills and experience required by workers in food services and drinking places differ by occupation and the type of establishment. Many entry-level positions, such as waiters and waitresses or food preparation workers, require little or no formal education or previous training. Similarly, work in limited-service eating places generally requires less experience than work in full-service restaurants.

Many fast-food worker or server jobs are held by young or part-time workers. For many youths, this is their first job; for others, part-time schedules allow flexible working arrangements. On-the-job training, typically under the close supervision of an experienced employee or manager, often lasts a few weeks or less. Some large chain operations require formal training sessions, many using online or video training programs, for new employees.

5.5 APPLICABLE OSHA REGULATIONS

Another way to gather an understanding of the hazards faced by retail workers is to see the types of violations that Occupational Safety and Health Administration

TABLE 5.7
Twenty-Five Most Frequent OSHA Violations for Accommodation Sector

Hotels, Rooming Houses, Camps, and Other Lodging Places (70)		
CFR Standard	**Number Cited**	**Description**
1910.1200	62	Hazard communication
1910.303	24	Electrical systems design, general requirements
1910.305	23	Electrical, wiring methods, components and equipment
1910.1030	21	Bloodborne pathogens
1910.134	19	Respiratory protection
1910.157	18	Portable fire extinguishers
1910.215	18	Abrasive wheel machinery
1910.132	15	Personal protective equipment, general requirements
1926.1101	15	Asbestos
1910.151	14	Medical services and first aid
1926.62	13	Lead
1910.23	9	Guarding floor and wall openings and holes
1910.37	9	Maintenance, safeguards, and operation features for exit routes
1910.213	8	Woodworking machinery requirements
1910.1001	8	Asbestos
1910.36	6	Design and construction requirements for exit routes
1910.178	6	Powered industrial trucks (forklifts)
1910.253	6	Oxygen-fuel gas welding and cutting
1926.451	6	Scaffolds, general requirements
5A1	6	General duty clause (section of OSHA Act)
1910.212	5	Machines, general requirements
1910.304	5	Electrical, wiring design and protection
1904.29	4	Forms
1910.22	4	Working/walking surfaces, general requirements
1910.67	4	Vehicle-mounted elevating/rotating work platforms

Note: Standards cited by federal OSHA for the accommodation services sector from October 2005 to September 2006 are included here.

(OSHA) have found during their inspections of retail establishments. These violations provide another way of targeting hazards that have the potential to cause injury, illness, and death of workers. As can be seen from the 25 most frequently cited violations for both the accommodation and food services sector, OSHA cites this industry under the general industry standard (29 CFR 1910) and the recordkeeping standard (29 CFR 1904) (see Tables 5.7 and 5.8).

Although the previous were the 25 most frequently issued violations, OSHA has cited other hazards with less frequency. Some of these are as follows:

- Hand protection
- Scaffolding
- Material handling

TABLE 5.8
Twenty-Five Most Frequent OSHA Violations for Food Services Sector

Eating and Drinking Places

CFR Standard	Number Cited	Description
1910.1200	284	Hazard communication
1910.305	92	Electrical, wiring methods, components and equipment
1910.132	19	Personal protective equipment, general requirements
1910.157	17	Portable fire extinguishers
1910.22	16	Working/walking surfaces, general requirements
1910.37	15	Maintenance, safeguards, and operation features for exit routes
1910.151	14	Medical services and first aid
1910.303	13	Electrical systems design, general requirements
1910.133	10	Eye and face protection
1910.212	7	Machines, general requirements
1910.304	7	Electrical, wiring design and protection
1910.101	4	Compressed gases, general requirements
1910.138	4	Hand protection
1910.147	4	The control of hazardous energy, lockout/tagout
1903.19	3	Abatement verification
1910.141	3	Sanitation
1910.334	3	Electrical, use of equipment
1926.451	3	Scaffolds, general requirements
1904.29	2	Forms
1910.38	2	Emergency action Plans
1910.103	2	Hydrogen
1910.253	2	Oxygen-fuel gas welding and cutting
5A1	2	General duty clause (section of OSHA Act)
1903.2	1	Posting of notice, availability of the act, regulations, and applicable standards
1904.2	1	Partial exemption for establishments in certain industries

Note: Standards cited by federal OSHA for the food services sector from October 2005 to September 2006 are included here.

- Head protection
- Eye and face protection
- Compressed gases
- Hazardous locations
- Asbestos
- Fall protection
- Automatic sprinklers
- Ladders
- Fixed ladders
- Fixed industrial stairs
- Lockout/tagout
- Safeguard for personnel protection
- Bakery equipment

With the hazards faced by this sector, it is imperative that safety and health be an integral part of doing business, with the specific purpose of protecting its employees.

REFERENCE

Bureau of Labor Statistics, United States Department of Labor, *Workplace Injuries and Illnesses in 2004*. Available at http://bls.gov.

6 Other Services

Automobile repair is one of the primary other services provided by the service industry.

The other services (81) sector comprises establishments engaged in providing services not specifically provided for elsewhere in the North American Industry Classification System (NAICS). Establishments in this sector are primarily engaged in activities such as equipment and machinery repairing, promoting or administering religious activities, grantmaking, advocacy, and providing dry cleaning and laundry services, personal care services, death care services, pet care services, photofinishing services, temporary parking services, and dating services.

Other services represent about 3.3% of all employment yet account for 12.9% of all establishments. Estimates show that the annual average employment in other services during 2005 was 5,386,000. The NAICS classification breakdown for the other services sector is as follows:

Other services (except public administration) (81)
   Repair and maintenance (811000)
         Automotive repair and maintenance (811100)
         Automotive mechanical and electrical repair and maintenance (811110)
         Automotive body, paint, interior and glass repair (811120)
         Other automotive repair and maintenance (811190)
      Electronic and precision equipment repair and maintenance (811200)
      Commercial and industrial machinery and equipment (except automotive and electronic) repair and maintenance (811300)
      Personal and household goods repair and maintenance (811400)

Personal and laundry services (811200)
   Personal care services (812100)
   Death services (812200)
   Dry cleaning and laundry services (812300)
   Other personal services (812900)
Religious, grantmaking, civic, professional, and similar organizations
   (813000)
   Religious organizations (813100)
   Grantmaking and giving services (813200)
   Social advocacy organizations (813300)
   Civic and social organizations (813400)
   Business, professional, labor, political and similar organizations (813900)
      Labor unions and similar labor organizations (813930)

In order to try to give some ideas as to the type of work and the hazards that exist in the other services sector, a description of two of the groups found in this sector follows.

6.1   MAINTENANCE AND REPAIR WORKERS—GENERAL

General maintenance and repair workers are employed in almost every industry. Many workers learn their skills informally on the job; others learn by working as helpers to other repairers or to construction workers such as carpenters, electricians, or machinery repairers.

Most craft workers specialize in one kind of work, such as plumbing or carpentry. General maintenance and repair workers, however, have skills in many different crafts. They repair and maintain machines, mechanical equipment, and buildings and work on plumbing, electrical, and air-conditioning and heating systems. They build partitions, make plaster or drywall repairs, and fix or paint roofs, windows, doors, floors, woodwork, and other parts of building structures. They also maintain and repair specialized equipment and machinery found in cafeterias, laundries, hospitals, stores, offices, and factories. Typical duties include troubleshooting and fixing faulty electrical switches, repairing air-conditioning motors, and unclogging drains. New buildings sometimes have computer-controlled systems that allow maintenance workers to make adjustments in building settings and monitor for problems from a central location. For example, they can remotely control light sensors that turn off lights automatically after a set amount of time or identify a broken ventilation fan that needs to be replaced.

General maintenance and repair workers inspect and diagnose problems and determine the best way to correct them, frequently checking blueprints, repair manuals, and parts catalogs. They obtain supplies and repair parts from distributors or storerooms. Using common hand and power tools such as screwdrivers, saws, drills, wrenches, and hammers, as well as specialized equipment and electronic testing devices, these workers replace or fix worn or broken parts, where necessary, or make adjustments to correct malfunctioning equipment and machines (see Figure 6.1).

General maintenance and repair workers also perform routine preventive maintenance and ensure that machines continue to run smoothly, building systems operate

FIGURE 6.1 Repair of electrical and electronics equipment is undertaken by a worker.

efficiently, and the physical condition of buildings does not deteriorate. Following a checklist, they may inspect drives, motors, and belts, check fluid levels, replace filters, and perform other maintenance actions. Maintenance and repair workers keep records of their work.

Employees in small establishments, where they are often the only maintenance worker, make all repairs, except for very large or difficult jobs. In larger establishments, their duties may be limited to the general maintenance of everything in a workshop or a particular area.

General maintenance and repair workers often carry out several different tasks in a single day, at any number of locations. They may work inside of a single building or in several different buildings. They may have to stand for long periods, lift heavy objects, and work in uncomfortably hot or cold environments, in awkward and cramped positions, or on ladders. They are subject to electrical shocks, burns, falls, cuts, and bruises. The nonfatal injuries and illnesses incidence rate was 3.2 per 100 full-time workers in other services and 4.8 per 100 full-time workers in all private industry.

Most general maintenance workers work a 40 h week. Some work evening, night, or weekend shifts or are on call for emergency repairs. Those employed in small establishments often operate with only limited supervision. Those working in larger establishments frequently are under the direct supervision of an experienced worker.

Many general maintenance and repair workers learn their skills informally on the job. They start as helpers, watching and learning from skilled maintenance workers. Helpers begin by doing simple jobs, such as fixing leaky faucets and replacing light bulbs and progress to more difficult tasks, such as overhauling machinery or building walls. Some learn their skills by working as helpers to other repair or construction workers, including carpenters, electricians, or machinery repairers.

6.2   BUILDING CLEANING WORKERS

This very large occupation requires few skills to enter and has one of the largest numbers of job openings of any occupation each year. Most job openings result from the need to replace the many workers who leave these jobs because of their limited opportunities for training or advancement, low pay, and high incidence of only part-time or temporary work.

Building cleaning workers—including janitors, maids, housekeeping cleaners, window washers, and rug shampooers—keep office buildings, hospitals, stores, apartment houses, hotels, and residences clean, sanitary, and in good condition. Some do only cleaning, while others have a wide range of duties.

Janitors and cleaners perform a variety of heavy cleaning duties, such as cleaning floors, shampooing rugs, washing walls and glass, and removing rubbish. They may fix leaky faucets, empty trashcans, do painting and carpentry, replenish bathroom supplies, mow lawns, and see that heating and air-conditioning equipment work properly. On a typical day, janitors may wet- or dry-mop floors, clean bathrooms, vacuum carpets, dust furniture, make minor repairs, and exterminate insects and rodents. They also clean snow or debris from sidewalks in front of buildings and notify management of the need for major repairs. While janitors typically perform most of the duties mentioned, cleaners tend to work for companies that specialize in one type of cleaning activity, such as washing windows.

Maids and housekeeping cleaners perform any combination of light cleaning duties to keep private households or commercial establishments such as hotels, restaurants, hospitals, and nursing homes clean and orderly. In hotels, aside from cleaning and maintaining the premises, maids and housekeeping cleaners may deliver ironing boards, cribs, and rollaway beds to guests' rooms. In hospitals, they also may wash bed frames, brush mattresses, make beds, and disinfect and sterilize equipment and supplies with germicides and sterilizing equipment (see Figure 6.2).

Janitors, maids, and cleaners use many kinds of equipment, tools, and cleaning materials. For one job they may need standard cleaning implements; another may require an electric polishing machine and a special cleaning solution. Improved building materials, chemical cleaners, and power equipment have made many tasks easier and less time consuming, but cleaning workers must learn the proper use of equipment and cleaners to avoid harming floors, fixtures, and themselves.

Cleaning supervisors coordinate, schedule, and supervise the activities of janitors and cleaners. They assign tasks and inspect building areas to see that work has been done properly; they also issue supplies and equipment and inventory stocks to ensure that supplies on hand are adequate. They also screen and hire job applicants; train new and experienced employees; and recommend promotions, transfers, or dismissals. Supervisors may prepare reports concerning the occupancy of rooms, hours worked, and department expenses. Some also perform cleaning duties.

Cleaners and servants in private households dust and polish furniture; sweep, mop, and wax floors; vacuum; and clean ovens, refrigerators, and bathrooms. They may also wash dishes, polish silver, and change and make beds. Some wash, fold, and iron clothes; a few wash windows. General house workers may also take clothes and laundry to the cleaners, buy groceries, and perform many other errands.

FIGURE 6.2 Janitorial services in company cafeteria.

Building cleaning workers in large office and residential buildings, and more recently in large hotels, often work in teams consisting of workers who specialize in vacuuming, picking up trash, and cleaning restrooms, among other things. Supervisors conduct inspections to ensure that the building is cleaned properly and the team is functioning efficiently. In hotels, one member of the team is responsible for reporting electronically to the supervisor when rooms are cleaned.

Because most office buildings are cleaned while they are empty, many cleaning workers work evening hours. Some, however, such as school and hospital custodians, work in the daytime. When there is a need for 24 h maintenance, janitors may be assigned to shifts. Most full-time building cleaners work about 40 h a week. Part-time cleaners usually work in the evenings and on weekends.

Building cleaning workers usually work inside heated, well-lighted buildings. However, they sometimes work outdoors, sweeping walkways, mowing lawns, or shoveling snow. Working with machines can be noisy, and some tasks, such as cleaning bathrooms and trash rooms, can be dirty and unpleasant. Janitors may suffer cuts, bruises, and burns from machines, hand tools, and chemicals. They spend most of their time on their feet, sometimes lifting or pushing heavy furniture or equipment. Many tasks, such as dusting or sweeping, require constant bending, stooping, and stretching. As a result, janitors may also suffer back injuries and sprains.

No special education is required for most janitorial or cleaning jobs, but beginners should know simple arithmetic and be able to follow instructions. High-school shop courses are helpful for jobs involving repair work. Most building cleaners learn their skills on the job. Beginners usually work with an experienced cleaner, doing routine cleaning.

TABLE 6.1

Occupational Death Cause by Percent

for Other Services

Cause	Other Services (%)
Highway	16
Homicides	20
Falls	11
Struck by	13

Source: From Bureau of Labor Statistics, United States Department of Labor. *Workplace Injuries and Illnesses in 2004.* Available at http://bls.gov.

6.3   PROFILE OF OTHER SERVICES WORKERS' DEATHS, INJURIES, AND ILLNESSES

6.3.1   DEATHS

There were 208 occupationally related deaths among other services workers in 2005. The other services sector accounted for 7.6% of the service industry deaths (2736). Table 6.1 shows the percent values from each major category of those deaths.

6.3.2   INJURIES

There were 31,350 reported injuries to other services workers in 2004. This was 3.7% of the total injuries for the service industries (850,930). In Tables 6.2 through 6.5,

TABLE 6.2

Nature of Injury by Number and Percent for Other Services

Nature of Injury	Number	Percent
Sprains/strains[a]	11,880	38
Fractures[a]	2,230	7
Cuts/punctures[a]	2,940	9
Bruises[a]	2,450	7.8
Heat burns	660	2
Chemical burns	190	0.6
Amputations	470	1.5
Carpal tunnel syndrome	360	1
Tendonitis	140	0.4
Multiple trauma[a]	1,460	4.7
Back pain	1,130	3.6

Source: From Bureau of Labor Statistics, United States Department of Labor. *Workplace Injuries and Illnesses in 2004.* Available at http://bls.gov.

[a] Five most frequently occurring conditions.

TABLE 6.3
Body Part Injured by Number and Percent for Other Services

Body Part Injured	Number	Percent
Head	2,600	8
Eyes	1,340	4
Neck	560	1
Trunk[a]	10,590	33
Back[a]	6,600	21
Shoulder	2,080	6.6
Upper extremities[a]	7,130	23
Finger	2,670	8.5
Hand	1,470	4.7
Wrist	1,320	4
Lower extremities[a]	6,510	21
Knee	2,550	8
Foot, toe	1,140	3.6
Body systems	410	1.3
Multiple body parts[a]	3,220	10

Source: From Bureau of Labor Statistics, United States Department of Labor. *Workplace Injuries and Illnesses in 2004.* Available at http://bls.gov.
[a] Five most frequently injured body parts.

TABLE 6.4
Source of Injury by Number and Percent for Other Services

Sources of Injuries	Number	Percent
Parts and materials[a]	3500	11
Worker motion/position[a]	5340	17
Floor, walkways, or ground surfaces[a]	6110	19
Handtools	1760	5.6
Vehicles[a]	3360	12
Health care patients	260	0.8
Chemicals and chemical products	570	1.8
Containers[a]	2540	8
Furniture and fixtures	830	2.6
Machinery	1820	5.8

Source: From Bureau of Labor Statistics, United States Department of Labor. *Workplace Injuries and Illnesses in 2004.* Available at http://bls.gov.
[a] Five most frequent sources of injury.

TABLE 6.5
Exposure/Accident Type by Number and Percent for Other Services

Type of Accidents	Number	Percent
Struck by an object[a]	4600	15
Struck against an object	1920	6
Caught in or compressed or crushed	1280	4
Fall to lower level	1830	5.8
Fall on same level[a]	4470	14
Slips or trips without a fall	97	0.3
Overexertion[a]	5970	19
Lifting[a]	3470	11
Repetitive motion	1210	3.9
Exposure to harmful substances or environment	1540	2.9
Transportation accident[a]	2060	6.5
Fires and explosions	80	0.3
Assaults/violent acts	690	2

Source: From Bureau of Labor Statistics, United States Department of Labor. *Workplace Injuries and Illnesses in 2004.* Available at http://bls.gov.

[a] Five most frequent exposures or types of accidents that led to an injury.

the distributions of the nature, body part, source, and exposure (accident type) of the 31,350 injuries are presented.

6.3.3   ILLNESSES

In the other services sector there were 3500 cases of occupationally related illnesses; this is 2.7% of the total for the service industry (see Table 6.6).

TABLE 6.6
Occupational Illnesses by Number of Cases and Percent for Other Services

Illness Type	Number	Percent
Skin diseases and disorders	800	23
Respiratory conditions	500	14
Poisoning	100	3
Hearing loss	0	0
All others	2100	60

Source: From Bureau of Labor Statistics, United States Department of Labor. *Workplace Injuries and Illnesses in 2004.* Available at http://bls.gov.

6.4 HAZARDS FACED BY OTHER SERVICES WORKERS

Performing all types of tasks, many of which other workers would not want to do, creates the dangers faced by other services workers. This includes all kinds of maintenance and cleaning tasks that are carried out behind the scenes with little interaction with the clients or patrons of the business.

The hazards covered in this book are the primary ones that affect other services workers found in facility and stores operations. In most cases, the most frequent hazards faced by other services' workers are

- Walking and working surfaces
- Electrocutions
- Material handling/lifting of containers
- Slips, trips, and falls
- Strains/sprains
- Trauma injuries
- Fall from elevation
- Fires
- Hand and power tools
- Hazardous chemicals
- Biological hazards
- Cleaning agents (Figure 6.3)
- Repetitive/cumulative trauma

FIGURE 6.3 Workers in a dry cleaner operation have a potential for exposure to hazardous chemicals.

6.5 OCCUPATIONS

Generally, these are not the most glamorous types of occupations. Many times workers are jack-of-all-trades or trade trained for specific jobs.

Some duties of general maintenance and repair workers are similar to those of carpenters; pipelayers, plumbers, pipe fitters, and steamfitters; electricians; and heating, air-conditioning, and refrigeration mechanics. Other duties are similar to those of coin, vending, and amusement machine servicers and repairers; electrical and electronics installers and repairers; electronic home entertainment equipment installers and repairers; radio and telecommunications equipment installers and repairers. Workers who specialize in one of the many job functions of janitors and cleaners include pest control workers; general maintenance and repair workers; and grounds maintenance workers.

The other services workers often perform tasks and jobs the construction workers do and many of the same hazards.

6.6 APPLICABLE OSHA REGULATIONS

Another way to gather an understanding of the hazards faced by other services workers is to see the types of violations that Occupational Safety and Health Administration (OSHA) have found during their inspections of retail establishments. These violations provide another way of targeting hazards that have the potential to cause injury, illness, and death of workers. As can be seen from the 25 most frequently cited violations, OSHA cites this industry under the general industry standard (29 CFR 1910) and the recordkeeping standard (29 CFR 1904) (see Tables 6.7 through 6.9).

TABLE 6.7
Twenty-Five Most Frequent OSHA Violations for Personal Services Subsector

CFR Standard	Number Cited	Description
1910.1200	73	Hazard communication
1910.1030	66	Bloodborne pathogens
1910.132	23	Personal protective equipment, general requirements
1910.141	19	Sanitation
1910.305	19	Electrical, wiring methods, components and equipment
1910.147	16	The control of hazardous energy, lockout/tagout
1910.134	14	Respiratory protection
1910.1048	14	Formaldehyde
1910.151	13	Medical services and first aid
1910.219	13	Mechanical power-transmission apparatus
1910.303	13	Electrical systems design, general requirements
1910.146	11	Permit-required confined spaces
1910.157	10	Portable fire extinguishers
1910.22	9	Working/walking surfaces, general requirements
1910.23	9	Guarding floor and wall openings and holes
1910.37	9	Maintenance, safeguards, and operational features of exit routes

TABLE 6.7 (continued)

Twenty-Five Most Frequent OSHA Violations for Personal Services Subsector

CFR Standard	Number Cited	Description
1910.178	9	Powered industrial trucks (forklifts)
1910.264	9	Laundry machinery and operations
1910.212	8	Machines, general requirements
1910.95	6	Occupational noise exposure
5A1	6	General duty clause (section of OSHA Act)
1910.133	5	Eye and face protection
1910.304	5	Electrical, wiring design, and protection
1910.36	4	Design and construction requirements for exit routes
1910.138	4	Hand protection
1910.215	4	Abrasive wheel machinery

Note: Standards cited by federal OSHA for the personal services subsector from October 2005 to September 2006 are included here.

TABLE 6.8

Twenty-Five Most Frequent OSHA Violations for Automobile Repair, Services, and Parking Subsector

CFR Standard	Number Cited	Description
1910.134	647	Respiratory protection
1910.1200	393	Hazard communication
1910.1000	123	Air contaminants
1910.132	118	Personal protective equipment, general requirements
1910.305	63	Electrical, wiring methods, components and equipment
1910.107	47	Spray finishing w/flammable/combustible materials
1910.151	47	Medical services and first aid
1910.303	47	Electrical systems design, general requirements
1910.157	40	Portable fire extinguishers
1910.22	30	Working/walking surfaces, general requirements
1910.215	29	Abrasive wheel machinery
1910.106	28	Flammable and combustible liquids
5A1	28	General duty clause (section of OSHA Act)
1910.133	24	Eye and face protection
1910.147	22	The control of hazardous energy, lockout/tagout
1910.138	21	Hand protection
1910.253	21	Oxygen-fuel gas welding and cutting
1910.23	20	Guarding floor and wall openings and holes
1910.141	16	Sanitation
1910.178	14	Powered industrial trucks (forklifts)
1910.146	13	Permit-required confined spaces

(continued)

TABLE 6.8 (continued)
Twenty-Five Most Frequent OSHA Violations for Automobile Repair, Services, and Parking Subsector

CFR Standard	Number Cited	Description
1910.212	13	Machines, general requirements
1910.1025	12	Lead
1910.37	11	Maintenance, safeguards, and operational features of exit routes
1910.242	10	Hand and portable-powered tools and equipment
1910.177	9	Servicing multipiece and single-piece rim wheels
1910.244	9	Other portable tools and equipment
1910.304	9	Electrical, wiring design and protection
1910.95	8	Occupational noise exposure
1910.219	8	Mechanical power-transmission apparatus
1910.24	7	Fixed industrial stairs
1910.334	7	Electrical, use of equipment
1904.2	6	Exemption for establishments partial in certain industries

Note: Standards cited by federal OSHA for the automobile repair, services, and parking subsector from October 2005 to September 2006 are included here.

TABLE 6.9
Twenty-Five Most Frequent OSHA Violations for Miscellaneous Repair Subsector

CFR Standard	Number Cited	Description
1910.146	94	Permit-required confined spaces
1910.1200	62	Hazard communication
1910.134	48	Respiratory protection
1910.1052	40	Methylene chloride
1910.178	28	Powered industrial trucks (forklifts)
1910.132	21	Personal protective equipment, general requirements
1910.215	16	Abrasive wheel machinery
1910.1027	16	Cadmium
1910.147	14	The control of hazardous energy, lockout/tagout
1910.253	14	Oxygen-fuel gas welding and cutting
1910.305	14	Electrical, wiring methods, components and equipment
1910.1025	14	Lead
1910.157	12	Portable fire extinguishers
1910.212	12	Machines, general requirements
5A1	9	General duty clause (section of OSHA Act)
1910.303	8	Electrical systems design, general requirements
1910.37	7	Maintenance, safeguards, and operational features of exit routes
1910.133	7	Eye and face protection
1910.22	6	Working/walking surfaces, general requirements

TABLE 6.9 (continued)
Twenty-Five Most Frequent OSHA Violations for Miscellaneous
Repair Subsector

CFR Standard	Number Cited	Description
1910.23	6	Guarding floor and wall openings and holes
1910.106	6	Flammable and combustible liquids
1910.107	6	Spray finishing w/flammable/combustible materials
1910.151	6	Medical services and first aid
1904.29	5	Forms
1910.177	4	Servicing multipiece and single-piece rim wheels
1910.213	4	Woodworking machinery requirements
1910.217	4	Mechanical power presses

Note: Standards cited by federal OSHA for the miscellaneous repair subsector from October 2005 to September 2006 are included here.

Although the previous were the 25 most frequently issued violations for these three subsectors (Personal services; automobile repair, services, and parking; miscellaneous repair services), OSHA has cited other hazards with less frequency. With the hazards faced by this sector, it is imperative that safety and health be an integral part of doing business, with the specific purpose of protecting its employees.

REFERENCE

Bureau of Labor Statistics, U.S. Department of Labor, *Workplace Injuries and Illnesses in 2004*. Available at http://bls.gov.

7 Managing Safety and Health People Service Sectors

Effective management will go a long way toward preventing incidents like this.

Managing the safety and health functions should be accomplished like any other aspect of doing business. First, management is ultimately accountable and responsible for its workforce's safety and health on the job. The key to establishing an effective safety and health effort is the implementation of the many facets of management techniques for safety and health, the least of which is planning and employing an organized approach. It is imperative that you must identify your problem areas (hazards) and make an honest effort to address them or control them in some way. This is especially true for the people service sectors.

It should be apparent to you that everyone has a role to play relevant to occupational safety and health. An atmosphere of genuine cooperation with regard to safety and health must prevail. The welfare of those working in your people service environment must be the common theme that everyone can support. You certainly want everyone to go to their homes after work with the same number of fingers and toes and not feeling as though they are ill. The same should be said for their clients, patients, customers, patrons, and students.

It is paramount that everyone be involved in ensuring that their workplace is safe, healthy, and free from hazards that might make them sick or cause them physical harm. This goal cannot be accomplished by management or supervisors on their own. It must involve those who perform the work. After all, who knows the most about the work they have been assigned to accomplish and especially the dangers of doing their job than the employees themselves.

You need to address all of the many facets of developing and implementing a functional safety and health effort. All safety and health management techniques need to be addressed as an integral part of the overall safety and health approach employed at your workplace. This includes the process, techniques, and the people side of managing safety and health.

It is imperative when addressing safety and health that you really mean what you say and you support your efforts by every action, word, and deed. If you drop the ball on any part of your support for safety and health, the perception will exist that you have only given lip service to safety and health and really do not care about your workforce.

Employees/workers need to feel that they are of value to their employers. How many companies do you know who say, "Our employees are our most valuable asset," but whose actions do not convey that message. Certainly, it is a sacrifice both in personal effort and in resources to make others believe that you really do care. Therefore, many in management drop the ball, trip, fall, and fail to really support the valued employee message on this most important issue. In surveys done recently, employees in workplaces stated that the most important aspect of their job to them was to be treated with respect and as a valued employee. These desires were evaluated as a statistically representative answer for this population. If any of us would venture a guess about what the results of such a survey would be at our own workplaces, surely they would be identical.

7.1   PRINCIPLES OF MANAGEMENT

Those managing safety and health should adhere to the following basic principles.

1. Management is ultimately responsible for occupational safety and health—Thus, the need for commitment, budgeting, and planning for safety falls upon management's shoulders.
2. Poor safety conditions and safety performance by the workforce result from management's failure to effectively manage workplace safety and health—Accidents and incidents result from management's inability to manage safety and health as they would any other company function.
3. Worker and supervisor involvement are critical to good workplace safety and health—A workforce that is not involved in safety and health has no ownership and thus feels no investment, responsibility, or accountability for it.
4. Workplace safety and health is not a dynamic fast evolving component of the workplace since it should go hand in hand with your normally evolving business—In safety and health there is little that is new since we know the causes of occupational injuries and illnesses as well as how

to intervene, mitigate, and prevent their occurrence. There should be no excuses for accidents and incidents since the philosophy should be that all are preventable.

5. You cannot have an effective safety and health program without specifically holding the first line supervisors accountable for their and their employee's safety and health performance (as well as other management personnel)— First line supervisors are the key to the success or failure of a safety and health program. All your planning, budgeting, and goal setting are for naught if they are not accountable and committed to safety and health.

6. Hazard identification and analysis are critical functions in assuring a safe and healthy work environment—If management and the workforce do not tolerate the existence of hazards and constantly ask the question, "How could this have happened?" they are better able to get to the basic causes of adverse workplace events.

7. Management's philosophy, actions, policies, and procedures regarding safety and health in the workplace put workers into situations where they must disregard good safety and health practices to perform their assigned task or work—Workers in most cases perform work in an unsafe or unhealthy manner when they have no choice or are forced to do so by existing conditions and expectations.

8. It is critical to obtain safe and healthy performance or behavior by effective communications and motivational procedures that are compatible with the culture of your workforce—If you do not understand what it is that fulfills the needs of your workforce, you will not be able to communicate or motivate them regarding safety and health outcomes no matter how good your management approach.

7.2 SAFETY AND HEALTH PROGRAM

The key to formalizing the management of safety and health in a workplace is the safety and health program. Many feel that written safety and health programs are just more paperwork, a deterrent to productivity, and nothing more than another bureaucratic way of mandating safety and health on the job. However, over the years, data and information have been mounting in support of the need to develop and implement written safety and health programs for all workplaces.

This perceived need for written programs must be tempered with a view to their practical development and implementation. A very small employer who employs one to four employees and no supervisors in all likelihood needs only a very basic written plan, along with any other written programs that are required as part of an OSHA regulation. However, in large workplace environments where the number of employees increases, the owner/employer becomes more removed from the hands-on aspects of what now may be multiple floors, office complexes, or different types of worksites in the multiple facilities.

Now you must find a way to convey support for safety to all those who work in the same office building. As with all other aspects of business, the employer must plan, set the policies, apply management principles, and assure adherence to the goals in order to facilitate the efficient and effective completion of projects or work. Again, job safety and health should be managed the same as any other part of the office building's business.

To effectively manage safety and health, an owner/employer must pay attention to some critical factors. These factors are the essence in managing safety and health on worksites. The questions that need to be answered regarding managing safety and health are

- What is the policy of the owner/employer regarding safety and health in the workplace?
- What are the safety and health goals of the owner/employer?
- Who is responsible for occupational safety and health?
- How are supervisors and employees held accountable for job safety and health?
- What are the safety and health rules for the work-building environment?
- What are the consequences of not following the safety rules?
- Are there set procedures to address safety and health issues that arise in the place of work?
- How are hazards identified?
- How are hazards controlled or prevented?
- What type of safety and health training occurs? And, who is trained?

Specific actions can be taken to address each of the previous questions. The written safety and health program is of primary importance in addressing these items. It seems apparent that to have an effective safety program, at a minimum, an owner/ employer must

- Have a demonstrated commitment to job safety and health
- Commit budgetary resources
- Train new personnel
- Ensure that supervisors are trained
- Have a written safety and health program
- Hold supervisors accountable for safety and health
- Respond to safety complaints and investigate accidents
- Conduct safety audits

Other refinements that will help in reducing workplace injuries and illnesses can always be part of the safety and health program. They are more worker involvement (for example, joint labor/management committees); incentive or recognition programs; getting outside help from a consultant or safety association; and setting safety and health goals.

An increase in occupational incidents that result in injury, illness, or damage to property is enough reason to develop and implement a written safety and health program.

7.2.1 REASONS FOR BUILDING A SAFETY AND HEALTH PROGRAM

The three major considerations involved in the development of a safety program are

- Humanitarian—Safe operation of workplaces is a moral obligation imposed by modern society. This obligation includes consideration for loss of life, human pain and suffering, family suffering and hardships, etc.
- Legal obligation—Federal and state governments have laws charging the employer with the responsibility for safe working conditions and adequate supervision of work practices. Employers are also responsible for paying the costs incurred for injuries suffered by their employees during their work activities.
- Economic—Prevention costs less than accidents. This fact has been proven consistently by the experiences of thousands of workplaces. The direct cost is represented by medical care, compensation, etc. The indirect cost of four to ten times the direct cost must be calculated, as well as the loss of wages to employees and the reflection of these losses on the entire community.

All three are good reasons to have a health and safety program. It is also important that these programs be formalized in writing, since a written program sets the foundation and provides a consistent approach to occupational health and safety for the company. There are other logical reasons for a written safety and health program. Some of them are

- It provides standard directions, policies, and procedures for all company personnel
- It states specifics regarding safety and health and clarifies misconceptions
- It delineates the goals and objectives regarding workplace safety and health
- It forces the company to define its view of safety and health
- It sets out in black and white the rules and procedures for safety and health that everyone in the company must follow
- It is a plan that shows how all aspects of the company's safety and health initiative work together
- It is a primary tool of communication of the standards set by the company regarding safety and health

Written safety and health programs have a real place in modern safety and health practices not to mention the potential benefits. If a decrease in occupational incidents that result in injury, illness, or damage to property is not reason enough to develop and implement a written safety and health program, the other benefits from having a formal safety and health program seem well worth the investment of time and resources. Some of these are

- Reduction of industrial insurance premiums/costs
- Reduction of indirect costs of accidents

- Fewer compliance inspections and penalties
- Avoidance of adverse publicity from deaths or major accidents
- Less litigation and fewer legal settlements
- Lower employee payroll deductions for industrial insurance
- Less pain and suffering by injured workers
- Fewer long-term or permanent disability cases
- Increased potential for retrospective rating refunds
- Increased acceptance of bids on more jobs
- Improved morale and loyalty from individual workers
- Increased productivity from work crews
- Increased pride in company personnel
- Greater potential of success for incentive programs

7.2.2 BUILDING A SAFETY AND HEALTH PROGRAM

The length of such a written plan is not as important as the content. It should be tailored to the people service sectors need and the health and safety of its workforce and those it serves. It could be a few pages or a multiple page document. In order to ensure a successful safety program, three conditions must exist. These are management leadership, safe working conditions, and safe work habits by all employees. The employer must

- Let the employees know that the management is interested in safety on the job by consistently enforcing and reinforcing safety regulations.
- Provide a safe working place for all employees; it pays dividends.
- Be familiar with federal and state laws applying to your operation.
- Investigate and report all OSHA recordable accidents and injuries. This information may be useful in determining areas where more work is needed to prevent such accidents in the future.
- Make training and information available to the employees, especially in such areas as first aid, equipment operation, and common safety policies.
- Develop a prescribed set of safety rules to follow, and ensure that all employees are aware of the rules.

7.2.3 OTHER REQUIRED WRITTEN PROGRAMS

Many of the OSHA regulations have requirements for written programs that coincide with the regulations. This may become a bothersome requirement to many within the workplace, but the failure to have these programs in place and written is a violation of the regulations and will result in a citation for the company. At times, it is difficult to determine which regulations require a written program but, in most cases, the requirements are well known. Some of the other OSHA regulations that require written programs are

- Process safety management of highly hazardous chemicals
- Bloodborne pathogens/exposure control plan

- Emergency action plan
- Respirator program
- Lockout/tagout/energy control program
- Hazard communications program
- Fall protection plan
- Confined space "permit entry" plan

The specific requirements for the content of written programs vary with the regulation.

7.2.4 OSHA GUIDELINES FOR A SAFETY AND HEALTH PROGRAM

Although federal regulations do not currently require employers to have a written safety and health program, the best way to satisfy OSHA requirements and reduce accidents is for employers to produce one. In addition, distributing a written safety and health program to employees can increase employee awareness of safety and health hazards while, at the same time, reduce the costs and risks associated with workplace injuries, illnesses, and fatalities.

Federal guidelines for safety and health programs suggest that an effective occupational safety and health program must include evidence of the following:

- Management commitment and employee involvement are complementary. Management commitment provides the motivation force and the resources for organizing and controlling activities within an organization. In an effective program, management regards worker safety and health as a fundamental value of the organization and applies its commitment to safety and health protection with as much vigor as to other organizational purposes. Employee involvement provides the means through which workers develop and express their own commitment to safety and health protection, for themselves and for their fellow workers.
- Worksite analysis involves a variety of worksite examinations, to identify not only existing hazards but also conditions and operations in which changes might occur to create hazards (Figure 7.1). Unawareness of a hazard, which stems from failure to examine the worksite, is a sure sign that safety and health policies and practices are ineffective. Effective management actively analyzes the work and worksite to anticipate and prevent harmful occurrences.
- Hazard prevention and control are triggered by a determination that a hazard or potential hazard exists. Where feasible, hazards are prevented by effective design of the job site or job. Where it is not feasible to eliminate them, they are controlled to prevent unsafe or unhealthful exposure. Elimination or control is accomplished in a timely manner, once a hazard or potential hazard is recognized.
- Safety and health training addresses the safety and health responsibilities of all personnel concerned with the site, whether salaried or hourly. It is often most effective when incorporated into other training about performance

FIGURE 7.1 Identifying hazards is a first step in prevention. (Courtesy of the U.S. Environmental Protection Agency.)

requirements and job practices. Its complexity depends on the size and complexity of the worksite and the nature of the hazards and potential hazards at the site. All training should be documented by the employer. You might use a form similar to the one found in Figure 7.2.

7.2.5 SAFETY AND HEALTH PROGRAM ELEMENTS

If a representative from Occupational Safety and Health Administration (OSHA) visits a workplace, they will evaluate the safety program using the elements listed above. The compliance officer will review the previous items to assess the effectiveness of the safety and health program. However, you are not limited only these elements of your safety and health program. You might want to address accountability and responsibility, emergency procedures, program evaluation, firefighting, or first aid and medical care. This is your program, designed to meet your specific needs. These can be addressed in add-on sections. You will find that your fines for OSHA violations can be reduced if you have a viable written safety and health program, which meets the minimum OSHA guidelines for safety and health programs.

Individual worker's training form*

Worker's Name——————————————   Soc. Sec.#—————————————

Clock Number—————————————————

Subject:	Date:	Length of Training	Instructor	Worker's Signature
New Hire Orientation				
Hazard Communications				
Workplace Violence				
Security				
Other				

* Keep this form in the employee's personnel file

FIGURE 7.2   Employee training documentation form.

The composition or components of your safety and health program may vary depending on the complexity of your operations. They should at least include

- Management's commitment to the safety and health policy
- Hazard identification and evaluation
- Hazard control and prevention
- Training

Of course, each of these may have many subparts that address the four elements in some detail. The safety and health program that you develop should be tailored to meet your specific needs. It is now up to you to develop and implement your own effective safety and health program. You can build a more comprehensive program or pare down the model to meet your specific needs.

In summary, "Management Commitment and Leadership" includes a policy statement that should be developed and signed by the Top official in the company. Safety and health goals and objectives should be included to assist with establishing workplace goals and objectives that demonstrate the company's commitment to safety. An enforcement policy is provided to outline disciplinary procedures for violations of the company's safety and health program. This safety and health plan, as well as the enforcement policy, should be communicated to everyone on the jobsite. Some of the key aspects found under the heading, "Management Commitment and Leadership," are

- Policy statement: goals established, issued, and communicated to employees
- Program revised annually
- Participation in safety meetings; inspections; safety items addressed in meetings
- Commitment of resources is adequate in the form of budgeted dollars

- Safety rules and procedures incorporated into jobsite operations
- Procedure for enforcement of the safety rules and procedures
- Statement that management is bound to adhere to safety rules

"Identification and assessment of hazards" includes those items that can assist you with identifying workplace hazards and determining the corrective action necessary to control them. Actions include jobsite safety inspections, accident investigations, and meetings of safety and health committees and project safety meetings. In order to accomplish the identification of hazards, the following should be reviewed using

- Periodic site safety inspections involving supervisors
- Preventative controls in place (PPE, maintenance, engineering controls)
- Actions taken to address hazards
- An established safety committee, where appropriate
- Documented technical references available
- Enforcing procedures implemented by management

The employer must carry out an initial assessment, and then reassess as often, thereafter, as necessary to ensure compliance. Worksite assessments involve a variety of worksite examinations to identify, not only existing hazards, but also conditions and operations where changes might occur and create hazards. Being aware of a hazard, which stems from failure to examine the worksite, is a sure sign that safety and health policies and practices are inadequate. Effective management actively analyzes the work and worksite, to anticipate and prevent harmful occurrences. Worksite analysis is intended to ensure that all hazards are identified. This can be accomplished by

- Conducting comprehensive baseline worksite surveys for safety and health and periodically carrying out a comprehensive updated survey
- Analyzing planned and new facilities, processes, materials, and equipment
- Performing routine job hazard analyses

"Hazard prevention and controls" are triggered by a determination that a hazard or potential hazard exists. Where feasible, hazards are prevented by effective design of the jobsite or job. Where it is not feasible to eliminate them, they are controlled to prevent unsafe and unhealthful exposure. Elimination of controls is to be accomplished in a timely manner. Once a hazard or potential hazard is recognized, employers must ensure that all current and potential hazards, however detected, are corrected or controlled in a timely manner. Procedures should be established using the following measures:

- Engineering techniques where feasible and appropriate
- Procedures for safe work, which are understood and followed by all affected parties, as a result of training, positive reinforcement, correction of unsafe performance, and, if necessary, enforcement through a clearly communicated disciplinary system

- Provision of personal protective equipment
- Administrative controls, such as reducing the duration of exposure

The employer must ensure that each employee is provided "information and training" on the safety and health program (Figure 7.3). Each employee exposed to a hazard must be provided information and training on that hazard. Note: Some OSHA standards impose additional, more specific requirements for information and training. The employer must provide general information and training on the following subjects:

- The nature of the hazards to which the employee is exposed and how to recognize them
- What is done to control these hazards
- What protective measures the employee must follow to prevent or minimize exposure to these hazards
- The provisions of applicable standards

The employer must provide specific information and training:

- New employees must be informed and properly trained, before their initial assignment to a job involving exposure to a hazard.
- The employer is not required to provide initial information and training for employees for whom the employer can demonstrate that adequate training has already been given.
- The employer must provide periodic information and training as often as necessary to ensure that employees are adequately informed and trained; and to be sure safety and health information and changes in workplace conditions, such as when a new or increased hazard exists, are communicated.

FIGURE 7.3 Workers cannot be expected to perform a job in a safe and healthy manner unless they have been trained how to.

Safety and health training addresses the safety and health responsibilities of all personnel concerned with the site, whether salaried or hourly. The employer must provide all employees who have program responsibilities with the information and training necessary to carry out their safety and health responsibilities.

You will find model written safety and health programs in *Industrial Safety and Health for Goods and Materials Services*, *Industrial Safety and Health for Infrastructure Services*, and *Industrial Safety and Health for Administrative Services*. These models should be taken and adapted to fit the needs of your business.

7.3   SUMMARY

The management of safety and health is well recognized as a vital component by those who have responsibility for workplace safety and health. It is not just a written proclamation or program, but a true and supported endeavor to provide a safe and healthy workplace for workers. It must be as well planned and organized as any other facet of the company's business. Managing safety and health is critical in providing the protections that a workforce within an office building is entitled to and deserves. This management process will probably progress much better if you have a person who by training and experience can take the responsibility for the development of the safety and health effort in your office environment. Thus, occupational safety and health management must be planned, organized, and implemented in a professional manner in order for it to be successful in ensuring that your workplace is safe, healthy, and secure.

8 Summary of 29 CFR 1910

Labor

29

PARTS 1900 TO 1910.999
Revised as of July 1, 1997

Example of the cover of the Code of Federal Regulations.

In this chapter you will find an overview to the general industry standard entitled 29 CFR 1910. A paragraph highlights the content of each subpart of this standard. You will also find a listing of the sections that are contained within each subpart. In addition, a checklist is included for each subpart. If you answer "yes" to any question, then some or all of the subpart would be applicable to your operation. Thus, a yes answer suggests that your workplace needs to be in compliance with the applicable sections of that subpart.

You can find a similar appendix for 29 CFR 1926 in the *Handbook of OSHA Construction Safety and Health (Second Edition)*, 2006, published by CRC Press/Lewis Publishers.

This chapter provides a tool that can be of assistance to any industry or business as they attempt to comply with the regulations that Occupational Safety and Health Administration (OSHA) have developed, promulgated, and enforced.

8.1 PART 1910—OCCUPATIONAL SAFETY AND HEALTH STANDARDS

8.1.1 Subpart A—General

This subpart explains the purpose and scope, definitions, petitions for issuance, amendment, and repel of a standard in 29 CFR 1910. The purpose of these standards is to make the workplace safer and healthier. The subpart explains the applicability of the OSHA standards relevant to the workplaces covered, the geographic location covered, and specific entities (i.e., federal agencies) not covered. It also lists regulations that have been incorporated in this standard by reference into 29 CFR 1910 as well as the requirements for nationally recognized testing laboratories.

Checklist

_____Do you want to see a standard issued, amended, or repealed?
_____Does this standard include your operation or business?
_____Are you interested in the requirements for nationally recognized testing laboratories?
_____Do you want to order a copy of a standard incorporated by reference into 29 CFR 1910?

Sections of Subpart A Regulation

1910.1 Purpose and scope
1910.2 Definitions
1910.3 Petitions for the issuance, amendment, or repeal of a standard
1910.4 Amendments to this part
1910.5 Applicability of standards
1910.6 Incorporation by reference
1910.7 Definition and requirements for a nationally recognized testing laboratory
1910.8 OMB control numbers under the Paperwork Reduction Act

8.1.2 Subpart B—Adoption and Extension of Established Federal Standards

Subpart B adopts and extends the applicability of established federal standards to every employer, employee, and place of employment covered by the act. Only standards relating to safety or health are adopted into this act. This also pertains to any facility engaging in construction, alterations, or repair, including painting and decorating.

The Construction Safety Act adopts as occupational safety and health standards under section 6 of the Act the standards that are prescribed in part 1926 of this chapter. Thus, the standards (substantive rules) published in Subpart C and the following subparts of part 1926 of this chapter are applied. This section does not incorporate subparts A and B of part 1926 of this chapter.

Adoption and extension of established safety and health standards for shipyard employment and the standards prescribed by part 1915 (formerly parts 1501–1503) of this title and in effect from April 28, 1971 (as revised) are adopted as occupational safety or health standards under section 6(a) of the Act and shall apply, according to the provisions thereof, to every employment and place of employment of every employee engaged in ship repair, shipbreaking, and shipbuilding, or a related employment.

Part 1918 of this chapter shall apply exclusively, according to the provisions thereof, to all employment of every employee engaged in longshoring operations, marine terminals, or related employment aboard any vessel. All cargo transfer accomplished with the use of shore-based material-handling devices shall be governed by part 1917 of this chapter.

Workplaces that expose workers to asbestos, tremolite, anthophyllite, and actinolite dust; vinyl chloride; acrylonitrile; inorganic arsenic; lead; benzene; ethylene oxide; 4,4'-methylenedianiline; formaldehyde; cadmium; 1,3-butadiene; and methylene chloride are covered by appropriate 1910 standards.

Checklist

_____Are construction activities taking place?
_____Is shipyard employment occurring?
_____Is longshoring taking place?
_____Are workers exposed to the hazardous chemicals mentioned here?

Sections of Subpart B Regulation

1910.11   Scope and purpose
1910.12   Construction work
1910.15   Shipyard employment
1910.16   Longshoring and marine terminals
1910.17   Effective dates
1910.18   Changes in established federal standards
1910.19   Special provisions for air contaminants

8.1.3   SUBPART C—[REMOVED AND RESERVED]

1910.20   [Redesignated as 1910.1020]

8.1.4   SUBPART D—WALKING–WORKING SURFACES

This subpart addresses the requirements for maintaining walking and working surfaces. Subpart D applies to all permanent places of employment. It contains regulations pertaining to housekeeping, aisles and passageways, guarding wall and floor openings, fixed stairs, portable wood and metal ladders, fixed ladders, scaffolding, and manually propelled mobile ladder stands and scaffolds from frame to suspended types as well as dockboards, forging machine areas, and veneering machine areas (see Figure 8.1).

FIGURE 8.1 Secured portable ladder.

Checklist

_____Do you use dockboards?

_____Do you have forging machines or veneering machines at your site?

_____Is attention paid to housekeeping?

_____Are there floor and wall openings or holes at your facility?

_____Do you have manually propelled mobile ladder stands and scaffolds?

_____Do your workers use scaffolds in the performance of their work?

_____Do you own scaffolds?

_____Do you enforce housekeeping?

_____Do you erect, tear down, or maintain scaffolds?

_____Are you responsible for training workers regarding scaffolds and their safety?

_____Are there scaffolds on your worksite?

_____Do your workers use ladders in performing their work?

_____Are your workers required to ascend and descend industrial stairs?

_____Does your company own ladders?

_____Do your workers have to climb fixed ladders?

Sections of Subpart D Regulation

    1910.21   Definitions

    1910.22   General requirements

    1910.23   Guarding floor and wall openings and holes

    1910.24   Fixed industrial stairs

1910.25   Portable wood ladders
1910.26   Portable metal ladders
1910.27   Fixed ladders
1910.28   Safety requirements for scaffolding
1910.29   Manually propelled mobile ladder stands and scaffolds (towers)
1910.30   Other working surfaces

8.1.5   SUBPART E—EXIT ROUTES, EMERGENCY ACTION PLANS, AND FIRE PREVENTION PLANS

This subpart deals specifically with providing a safe continuous and unobstructed means of exit and the design, construction, and maintenance to assure an open travelway from any point in a building or structure to a safe exit. The standard addresses exits by describing the make up of an exit, specific physical requirements for an exit, and the number of exits required. It requires the employer to comply with the National Fire Protection Association's life safety code regarding safe exits. This subpart also contains the requirements essential to provide a safe means of exit from fire and like emergencies. The subpart sets forth the requirements for emergency action plans and fire prevention plans. Emergency action and fire prevention plans, which can assure adequate escape procedures, evacuation routes, alarm systems, and other emergency actions, are required (see Figure 8.2).

FIGURE 8.2   Well-designed emergency exit.

Checklist

_____Are exits designed to meet the NFPA life safety code?

_____Do exits meet the requirements of the regulation for construction?

_____Are exits maintained to meet their intended purpose?

_____Do you have an emergency action or escape plan or procedure?

_____Do you have a fire prevention plan for your facility?

_____Are all exits unlocked and free from impediments?

_____Is there a safe means of exit for all your workers?

_____Are all exits designed to be visible and allow for a safety exit from your facility?

Sections of Subpart E Regulation

1910.33	Table of contents
1910.34	Coverage and definitions
1910.35	Compliance with the NFPA 101–2000, life safety code
1910.36	Design and construction requirements for exit routes
1910.37	Maintenance, safeguards, and operational features for exit routes
1910.38	Emergency action plans
1910.39	Fire prevention plans

Subpart E Appendix—Exit routes, emergency action plans and fire prevention plans

8.1.6   Subpart F—Powered Platforms, Manlifts, and Vehicle-Mounted Work Platforms

This subpart covers powered platform installations permanently dedicated to interior or exterior building maintenance of a specific structure or group of structures. It does not apply to suspended˙ self-powered platforms used to service buildings as well as the guidelines for personal fall arrest systems. This subpart applies to all permanent installations completed after July 23, 1990 and contains information on powered platforms for building maintenance. Building maintenance covers a wide array of activities from window cleaning to engineering design of equipment as well as expressing the need to train workers. In addition, this section specifically addresses the requirements for vehicle-mounted elevating and rotating work platforms and manlifts.

Checklist

_____Do you provide fall protection for your workforce?

_____Are powered platforms used for building maintenance?

_____Do you have vehicle-mounted elevated and rotating work platforms?

_____Do you have manlifts?

Sections of Subpart F Regulation

1910.66	Powered platforms for building maintenance
1910.67	Vehicle-mounted elevating and rotating work platforms
1910.68	Manlifts

8.1.7 Subpart G—Occupational Health and Environmental Control

The standards in Subpart G deal with air quality, noise exposure exceeding 85 dB, and nonionizing radiation exposure in the workplace. Ventilation is specific for facilities that use abrasive blasting; facilities that have spray booths, or open surface tanks used for cleaning, and facilities with grinding, polishing, and buffing operations.

Checklist

_____Do you have ventilation issues caused by abrasive blasting?

_____Do you have ventilation issues caused by spray booths?

_____Do you have ventilation issues caused by open surface tanks?

_____Do you have ventilation issues caused by grinding, polishing, and buffing operations?

_____Do you have noise exposure in excess of the 85 dBA level?

_____Do you have a source of nonionizing radiation?

_____Does you company use any chemicals that could be considered hazardous?

_____Does your company have a medical officer for examinations, advice, or consultation?

_____Have you had injuries or illnesses that require first aid?

_____Have you had to do environmental or air monitoring?

_____Do you provide drinking water to workers?

_____Do you provide for toilets and washing facilities?

_____Do you have high-noise worksites or tasks?

_____Do you have sources of ionizing or nonionizing (lasers) radiation at your worksites?

_____Do you do contracting jobs where chemical processes involving highly hazardous chemicals take place?

_____Do you use some form of ventilation to remove airborne contaminants?

_____Do you do hazardous waste remediation work?

_____Do you do night work or work in areas with limited light?

Sections of Subpart G Regulation

1910.94 Ventilation
1910.95 Occupational noise exposure
1910.96 [Redesignated as 1910.1096]
1910.97 Nonionizing radiation
1910.98 Effective dates

8.1.8 Subpart H—Hazardous Materials

Subpart H contains information on compressed gases, acetylene, hydrogen, oxygen, nitrous oxide, flammable and combustible liquids, spray finishing using flammable and combustible materials, dip tanks using flammable or combustible liquids, explosive and blasting agents, storage and handling of liquid petroleum gases, and storage and handling of anhydrous ammonia (see Figure 8.3). This section also covers process-safety management requirements of highly hazardous chemicals

FIGURE 8.3   Safe storage of compressed-gas cylinder.

and hazardous waste operations and emergency response. The final part of the regulation deals with dipping and coating processes.

Checklist

_____Do you use compressed gases?

_____Do you have acetylene, hydrogen, oxygen, or nitrous oxide on the premises?

_____Do you handle, use, or store flammable or combustible liquids?

_____Do you have spray finishing operations with flammable or combustible liquids?

_____Do you have dip tanks using flammable or combustible liquids?

_____Do you have highly hazardous chemicals or chemical processes?

_____Do you have workers trained to remediate hazardous chemicals or respond to HAZMAT situations?

_____Do you have dipping and coating operations?

_____Does your company have equipment used for explosives and blasting?

_____Do any of your workers perform explosive handling and blasting operations?

_____Do you have blasting materials on your jobsites?

_____Do you employ individuals who are qualified blasters?

_____Do you have a contract blaster doing your blasting operations?

_____Does your company contract to carry out blasting activities?

_____Does your company or workers transport explosives or blasting materials?

_____Do blasting activities occur on your jobsites or projects?

Sections of Subpart H Regulation

1910.101   Compressed gases (general requirements)

1910.102   Acetylene

1910.103　　Hydrogen
1910.104　　Oxygen
1910.105　　Nitrous oxide
1910.106　　Flammable and combustible liquids
1910.107　　Spray finishing using flammable and combustible materials
1910.108　　Dip tanks containing flammable or combustible liquids
1910.109　　Explosives and blasting agents
1910.110　　Storage and handling of liquefied petroleum gases
1910.111　　Storage and handling of anhydrous ammonia
1910.112　　[Reserved]
1910.113　　[Reserved]
1910.119　　Process-safety management of highly hazardous chemicals
1910.120　　Hazardous waste operations and emergency response
1910.121　　[Reserved]
1910.122　　Table of contents
1910.123　　Dipping and coating operations: Coverage and definitions
1910.124　　General requirements for dipping and coating operations
1910.125　　Additional requirements for dipping and coating operations that use flammable or combustible liquids
1910.126　　Additional requirements for special dipping and coating applications

8.1.9　Subpart I—Personal Protective Equipment

Subpart I requires employers to provide employees with proper personal protective equipment (PPE) for the work performed. As part of this requirement, the employer must conduct a hazard survey of the work to determine the control measures to use where hazards cannot be eliminated. This serves as a resource in guiding the selection of the appropriate PPE. This includes PPE for eyes, face, head, and extremities. Other types of equipment that may be required are protective clothing and equipment as well as respiratory devices All PPE is to be maintained in a sanitary condition.

Not only are the employers required to provide the needed PPE, but also they are required to train workers how to use and wear their PPE. Equipment for emergency use should be stored in an accessible location known to all workers. The requirements for respirators and their use is the most extensive part of the subpart.

This subpart provides the standard for quality and selection of PPE such as eye/face protection, head protection, respiratory protection, foot protection, and hand/arm protection.

Checklist

_____Do you require personal protective equipment to be used?
_____Do you have the potential for falling, flying, or electrical hazards?
_____Do you require head protection?
_____Are there opportunities for heavy material to fall onto the workers' feet?
_____Do you provide hand and arm protection, i.e., gloves?
_____Does your workforce come into contact with electricity and need protective equipment?

_____Do you have the potential at any time for workers to suffer eye injuries?

_____Do environment or air contaminants require the use of respirators by the workers?

_____Do your workers need eye and face protection?

_____Are your workers potentially exposed to tuberculosis?

Sections of Subpart I Regulation .

    1910.132   General requirements
    1910.133   Eye and face protection
    1910.134   Respiratory protection
    1910.135   Head protection
    1910.136   Foot protection
    1910.137   Electrical protective devices
    1910.138   Hand protection
    1910.139   Respiratory protection for *M. tuberculosis*

8.1.10 Subpart J—General Environmental Controls

This section specifically applies to places of employment where such items as sanitary facilities, for example, toilet facilities, washing facilities, sanitary food storage, and food handling, are required. It also addresses temporary labor camps, safety colors for marking physical hazards, and requirements for accident prevention signs and tags. Two additional items specifically addressed by this section and of considerable importance are permit-required confined spaces and the control of hazardous energy (lockout/tagout).

Checklist

_____Do you provide sanitary facilities for your workforce?

_____Do you have temporary labor camps?

_____Do you have warning or accident prevention signs or tags posted in your workplace?

_____Do you use the appropriate colors to mark physical hazards?

_____Do you have confined spaces in your workplace?

_____Does you workforce enter confined spaces where permits are needed?

_____Do you have a lockout/tagout program in place?

_____Do your require lockout/tagout procedures to be followed?

Sections of Subpart J Regulation

    1910.141   Sanitation
    1910.142   Temporary labor camps
    1910.143   Nonwater carriage disposal systems [Reserved]
    1910.144   Safety color code for marking physical hazards
    1910.145   Specifications for accident prevention signs and tags
    1910.146   Permit-required confined spaces
    1910.147   The control of hazardous energy (lockout/tagout)

8.1.11 Subpart K—Medical and First Aid

The purpose of medical and first aid is to provide the employee with readily available medical consultation. If medical personnel are not readily available, then personnel adequately trained to administer first aid are to be present. These individuals should be provided protection and PPE to prevent exposure to bloodborne pathogens. The employer is required to provide fully equipped first-aid kits and they are to be maintained in suitable numbers to meet the needs of the workforce.

Checklist

_____Are there qualified medical personnel at the facility?
_____Do you have personnel trained in first aid available?
_____Do you have first-aid kits available?
_____Do you keep first-aid kits adequately stocked?

Sections of Subpart K Regulation

1910.151   Medical services and first aid
1910.152   [Reserved]

8.1.12 Subpart L—Fire Protection

Subpart L is concerned with fire protection and fire prevention. This subpart contains requirements for fire brigades, all portable extinguishers, fixed-fire suppression systems and fire detection systems, and alarm systems. It contains training requirements for the organization and personnel. It describes requirements for training and protective equipment for fire brigades.

In addition, this subpart establishes the requirements for the placement, use, maintenance, and testing of portable fire extinguishers provided for use by employees, as well as the requirements for all automatic sprinkler systems installed to meet a particular OSHA standard. Firefighting equipment is to be available and readily accessible. Workers are to be trained annually on the use of fire extinguishers.

The fire detection system should be in a labeled specific location. Lastly, a unique alarm system must be established at the worksite, which will alert employees to a fire.

Checklist

_____Does your worksite have a fire hazard potential?
_____Do you have a fire prevention program?
_____Do you use fire extinguishers at your site?
_____Do you train workers in fire prevention and firefighting?
_____Do you have a fire brigade?
_____Do you have a fire detection system or fire alarm system?
_____Are your employees expected to fight fires?

Sections of Subpart L Regulation

1910.155   Scope, application, and definitions applicable to this subpart
1910.156   Fire brigades
PORTABLE FIRE SUPPRESSION EQUIPMENT
1910.157   Portable fire extinguishers
1910.158   Standpipe and hose systems
FIXED FIRE SUPPRESSION EQUIPMENT
1910.159   Automatic sprinkler systems
1910.160   Fixed extinguishing systems, general
1910.161   Fixed extinguishing systems, dry chemical
1910.162   Fixed extinguishing systems, gaseous agent
1910.163   Fixed extinguishing systems, water spray and foam
OTHER FIRE PROTECTIVE SYSTEMS
1910.164   Fire detection systems
1910.165   Employee alarm systems
APPENDICES TO SUBPART L
APPENDIX A TO SUBPART L—FIRE PROTECTION
APPENDIX B TO SUBPART L—NATIONAL CONSENSUS STANDARDS
APPENDIX C TO SUBPART L—FIRE PROTECTION REFERENCES FOR
     FURTHER INFORMATION
APPENDIX D TO SUBPART L—AVAILABILITY OF PUBLICATIONS
     INCORPORATED BY REFERENCE IN SECTION 1910.156 FIRE
     BRIGADES
APPENDIX E TO SUBPART L—TEST METHODS FOR PROTECTIVE
     CLOTHING

8.1.13   Subpart M—Compressed-Gas and Compressed-Air Equipment

This subpart applies to compressed-air receivers and other equipment used in providing and using compressed air for performing operations such as cleaning, drilling, hoisting, and chipping. However, this section does not deal with the special problems created by using compressed air to convey materials, nor the problems created when work is performed in compressed-air environments such as in tunnels and caissons. This section is not intended to apply to compressed-air machinery and equipment used or transportation vehicles such as steam railroad cars, electric railway cars, and automotive equipment.

Checklist
_____Do you use a compressed-air receiver?
_____Do you have equipment that provides compressed air?

Sections of Subpart M Regulation

1910.166   [Reserved]
1910.167   [Reserved]
1910.168   [Reserved]
1910.169   Air receivers

8.1.14 Subpart N—Materials Handling and Storage

Subpart N details the storage of materials and how to stack, rack, and secure them against falling or sliding. Materials should not create a hazard due to storage in aisles or passageways. Housekeeping is an important component of handling and storing of materials.

Subpart N provides provisions for cranes, derricks, hoists, helicopters, conveyors, and aerial lifts. This subpart delimits many common safety requirements for material-handling equipment and reinforces the need to follow the manufacturer's requirements regarding load capacities, speed limits, special hazards, and unique equipment characteristics. A competent person must inspect all cranes and derricks before daily use and a thorough inspection must be accomplished annually by an OSHA-recognized qualified person. A record must be maintained of that inspection for each piece of hoisting equipment.

The industrial trucks section covers the classifications of trucks and designated areas where a truck can be used. It also describes the required inspections and maintenance actions for those vehicles. Safe operation procedures are also covered in this section.

Procedures for keeping and using slings are also covered in this section. It describes the proper sizes for loads as well as safe hookup procedures and inspection requirements are stated and required markings are discussed. The rigging of materials for handling is a critical component of Subpart H. This includes the safe use of slings made from wire rope, chains, synthetic fiber ropes or webs, and natural fiber ropes. Specifications for the use of rigging are found in this subpart regarding carrying capacity, inspection for defects, and safe operating procedures.

This subpart applies to the use of helicopters for lifting purposes. Helicopters must comply with the Federal Aviation Administration (FAA) regulations. The pilot of the helicopter has the primary responsibility for the load's weight, size, and rigging. Static charge must be eliminated prior to workers touching the load. Visibility is critical to the pilot in maintaining visual contact with ground crew members so that constant communications can be maintained.

All hoists are to comply with the manufacturer's specifications. If these do not exist, then as with cranes and derricks, the limitations are based on the determination of a professional engineer. In the operation of a hoist, there should be a signaling system, specified line speed, and a sign stating "No Riders". Permanently enclosed hoist cars are to be used to hoist personnel and these cars must be able to stop at any time by safety breaks or a similar system. All hoists are to be tested, inspected, and maintained on an ongoing basis and at least every 3 months. In addition, requirements exist in this subpart for base-mounted drum hoists and overhead hoists.

The servicing of single- and multipiece rims is also covered in this subpart.

Checklist

_____Does your company own or use cranes or derricks?
_____Does your company employ helicopters for lifting purposes?
_____Do you use material hoists on your worksite?
_____Do your workers work around cranes, derricks, helicopters, or hoists?
_____Does your company use cranes or hoists for lifting personnel?
_____Do you rent cranes, derricks, hoists, or other lifting devices?

_____Do you operate powered industrial trucks (forklifts) at your facility?

_____Do you rig loads of handling?

_____Do you use slings for rigging?

_____Do you have single- and multiple piece rims at your site?

_____Do you have materials stored on the worksite?

_____Do you have waste materials on the jobsite?

_____Do your workers use rigging to handle materials?

_____Do your workers know the limitations of the use of wire ropes, chains, etc.?

_____Does your company have responsibility for housekeeping?

Sections of Subpart N Regulation

1910.176   Handling material—general
1910.177   Servicing multipiece and single-piece rim wheels
1910.178   Powered industrial trucks
1910.179   Overhead and gantry cranes
1910.180   Crawler locomotive and truck cranes
1910.181   Derricks
1910.183   Helicopters
1910.184   Slings

APPENDIX A to 1910.178—Stability of powered industrial trucks (nonmandatory Appendix to paragraph (l) of this section.

8.1.15   Subpart O—Machinery and Machine Guarding

Subpart O covers the machine guarding for any equipment that exposes employees to a hazard during use due to exposed moving or rotating parts; generally, this covers any device that has an exposed point of operation. This subpart covers guards for woodworking machinery, abrasive wheel machinery, cooperage machinery, mills and calenders, mechanical presses, forging machines, and mechanical power-transmission apparatus (see Figure 8.4).

The woodworking section covers the parts that must be guarded and the types of guards that must be used, while the abrasive-wheel section describes the amount of wheel that can be exposed for the various types of abrasive grinding equipment and other precautions to take.

Mechanical power presses are required to have switches and brakes to protect the operator. Many presses are to be protected by mechanical guards and by other means. This section describes the actions that are to be taken to ensure that operations are as safe as possible. Many types of guarding systems can be used on presses.

The safe operation, inspection, and maintenance of forging machines as well as the best practices for guarding these pieces of equipment are discussed in this section. Special guarding needs are discussed for certain processes.

The mechanical power-transmission apparatus in this subpart covers all belts, pulleys, and conveyors that are used in industry. It describes those that need to be guarded. For specific applications, guidance is given for preferred operations.

FIGURE 8.4 Proper guarding could have prevented this accident. (Courtesy of Mine Safety and Health Administration.)

Checklist

_____Do you have power presses?
_____Do your workers use woodworking machinery?
_____Does your facility operate any abrasive-wheel machinery?
_____Are you considered to be a rubber or plastics industry that has mill and calenders?
_____Do you operate mechanical power presses?
_____Do you have a forging or die shop?
_____Are there power-transmission belts, pulleys, etc., present?

Sections of Subpart O Regulation

1910.211   Definitions
1910.212   General requirements for all machines
1910.213   Woodworking machinery requirements
1910.214   Cooperage machinery
1910.215   Abrasive wheel machinery
1910.216   Mills and calenders in the rubber and plastics industries
1910.217   Mechanical power presses
1910.218   Forging machines
1910.219   Mechanical power-transmission apparatus

8.1.16   SUBPART P—HAND AND PORTABLE POWERED TOOLS AND OTHER HANDHELD EQUIPMENT

The Subpart P regulation is dedicated to the safe use of both power and hand tools including employer and employee owned tools. The subpart requires that hand tools be safe and free from defects. It also cautions against misuse of tools.

This subpart addresses the need for properly guarded power tools. It discusses the areas where guarding is required and the types of guards that should be used, as well as the proper protective equipment to be used, when tools create such hazards as flying materials. The power tools that are covered by the regulation include electrical, pneumatic, fuel, hydraulic, and powder-actuated powered tools. These tools are to be secured if maintained in a fixed place, and all electrically powered equipment must be effectively grounded. Special attention is given to abrasive wheels and tools. Some special requirements exist for powder-actuated tools.

Jacks and their use are covered regarding the blocking and securing of objects that are lifted. This includes jack maintenance and inspection. It also pertains to riding and walk-behind lawn mowers, and other internal-combustion-engine-powered machines are included in this section.

Checklist

_____Do your workers use hand or power tools?
_____Do your workers use woodworking tools?
_____Do your workers use abrasive wheels or tools?
_____Do you supply tools to workers?
_____Do your workers use jacks?
_____Do your workers use walk-behind or riding mowers?

Sections of Subpart P Regulation

   1910.241   Definitions
   1910.242   Hand and portable powered tools and equipment, general
   1910.243   Guarding of portable powered tools
   1910.244   Other portable tools and equipment

8.1.17 Subpart Q—Welding, Cutting, and Brazing

Subpart Q covers the use and installation of arc or gas welding, cutting, and brazing equipment. It covers the different types of welding and ties the specific safety needs of each. This subpart also regulates the use of oxygen-fuel gas welding and cutting, arc welding and cutting, and resistance welding. Subpart Q covers the procedures and precautions associated with gas welding, cutting, arc welding, fire prevention, compressed-gas cylinders, and welding materials. Special attention is given to the transporting, moving, and storing of compressed-gas cylinders, as well as apparatuses such as hoses, torches, and regulators used for welding. Defective gas cylinders should not be used. All cylinders should be marked and labeled with 1 in. letters. Hoses should be identifiable and designed such that they cannot be misconnected to the wrong cylinder regulators. Prework inspections are an important component of this subpart.

Arc welding and its unique precautions are covered by this regulation. This includes grounding, care of cables, and care of electrode holders. As with all welding and cutting operations, appropriate PPE and safety are addressed in this subpart.

Fire prevention is an important part of welding and cutting and such work is not to be performed near flammable vapors, fumes, or heavy dust concentrations. Firefighting equipment must be readily accessible and in good working order.

Checklist

_____Do your workers perform welding and cutting tasks?

_____Do you have compressed-gas cylinders on your jobsite?

_____Do you have adequate firefighting equipment?

_____Is there a need for ventilation?

_____Do your welders wear personal protective equipment?

_____Does your company weld or cut in confined spaces?

_____Do your workers have to weld or cut on toxic materials?

_____Do your workers perform resistance welding?

Sections of Subpart Q Regulation

1910.251   Definitions

1910.252   General requirements

1910.253   Oxygen-fuel gas welding and cutting

1910.254   Arc welding and cutting

1910.255   Resistance welding

8.1.18   SUBPART R—SPECIAL INDUSTRIES

Subpart R deals with industries singled out by OSHA that need to be addressed in industry-specific standards. These industries include pulp, paper and paperboard mills, textile mills, bakeries, laundries, and sawmills. It also includes industries such as pulpwood logging, telecommunications, electric-power generation, transmission, and distribution, and grain-handling facilities.

Checklist

_____Does the work involve the manufacturing of pulp, paper, and paperboard?

_____Does the work involve operation and maintenance of textile mills and machinery?

_____Does the work involve the operation and maintenance of machinery and equipment used within a bakery?

_____Does the laundry equipment that is used have point of operation hazards?

_____Is the work conducted at a sawmill?

_____Does the work involve the normal operations included in logging operations?

_____Does the work involve processes in telecommunications centers and at telecommunications field installations?

_____Does the procedure involve working with the operation and maintenance of electric-power generation, transmission, and distribution lines and equipment?

_____Does the process require the operation of grain elevators, grain storage, and processing facilities?

Sections of Subpart R Regulation

1910.261   Pulp, paper, and paperboard mills

1910.262   Textiles

1910.263   Bakery equipment

1910.264   Laundry machinery and operations

1910.265   Sawmills

1910.266   Logging operations
1910.267   [Reserved]
1910.268   Telecommunications
1910.269   Electric power generation, transmission, and distribution
1910.272   Grain-handling facilities

8.1.19 SUBPART S—ELECTRICAL

Subpart S relates to the installation and use of electrical power on worksites, including both permanent and temporary. The two areas of emphasis within this subpart are installation safety requirements and safety-related work practices.

Installation safety requirements sections of Subpart S require that all electrical parts be inspected for durability, quality, and appropriateness. An installation that follows the National Electric Code (NEC) is considered in compliance with OSHA. Grounding is an important part of this regulation and the use of ground-fault circuit interrupters (GFCIs) or assured grounding is required. Emphasis is placed on temporary and portable lighting, as well as the use of extension cords. All listed, labeled, and certified equipment must be installed according to instructions from the manufacturer. This subpart includes special purpose equipment installation such as cranes and monorail hoists, electric welders, and x-ray equipment. It discusses work in high hazard locations as well as special systems such as remote control and power-limited circuits.

Safety-related work practices include workers not working on energized circuits. This includes precautions for working on hidden underground power sources. This subpart addresses the use of barriers to protect workers from electrical sources. In addition, working around electrically energized equipment and power lines is explained as well as the procedures for lockout/tagout of energized circuits to protect workers.

The primary purpose of this subpart is to protect workers from coming into contact with energized electrical power sources (see Figure 8.5).

FIGURE 8.5   Warning signs convey the dangers posed by electricity.

Checklist

_____Do you employee electricians?

_____Do your employees perform electrical installations?

_____Do your workers work around energized electrical circuits?

_____Do you follow a lockout/tagout procedure?

_____Do you use temporary lighting and extension cords?

_____Do your workers use GFCIs?

_____Do you have workers working in hazardous environments?

_____Do your workers use electrically powered tools?

_____Are there energized power lines on your jobsite?

_____Do your workers work around energized power lines?

_____Is there special electrically powered equipment on your worksite?

Sections of Subpart S Regulation

GENERAL

1910.301   Introduction

DESIGN SAFETY STANDARDS FOR ELECTRICAL SYSTEMS

1910.302   Electric utilization systems

1910.303   General requirements

1910.304   Wiring design and protection

1910.305   Wiring methods, components, and equipment for general use

1910.306   Specific purpose equipment and installations

1910.307   Hazardous (classified) locations

1910.308   Special systems

1910.309–1910.330 [Reserved]

SAFETY-RELATED WORK PRACTICES

1910.331   Scope

1910.332   Training

1910.333   Selection and use of work practices

1910.334   Use of equipment

1910.335   Safeguards for personnel protection

1910.336–1910.360 [Reserved]

SAFETY-RELATED MAINTENANCE REQUIREMENTS

1910.361–1910.380 [Reserved]

SAFETY REQUIREMENTS FOR SPECIAL EQUIPMENT

1910.381–1910.398 [Reserved]

DEFINITIONS

1910.399   Definitions applicable to this subpart

APPENDIX A TO SUBPART S—REFERENCE DOCUMENTS

APPENDIX B TO SUBPART S—EXPLANATORY DATA [RESERVED]

APPENDIX C TO SUBPART S—TABLES, NOTES, AND CHARTS [RESERVED]

8.1.20 SUBPART T—COMMERCIAL DIVING OPERATIONS

Subpart T applies to dives and diving support operations that take place within all waters in the United States, trust territories, DC, Commonwealth of Puerto Rico, other U.S. protected islands, etc. It does not apply to instructional diving and search and rescue. This subpart describes requirements, qualifications, and training certifications for divers and dive teams, as well as the need to use specific safe practices for pre, during, and postdives. It also includes emergency care procedures such as recompression and evacuation.

This subpart delineates the criteria and procedures for different types of diving operations such as scuba, surface supplied air, and mixed gas diving. The margin for error and risk are high; thus, all diving procedures within this regulation are very precise and require more than superficial knowledge and experience with diving operations.

The care and maintenance of all equipment involved, whether cylinders, decompression chambers, oxygen safety, or other diving equipment, require a unique expertise. This subpart makes all diving and diving operation procedures very exacting and requires recordkeeping of all dives and injuries.

Checklist

_____Does your company employ any divers?
_____Does your company oversee any diving operations?
_____Does your company own any diving equipment?
_____Do you have divers or diving operations on or at your workplace that belong to other contractors?

Sections of Subpart T Regulation

GENERAL
1910.401   Scope and application
1910.402   Definitions

PERSONNEL REQUIREMENTS
1910.410   Qualifications of dive team

GENERAL OPERATIONS PROCEDURES
1910.420   Safe practices manual
1910.421   Pre-dive procedures
1910.422   Procedures during dive
1910.423   Postdive procedures

SPECIFIC OPERATIONS PROCEDURES
1910.424   SCUBA diving
1910.425   Surface-supplied air diving
1910.426   Mixed-gas diving
1910.427   Liveboating

EQUIPMENT PROCEDURES AND REQUIREMENTS
1910.430   Equipment

RECORDKEEPING
1910.440   Recordkeeping requirements
1910.441   Effective date

APPENDIX A TO SUBPART T—EXAMPLES OF CONDITIONS WHICH
    MAY RESTRICT OR LIMIT EXPOSURE TO HYPERBARIC CONDITIONS
APPENDIX B TO SUBPART T—GUIDELINES FOR SCIENTIFIC DIVING

8.1.21   SUBPARTS U–Y [RESERVED]

1910.442–1910.999 [Reserved]

8.1.22   SUBPART Z—TOXIC AND HAZARDOUS SUBSTANCES

Subpart Z provides specific regulations for a select group of toxic or hazardous chemicals. The regulations set specific exposure limits, detail acceptable work procedures, delineate workplace/environmental sampling requirements, set specific PPE requirements, and denote the need for regulated work areas. This subpart also has the permissible exposure limits (PELs) for more than 500 hazardous chemicals. Subpart Z discusses, in some detail, working with and around potential cancer-causing chemicals. With many of the chemicals unique training requirements, as well as medical monitoring and surveillance exist. Requirements exist for posting and labels that warn of the dangers from exposure to specific chemicals. In many cases, precise decontamination is required, along with hygiene procedures to minimize potential contamination to workers or the spread of contamination. These regulations communicate the hazards involved and discuss the target organs, signs, and symptoms that accompany an occupational illness from one of these hazardous or toxic chemicals.

Because each of these chemicals have unique properties, adverse effects, handling procedures, signs and symptoms of overexposure, and regulatory requirements, the regulations specific to each chemical must be consulted and complied with.

This subpart also covers hazard communication, bloodborne pathogens, ionizing radiation, placarding, and laboratories' chemical safety.

Checklist

_____Does your company use any of the chemicals listed in sections 1000 through 1052?

_____Do any of the chemical mixtures you use on your jobsites contain any chemicals in sections 1000 through 1052?

_____Do your workers do asbestos or lead abatement work?

_____Do you have any sources of ionizing radiation?

_____Do your workers perform hazardous waste remediation work?

_____Do other contractors use any of the chemicals in sections 1000 through 1052 that might expose your own workers inadvertently?

_____Do you have a hazard communication program?

_____Do you provide training to your workers on any of the chemicals listed in section 1000 through 1052?

_____Does any of your work take your workers onto or into worksites where exposure to any of the chemicals in 1000 through 1052 could occur?

_____Do you have laboratories where hazardous chemicals exist or are used?

Sections of Subpart Z Regulation

1910.1000	Air contaminants
1910.1001	Asbestos
1910.1002	Coal tar pitch volatiles; interpretation of term
1910.1003	13 Carcinogens (4-nitrobiphenyl, etc.)
1910.1004	Alpha-naphthylamine
1910.1005	[Reserved]
1910.1006	Methyl chloromethyl ether
1910.1007	3,3'-Dichlorobenzidine (and its salts)
1910.1008	Bis-chloromethyl ether
1910.1009	Beta-naphthylamine
1910.1010	Benzidine
1910.1011	4-Aminodiphenyl
1910.1012	Ethyleneimine
1910.1013	Beta-propiolactone
1910.1014	2-Acetylaminofluorene
1910.1015	4-Dimethylaminoazobenzene
1910.1016	N-Nitrosodimethylamine
1910.1017	Vinyl chloride
1910.1018	Inorganic arsenic
1910.1020	Access to employee exposure and medical records
1910.1025	Lead
1910.1026	Chromium(VI)
1910.1027	Cadmium
1910.1028	Benzene
1910.1029	Coke oven emissions
1910.1030	Bloodborne pathogens
1910.1043	Cotton dust
1910.1044	1,2-Dibromo-3-chloropropane
1910.1045	Acrylonitrile
1910.1047	Ethylene oxide
1910.1048	Formaldehyde
1910.1050	Methylenedianiline
1910.1051	1,3-Butadiene
1910.1052	Methylene chloride
1910.1096	Ionizing radiation
1910.1200	Hazard communication
1910.1201	Retention of DOT markings, placards, and labels
1910.1450	Occupational exposure to hazardous chemicals in laboratories

9 Safety Hazards

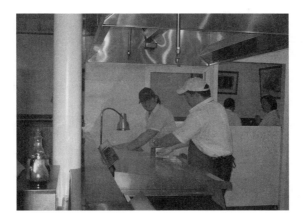

This hot grill poses a burn hazard to these workers.

9.1 HAZARD IDENTIFICATION

Hazard identification is used to examine the workplace for hazards with the potential to cause accidents. Hazard identification is often a worker-oriented process. The workers are trained in hazard identification and asked to recognize and report hazards for evaluation and assessment. Management is not as close to the actual work performed as are those performing the work. Even supervisors can use extra pairs of eyes looking for areas of concern.

Workers already have hazard concerns and have often devised ways to mitigate the hazards, thus preventing injuries and accidents. This type of information is invaluable when removing and reducing workplace hazards.

This approach to hazard identification does not require that it be conducted by someone with special training and can usually be accomplished by the use of a short fill-in-the-blank questionnaire. This hazard identification technique works well where management is open and genuinely concerned about the safety and health of its workforce. The most time-consuming portion of hazard identification is analyzing the information that has been relayed to management and the response to potential hazards identified. Empowering workers to identify hazards, making recommendations on abatement of the hazard, and suggesting how management can respond to these potential hazards are essential. Only three responses are required:

Hazard identification form

Worker's name (optional):_____

Date:_____

Jobsite:_____

Job Titles:_____

_____

1. Describe the hazard that exists.

_____

2. What are your recommendations for reducing or removing the hazard?

_____

3. What suggestions do you have for management for handling the hazard?

_____

4. Manager's or supervisor's response to hazard concern identified.

_____

Note: Use a separate form for each hazard identified.

FIGURE 9.1   Hazard identification form.

1. Identify the hazard
2. Explain how the hazards could be abated
3. Suggest what the company could do

Use a form similar to the one found in Figure 9.1.

The information obtained by hazard identification provides the foundation for making decisions on which jobs should be altered in order for the worker to perform the work safer and expeditiously. In addition, this process allows workers to become more involved in their own destiny. For some time, involvement has been recognized as a key motivator of workers.

It is important to remember that a worker may perceive something as a hazard, when in fact it may not be a true hazard; the risk may not match the ranking that the worker placed on it. Also, even if hazards exist, you need to prioritize them. You need to prioritize them according to the ones that can be handled quickly, which may take time or cost money above your budget. If the correction will cause a large capital expense and the risk is real but does not exhibit an extreme danger to life and health, you might need to wait until next year's budget cycle. An example of this would be when workers complained of smell and dust created by a chemical process. The dust was not above accepted exposure limits and the smell was not overwhelming. Therefore, the company elected to install a new ventilation system, but not until the next year because of budgetary constraints, and the use of PPE would be required until the hazard can be removed. Hazard identification is a process that is controlled by management. Management must assess the outcome of the hazard identification process and determine whether immediate action is necessary or if, in fact, there is an actual hazard involved. When you do not view a reported hazard as an actual hazard, it is critical to the ongoing process to inform the worker that you do not view it as a true hazard and explain why. This will ensure the continued cooperation of workers in hazard identification.

The expected benefits are a decrease in the incidents of injuries, a decrease in lost workdays and absenteeism, a decrease in workers' compensation cost, better productivity, and an increase in cooperation and communication. The baseline for determining the benefits of hazard identification can be formulated from existing company data on occupational injuries/illnesses, workers' compensation, attendance, profit, and production.

9.2 WORKSITE HAZARD ANALYSIS

Worksite analysis is the process of identifying hazards related to a project, process, or activities at the worksite. Identify the workplace hazards before determining how to protect employees. In performing worksite analyses, consider not only hazards that currently exist in the workplace but also those hazards that could occur because of changes in operations or procedures or because of other factors, such as concurrent work activities. First, perform safety hazard analyses of all activities or projects before the start of work, determine the hazards involved with each phase of the work process, and perform regular safety and health site inspections.

Secondly, require supervisors and employees to inspect their workplace prior to the start of each workshift or new activity, investigate accidents and near misses, and analyze trends in accident and injury data.

When performing a worksite analysis, all hazards should be identified. This means conducting comprehensive baseline worksite surveys for safety and health and periodic comprehensive updated surveys. You must analyze planned and new facilities, processes, materials, and equipment, as well as perform routine job hazard analyses. This also means that regular site safety and health inspections need to be conducted so that new or previously missed hazards and failures in hazard controls are identified.

A job safety assessment or analysis should be performed at the start of any task or operation. The designated competent or authorized person should evaluate the task or operation to identify potential hazards and determine the necessary controls. This assessment should focus on actual worksite conditions or procedures that differ from or were not anticipated in the related project or phase hazard analysis. In addition, the authorized person shall ensure that each employee involved in the task or operation is aware of the hazards related to the task or operation and of the measures or procedures that they must use to protect themselves. Note: the job safety assessment is not intended to be a formal, documented analysis, but instead is more of a quick check of actual site conditions and a review of planned procedures and precautions.

9.3 TRAINING ON HAZARD IDENTIFICATION

Supervisors and workers must be trained to identify hazards in order to prevent accidents, and identify existing and potential hazards that can prevail in the workplace. When looking at specific jobs, identify the hazards by breaking the job down into a step-by-step sequence and identify potential hazards associated with each step. Consider the following:

- Is there a danger of striking against, being struck by, or otherwise making injurious contact with an object? (For example, can tools or materials be dropped from overhead striking workers below.)
- Can the employee be caught in, on, or between objects? (For example, an unguarded v-belt, gears, or reciprocating machinery.)
- Can the employee slip, trip, or fall on the same level, or to another level? (For example, slipping in an oil-changing area of a garage, tripping on material left on stairways, or falling from a scaffold.)
- Can the employees strain themselves by pushing, pulling, or lifting? (For example, pushing a load into place or pulling a load on a hand truck.)
- Does the environment have hazardous toxic gas, vapors, mist, fumes, dust, heat, or ionizing or nonionizing radiation? (For example, arc welding on galvanized sheet metal produces toxic fumes and nonionizing radiation.)

During training, practice identifying all hazards or potential hazards. With the identification of the hazards, take steps to prevent the accidents or incidents from occurring. If you know the hazard, it is easier to develop interventions that mitigate the risk potential. These interventions may be in the form of safe operating procedures. Training workers to identify potential or real hazards quickly will definitely reduce the number of accidents/incidents that could occur. Once hazards are identified, you must have a system of reporting these hazards, which is accompanied by real-time response by supervision and management.

9.4 WORKSITE HAZARD IDENTIFICATION

You must identify the workplace hazards before you can determine how to protect your employees. In performing worksite hazard identification, you must consider not only hazards that currently exist in the workplace but also those hazards that could occur because of changes in operations or procedures or because of other factors such as concurrent work activities. For this element, you should

- Perform hazard identification of all worksites before the start of work
- Perform regular safety and health inspections
- Require supervisors and employees to inspect their workplace before the start of each workshift or new activity
- Investigate accidents and near misses
- Analyze trends in accident and injury data

To ensure that all hazards are identified, conduct comprehensive baseline worksite surveys for safety and health and periodic comprehensive updated surveys. Analyze planned and new facilities, processes, materials, and equipment and perform routine job hazard analyses.

Provide for regular site safety and health inspections, so that new or previously missed hazards and failures in hazard controls are identified. To ensure that employee insight and experience in safety and health protection are utilized and employee concerns are addressed, provide a reliable system for employees, without fear of reprisal, to notify management personnel about conditions that appear hazardous and to receive timely and appropriate responses; and encourage employees to use the system. Provide for investigation of accidents and near-miss incidents, so that their causes and means for their prevention are identified. Analyze injury and illness trends over time, so that patterns with common causes can be identified and prevented. Only through effective hazard assessment and control/prevention application can safety hazard be mitigated.

9.5 HAZARD ASSESSMENT GUIDE

Conducting a hazard assessment to determine if there are any safety hazards present or likely to be present, which require the use of controls, is a critical step in prevention. The assessment must match the particular hazard with appropriate controls. The following is a recommended procedure for conducting a hazard assessment:

- Review injury and accident data: Two sources of injury data that can provide helpful information for assessing hazards are the OSHA form 300 log and workers' compensation claims.
- Inform employees and supervisors of the process: Involve the employees and supervisors from each work area that is assessed. Review the job procedures, potential hazards, and existing controls. Discuss the reasons for the survey and the procedures used for the assessment. Point out that the assessment is not a review of their job performance.
- Conduct a walk-through survey: Conduct a walk-through survey of the work areas that may need new or improved safety hazard approaches. The purpose of the survey is to identify sources of hazards to workers and coworkers. Observe the following: layout of the workplace, location of the workers, work operations, hazards, and places where controls and personal protective equipment (PPE) are currently used, including any control devices and reasons for their use.
- Consideration should be given to the following basic hazard categories that have the potential to cause errant release of energy or agents:
  - Air contaminants
  - Biological agents
  - Chemical exposure (inhalation, ingestion, skin contact, eye contact, or injection)
  - Compression (roll-over or pinching objects)
  - Dust
  - Electrical hazard
  - Extreme cold
  - Fire and explosion hazards
  - Heat

o   Impact (falling/flying objects)
o   Lifting and other manual handling operations
o   Light (optical) radiation (welding, brazing, cutting, furnaces, etc.)
o   Materials falling from height, rolling, shifting, or caving in
o   Moving parts of machinery, tools, and equipment
o   Noise
o   Penetration (sharp objects piercing foot/hand)
o   Pressure systems (such as steam boilers and pipes)
o   Radiation
o   Slipping/tripping hazards
o   Vehicles (such as forklifts and trucks)
o   Vibration
o   Violence
o   Water (potential for drowning or fungal infections caused by wetness)
o   Work at height (such as work done on scaffolds or ladders)

9.5.1   ORGANIZE THE DATA

Following the walk-through survey, organize the data and information for use in the hazard assessment. The objective is to prepare for an analysis of the hazards in the environment to enable proper selection of controls and PPE.

9.5.2   ANALYZE THE DATA

Having gathered and organized the data, an estimate of the potential for injuries and illnesses should be made. Each of the basic hazards should be reviewed from the walk-through survey and determination made as to the type, level of risk, and seriousness of potential injury from each of the hazards found in the area. The possibility of exposure to several hazards simultaneously should be considered.

9.5.3   SELECTION OF CONTROLS

If the safety hazards cannot be removed, isolated, or a less hazardous procedure or process implemented, the development of engineering (guards) and administrative (enhanced training) controls should be considered. When all else fails and no other options exist, then appropriate PPE for the exposure should be selected and worn.

After completion of the hazard assessment, the general suggested process for the selection of PPE is to

*   Become familiar with the potential hazards, the PPE available, and its features (splash protection, impact protection, etc.) to prevent injuries and illnesses
*   Compare the hazards associated with the work environment and the capabilities of the available PPE (such as shaded lenses for welding or flying objects during a grinding operation)
*   Select the PPE that ensures a level of protection greater than the minimum required to protect employees from the hazards
*   Fit the user with the devise and provide instruction on care, use, and limitations of PPE

Note: PPE alone should not be relied upon to provide protection against hazards but should be used in conjunction with engineering controls, administrative controls, and procedural controls.

9.5.4   FITTING THE PPE

Careful consideration must be given to comfort and fit. The right size should be selected to encourage continued use of the devise. Adjustments should be made on an individual basis for a comfortable fit while still maintaining the PPE in proper position (see Figure 9.2).

9.5.5   REASSESSMENT OF THE HAZARDS

Reassessment of the workplace is necessary to ensure continued protection of the workers by identifying and evaluating

- New equipment and processes
- Accident records
- The suitability of previously selected PPE

Fully encapsulating suit

Apron, gloves, hardhat,
faceshield, boot covers

FIGURE 9.2   Hazard assessment determines what personal protective equipment is needed.

9.6 CONTROLLING HAZARDS

Ideally, hazards should be controlled by applying modern management principles. Use a comprehensive, proactive system to control hazards rather than a reactive, piecemeal response to each concern as it arises. To be proactive, an employer should

- Establish a health and safety policy, in consultation with the committee, to communicate to employees that the employer is committed to health and safety. Build health and safety into all aspects of the organization such as tendering, purchasing, hiring, and so on. Ensure everyone understands that health and safety is as important as any other area of the organization.
- Communicate the health and safety policy through the management structure. Ensure everyone understands his or her duties.
- Train managers, supervisors, and workers to carry out their responsibilities under the policy.
- Equip managers and supervisors to apply modern management and supervisory practices in their safety responsibilities.
- Administer the policy in the same way that other policies are managed.

9.6.1 MANAGING THE SAFETY CONTROLS

In order to control safety hazards and manage them effectively, employers need to apply a comprehensive approach. Part of that approach should include

- Review chemical and biological substances in the workplace.
- List all chemical and biological substances of concern to workers in the workplace (see Figure 9.3).
- Identify things such as work areas and production processes where products are stored, handled, or produced.
- Determine which substances are controlled products.
- Obtain current material safety data sheets (MSDSs) for controlled products and make sure each MSDS is less than 3 years old.
- Ensure that a mechanism is in place to update MSDSs regularly and as new information becomes available.
- Ensure that transport, handling, storage, and disposal practices, etc. for each product meet standards set by the product MSDSs.
- Make sure that MSDSs are readily available to workers throughout the organization.
- Assess the risk associated with handling, use, storage, and disposal of the product. For example, consider its flammability, toxicity, corrosiveness, reactivity, and explosiveness. Include emergency response requirements in your assessment. For example, what happens if there is a spill, fire, explosion, or other mishap? How will the emergency be handled? What must be done to care for the injured and protect others at risk?
- Determine what measures need to be taken to control those risks and apply the technical steps discussed in this chapter.

FIGURE 9.3   With the wide variety of chemicals used in the workplace, special care should be taken.

- Establish a system to ensure containers are adequately labeled.
- Ensure suppliers attach appropriate supplier labels to their controlled products. Labels should include the hazards of the product, and the precautionary and first-aid measures.
- Replace damaged or missing labels.
- Provide appropriate workplace labels for containers holding a product taken from its original container.
- Develop written work procedures based on the control measures.
- Arrange for the training of workers.
- Train workers and supervisors handling controlled products how to identify and control the hazards of the products they use.
- Update workers and supervisors when new information becomes available.

9.6.2 TECHNICAL ASPECTS OF HAZARD CONTROL

As a first step in hazard control, determine if the hazards can be controlled at their source (where the problem is created) through applied engineering. If this does not work, try to put controls between the source and the worker. The closer a control is to the source of the hazard, the better. If this is not possible, hazards must be controlled at the level of the worker. For example, workers can be required to use a specific work procedure to prevent harm.

One type of hazard control may not be completely effective. A combination of several different types of hazard controls often works well. Whatever method is used, an attempt should be made to try to find the root cause of each hazard and not simply control the symptoms. For example, it might be better to redesign a work process than simply improve a work procedure. It is better to replace, redesign, isolate, or quiet a noisy machine than to issue nearby workers with hearing protectors.

9.7 SOURCE CONTROL

9.7.1 ELIMINATION

First, try eliminating the hazard. Getting rid of a hazardous job, tool, process, machine, or substance may be the best way of protecting workers. For example, a salvage firm might decide to stop buying and cutting up scrapped bulk fuel tanks (due to explosion hazards).

9.7.2 SUBSTITUTION

If elimination is not practical, try replacing hazardous substances with something less dangerous. For example, a hazardous chemical can be replaced with a less hazardous one. A safer work practice can be used. Be sure to also identify, assess, and control the hazards of substitutes.

9.7.3 REDESIGN

Sometimes engineering can be used to redesign the layout of the workplace, workstations, work processes, and jobs to prevent ergonomic hazards. For example, containers can be redesigned to be easier to hold and lift. Engineering may be able to improve workplace lighting, ventilation, temperature, process controls, and so forth.

9.7.4 ISOLATION

Isolating, containing, or enclosing the hazard is often used to control chemical hazards and biohazards. For example, negative-pressure glove boxes are used in medical labs to isolate biohazards.

9.7.5 AUTOMATION

Dangerous processes can sometimes be automated or mechanized. For example, spot welding operations in car plants can be handled by computer-controlled robots. Care must be taken to protect workers from robotic hazards.

9.8 CONTROL ALONG THE PATH FROM THE HAZARD TO THE WORKER

Hazards that cannot be isolated, replaced, enclosed, or automated can sometimes be removed, blocked, absorbed, or diluted before they reach workers. Usually, the further a control keeps hazards away from workers, the more effective it is.

9.8.1 Barriers

A hazard can be blocked. For example, proper equipment guarding can protect workers from contacting moving parts. Screens and barriers can block welding flash from reaching workers. Machinery lockout systems can protect maintenance workers from physical agents such as electricity, heat, pressure, and radiation.

9.8.2 Absorption

Baffles can block or absorb noise. Local exhaust ventilation can remove toxic gases, dusts, and fumes where they are produced.

9.8.3 Dilution

Some hazards can be diluted or dissipated. For example, general (dilution) ventilation might dilute the concentration of a hazardous gas with clean, tempered air from the outside. Dilution ventilation is often quite suitable for less toxic products. However, it is not effective for substances that are harmful in low concentrations. It may also spread dusts through the workplace rather than completely removing them.

9.9 CONTROL AT THE LEVEL OF THE WORKER

Control at the level of the worker usually does not remove the risk posed by a hazard. It only reduces the risk of the hazard injuring the worker and lessens the potential seriousness of an injury. Therefore, most safety experts consider control at the level of the worker to be the least effective means of protecting workers.

9.9.1 Administrative Controls

These include introducing new policies, improving work procedures, and requiring workers to use specific PPE and hygiene practices. For example, job rotations and scheduling can reduce the time that workers are exposed to a hazard. Workers can be rotated through jobs requiring repetitive tendon and muscle movements to prevent cumulative trauma injuries. Noisy processes can be scheduled when few workers are in the workplace. Standardized written work procedures can ensure that work is done safely. Employees can be required to use shower and change facilities to prevent absorption of chemical contaminants. The employer is responsible for enforcing administrative controls.

9.9.2 Work Procedures, Training, and Supervision

Supervisors can be trained to apply modern safety management and supervisory practices. Workers can be trained to use standardized safe work practices. The employer should periodically review and update operating procedures and worker training. Refresher training should be offered periodically. The employer is expected to ensure that employees follow safe work practices.

9.9.3 EMERGENCY PLANNING

Written plans should be in place to handle fires, chemical spills, and other emergencies. Workers should be trained to follow these procedures and use appropriate equipment. Refresher training should be provided regularly.

9.9.4 HOUSEKEEPING, REPAIR, AND MAINTENANCE PROGRAMS

Housekeeping includes cleaning, waste disposal, and spill cleanup. Tools, equipment, and machinery are less likely to cause injury if they are kept clean and well maintained.

9.9.5 HYGIENE PRACTICES AND FACILITIES

Hygiene practices can reduce the risk of toxic materials being absorbed by workers or carried home to their families. Street clothing should be kept in separate lockers to avoid contamination from work clothing. Eating areas can be segregated from work areas. Eating, drinking, and smoking should be forbidden in toxic work areas. Where applicable, workers may be required to shower and change clothes at the end of the shift.

9.9.6 SELECTING CONTROLS

Selecting a control often involves evaluating and selecting temporary and permanent controls, implementing temporary measures until permanent (engineering) controls can be put in place, and implementing permanent controls when reasonably practicable. For example, suppose a noise hazard is identified. Temporary measures might require workers to use hearing protection. Long-term, permanent controls might use engineering to remove or isolate the noise source.

9.10 PERSONAL PROTECTIVE EQUIPMENT

PPE and clothing are used when other controls are not feasible (for example, to protect workers from noise exposure when using chain saws), where additional protection is needed, and where the task or process is temporary (such as periodic maintenance work).

PPE is much less effective than engineering controls since it does not eliminate the hazard. It must be used properly and consistently to be effective. Awkward or bulky PPE may prevent a worker from working safely. In some cases, PPE can even create hazards, such as heat stress.

The employer must require workers to use PPE wherever its use is prescribed by the regulations or organizational work procedures. Workers must be trained to use, store, and maintain their equipment properly. The employer, supervisor, and workers must understand the limitations of their PPE.

A more detailed coverage of PPE can be found in Chapter 21.

9.10.1   Eye and Face Protection

Refer to the eye and face protection manuals for guidance on the proper selection of PPE for eye and face protection. Some occupations for which eye and face protection should be routinely considered are carpenters, electricians, machinists, lathe operators, mechanics, plumbers, health care workers, equipment operators, maintenance personnel, mechanics, welders, and laborers.

9.10.2   Head Protection

Refer to the head protection manuals for guidance on proper selection of PPE for head protection. Some examples of the occupations for which head protection should be routinely considered are carpenters, electricians, mechanics, plumbers, packers, welders, laborers, freight handlers, and timber cutting and warehouse laborers.

9.10.3   Foot Protection

Refer to the foot protection manuals for guidance on proper selection of PPE for foot protection. Some examples of the occupations for which foot protection should be routinely considered are shipping and receiving clerks, stock clerks, carpenters, electricians, machinists, mechanics, plumbers, welders, gardeners, and groundskeepers.

9.10.4   Hand Protection

No one type or style of glove can provide protection against all potential hand hazards. Therefore, it is important to select the most appropriate glove for a particular application and determine how long it can be worn and whether it can be reused. It is important to know the performance characteristics of gloves relative to the specific hazard. Documentation in the form of glove charts from the manufacturer should be requested. The work activities of the employee should be analyzed to determine the degree of dexterity required, the duration, frequency, degree of exposure and physical stresses that will be applied. Consider the following factors in glove selection for chemical hazards:

- Toxic properties of the chemical must be determined in relation to skin absorption.
- MSDSs are an excellent source of information.
- For mixtures and formulated chemicals, a glove must be selected on the basis of the chemical component with the shortest breakthrough time.
- Employees must be able to remove the gloves in such a manner as to prevent skin contamination.

Some of the occupations that should use gloves to prevent injury and contamination are nurses, laboratory technicians, janitors, maintenance personnel, and those performing lifting and carrying tasks.

9.10.5 UPPER/LOWER BODY PROTECTION

Refer to the upper/lower body protection manuals for guidance on the proper selection of PPE for upper or lower body protection. Some occupations for which body protection should be routinely considered include lab technicians and researchers, fire control workers, nurses, health care workers, and waste collectors.

9.10.6 CLEANING AND MAINTENANCE

All PPE must be kept clean and properly maintained. Cleaning is particularly important for eye and face protection where dirty or fogged lenses could impair vision. All PPE should be cleaned, inspected, and maintained at regular intervals so that PPE can provide the requisite protection. Contaminated PPE that cannot be decontaminated must be disposed of in a manner that protects employees from exposure to hazards.

9.11 EVALUATING THE EFFECTIVENESS OF CONTROLS

Sometimes hazard controls do not work as well as expected. Therefore, the committee or representative should monitor the effectiveness of the corrective action taken by the employer during inspections and other activities. Ask these questions:

- Have the controls solved the problem?
- Is the risk posed by the original hazard contained?
- Have any new hazards been created?
- Are new hazards appropriately controlled?
- Are monitoring processes adequate?
- Have workers been adequately informed about the situation?
- Have orientation and training programs been modified to deal with the new situation?
- Are any other measures required?
- Has the effectiveness of hazard controls been documented in your committee minutes?

9.12 SUMMARY

It is important to identify existing or potential safety hazards and take steps to remove or limit their effects on the workforce. This can be accomplished by using many approaches to control, prevent, or remove safety hazards that could cause injury, illness, and death in the workplace.

All identified hazardous conditions should be eliminated or controlled immediately. Where this is not possible, interim control measures are to be implemented immediately to protect workers. Warning signs must be posted at the location of the hazard; all affected employees need to be informed of the location of the hazard and of the required interim controls, and permanent control measures must be implemented as soon as possible.

Once hazards have been identified, assessed, and controlled, the employer and worker representatives should work together to develop training programs for workers, emergency response procedures, and health and safety requirements for contractors. Someone needs to be responsible to monitor these activities to ensure that they are effective.

The employer is responsible for ensuring that workplace hazards are identified, assessed, and appropriately controlled. Workers must be told about the hazards they face and taught how to control them.

The employer is expected to consult and involve occupational safety health professionals or worker representatives in the hazard control process. Helping the employer identify, assess, and control hazards is one of the most important roles of the responsible party in the internal responsibility system. Hazards are broadly divided into two groups: hazards that cause illness (health hazards) and those that cause injury (safety hazards). Hazards can be identified by asking what harm could result if a dangerous tool, process, machine, piece of equipment, and so forth, failed. Health and safety hazards can be controlled at the source, along the path, or at the level of the worker. Once controls are in place, they must be checked periodically to make sure that they are still working properly. Someone should be responsible to audit the hazard controls in the internal responsibility system and help the employer keep them effective.

Where a supervisor or foreman is not sure how to correct an identified hazard or is not sure if a specific condition presents a hazard, the supervisor or foreman should seek technical assistance from the designated competent person, safety and health officer, or technical authority. Some of the other techniques used to control hazards are as follows.

9.12.1 Job Safety Assessment

Prior to the start of any task or operation, the designated competent or company authorized person should evaluate the task or operation to identify potential hazards and to determine the necessary controls. This assessment shall focus on actual worksite conditions or procedures that differ from or were not anticipated to the related other hazard analysis. In addition, the competent person shall ensure that each employee involved in the task or operation is aware of the hazards related to the task or operation and of the measures or procedures that they must use to protect themselves. Note that the job safety assessment is not intended to be a formal documented analysis, but instead is more of a quick check of actual site conditions and a review of planned procedures and precautions.

9.12.2 Controls

Controls come in all forms, from engineering, administrative, to PPE. The best controls are those that can be placed on equipment before involving people and thus either preclude or guard the workforce from the hazard. The use of administrative controls relies on individuals following policies and procedures to control hazards and exposure to hazards. However, as we all know this certainly provides no guarantees that the protective policies and procedures will be adhered to unless effective supervision and enforcement exist. Again, this relies on the company having a strong commitment to occupational safety and health. The use of PPE will not control hazards unless the individuals who are exposed to the hazards wear the appropriate PPE. The use of PPE is usually considered the control of last resort since it has always been difficult for companies to ensure that exposed individuals are indeed wearing the required PPE.

9.12.3 ACCIDENT REPORTING

All incidents and accidents resulting in injury or causing illness to employees and events (near-miss accidents) shall be reported to

- Establish a written record of factors that cause injuries and illnesses and occurrences (near misses) that might have resulted in injury or illness, but did not, as well as property and vehicle damage
- Maintain a capability to promptly investigate incidents and events to initiate and support corrective or preventive action
- Provide statistical information for use in analyzing all phases of incidents and events
- Provide the means for complying with the reporting requirements for occupational injuries and illnesses

Incident reporting system requirements apply to all incidences involving company employees, onsite vendors, contractor employees, and visitors, which results in (or might have resulted in) personal injury, illness, and property and vehicle damage.

Injuries and illnesses that require reporting include those injuries and illnesses occurring on the job that result in any of the following: lost work time, restrictions in performing job duties, requirement for first aid or outside medical attention, permanent physical bodily damages, or death. Examples of reportable injuries and illnesses include, but are not limited to, heat exhaustion from working in hot environments, strained back muscles from moving equipment, acid burns on fingers, etc.

Other incidents requiring reporting include those incidents occurring on the job that result in any of the following: injury or illness, damage to a vehicle, fire/explosion, property damage of more than $100, or chemical releases requiring evacuation of at least the immediate spill area.

Examples of nonreportable injuries and illnesses include small paper cuts, common colds, and small bruises not resulting in work restrictions or requiring first aid or medical attention.

Events (near misses) or other incidents (near misses) that, strictly by chance, do not result in actual or observable injury, illness, death, or property damage are required to be reported. The information obtained from such reporting can be extremely useful in identifying and mitigating problems before they result in actual personal or property damage. Examples of near-miss incidences required to be reported include the falling of a compressed-gas cylinder, overexposures to chemical, biological, or physical agents (not resulting in an immediately observable manifestation of illness or injury), and slipping and falling on a wet surface without injury.

9.12.4 INCIDENT REPORTING PROCEDURES

The following procedures are to be followed by all employees to effectively report occupational injuries and illnesses and other incidents or events. Serious injury or illness posing a life-threatening situation should be reported immediately to the local emergency response medical services (call 911).

Injuries and illnesses shall be reported, by the injured employee, to his or her supervisor in person or by phone soon after any life-threatening situation has been addressed. If the injured employee is unable to report immediately, then the incident should be reported as soon as possible.

Upon notification of an occupational injury or illness, the supervisor should complete the incident/accident report and, if possible, send it with the injured employee to the medical professional involved. The incident/accident report form must be completed and forwarded to the company's medical department even if the employee receives medical treatment at the hospital or from a private physician.

Incidents not involving injury or illness, but resulting in property damage, must also be reported within 24 h of the incident. In cases of a fire or explosion that cannot be controlled by one person, vehicular accident resulting in injury or more than $500 worth of damage, or a chemical release requiring a building evacuation, the involved party must immediately report the incident to the emergency response services in the area (911—police, fire, etc.)

All near-miss incidences also must be reported on the incident/accident report form within 24 h of occurrence. In place of indicating the result of the incident (i.e., actual personal or property damage), the reporting person shall indicate the avoided injury or damage.

Events, hazardous working conditions or situations, and incidents involving contractor personnel must be reported to the supervisor or safety professional immediately.

The safety department will record and maintain injury and illness data on the OSHA 301 "Injury and Illness Incident Report." The OSHA 300 log "Log of Work-Related Injuries and Illnesses" and the OSHA 300A "Summary of Work-Related Injuries and Illnesses" must be posted from February 1 to April 30.

9.12.5 TRAINING

To ensure that all employees understand the incident reporting requirements and are aware of their own and other's responsibilities, annual training sessions will be held with all employees to review procedures and responsibilities. New employee orientation training should include information on incident reporting and procedures.

9.12.6 PROGRAM AUDITS

The effectiveness of a program can only be accomplished if the program is implemented. Therefore, periodic reviews and audits shall be conducted to confirm that all employees are familiar with the incident reporting requirements.

The identification of hazards and the controlling of hazards within the workplace are the responsibilities of the employer and their management. Since employers are in control of the workplace, they have the right to set and enforce their own occupational safety and health requirements.

10 Health Hazards

Ergonomic-related health issues can arise from poorly designed material-handling tasks.

A health hazard is any agent, situation, or condition that can cause an occupational illness. There are five types:

- Chemical hazards, such as battery acid and solvents
- Biological hazards, such as bacteria, viruses, dusts, and molds. Biological hazards are often called biohazards
- Physical agents (energy sources) strong enough to harm the body, such as electric currents, heat, light, vibration, noise, and radiation
- Work design (ergonomic) hazards
- Workplace stress

A health hazard may produce serious and immediate (acute) affects and symptoms. It may cause long-term (chronic) problems or may have long periods between exposure and the occurrence of the disease or illness (latency period). All or part of the body may be affected. Someone with an occupational illness may not recognize the symptoms immediately. For example, noise-induced hearing loss is often difficult for victims to detect until it is advanced.

10.1 IDENTIFYING HEALTH HAZARDS

Finding health hazards is an investigative process that requires a systematic approach and many facets and information need to be reviewed:

- Prepare a list of known health hazards in the workplace based on records and events.
- Review the total facility, floor plans, and work process diagrams to identify health hazard sources and locations.
- Interview workers, supervisors, and managers to identify known and suspected health hazards not already on the list.
- Make use of the five senses and use an industrial hygienist if validations of your observations are needed. The industrial hygienist can perform accurate sampling as well as offer expert advice.

10.1.1 Prepare a List of Known Health Hazards in the Workplace

As a first step, someone need to be assigned responsibility to help the employer prepare a current list of chemical and biological substances, physical agents, work design hazards, and stress-related problems at the workplace. This will require the employer to provide the responsible party with an updated copy of existing lists of chemical and biological substances.

Check current product labels and material safety data sheets (MSDSs). Each chemical and biological substance controlled under the auspices of the hazard communication standard in the workplace must have appropriate container labels. Current MSDSs must be readily available to the responsible party. Look at container labels and MSDSs for hazard warnings and symbols (such as the skull and cross bones). Identify substances controlled under Occupational Safety and Health Administration (OSHA). This process should include

- Examining products exempted from OSHA controls, but of concern to workers (such as pesticides).
- Conducting inspections to identify defects, such as substance containers and pipes that are not properly labeled.
- Reading inspection and accident reports, complaint files, shop plans, first-aid logs, OSHA logs, product literature, and other documents.
- Monitoring the workplace (measuring noise, temperatures, concentrations of airborne chemicals and so forth). This usually occurs under the guidance of an industrial hygienist or qualified safety and health personnel. The results must be provided to the responsible party including any measurements of biological and chemical substances taken in the workplace.

10.1.2   REVIEW FLOOR PLANS AND WORK PROCESS DIAGRAMS

A diagram of the facility or floor plans may show, for example, that certain points in the production line release chemicals into the air or that inadequate exhaust or ventilation exists. Put monitoring equipment at these locations to determine what hazardous substances are present and in what quantities. Check for work design problems that may cause back injury and other ergonomic hazards. Look for tasks associated with accidents, complaints, and ill health.

10.1.3   INTERVIEW WORKERS, SUPERVISORS, AND MANAGERS

Interview workers, supervisors, and managers during the inspection process and ask them what hazards they work with and what work-related health problems they know about. However, deal with the concerns of workers at any time, not just during inspections.

Talk to the vendor or supplier if you need more information about a specific product, tool, or piece of equipment. Contact OSHA if you need more information about specific hazards.

10.1.4   USE YOUR FIVE SENSES

Some substances and physical agents can be detected with your five senses. For example, dusts and fumes sometimes form a haze. Vibration and temperature can be felt. An abnormal taste may be a sign of airborne chemicals. Some substances have a distinct color, visual appearance, or odor. The human nose can quickly become overcome by smells and cannot detect amounts or the nature of the hazard. Therefore, it is not to be trusted.

Odor is a common warning property. Be careful to check the substance's odor threshold in the physical properties section of its MSDS. Use odor to detect a substance only if it can be smelled at levels below hazardous concentrations.

Unfortunately, many hazardous agents and conditions cannot be detected with the senses. Others, such as hydrogen sulfide (H_2S) gas, are often dangerous when strong enough to be detected in this way. Using your senses is not always a safe way of detecting hazards.

10.1.5   QUICK HEALTH HAZARD IDENTIFICATION CHECKLIST

Each of these questions needs to be answered:

- What chemical substances are produced, used, handled, stored, or shipped in the workplace?
- Are any vapors, gases, dusts, mists, or fumes present (including chemical by-products of work processes)?
- Are biological substances (such as bacteria, viruses, parasites, dusts, molds, and fungi) present in the workplace, the ventilation systems, and other components of the physical plant?
- Are physical agents (energy sources strong enough to harm the body, such as electric currents, heat, light, vibration, noise, and radiation) present?

- Are temperature extremes present?
- Do ergonomic hazards exist—such as work requiring lifting, awkward posture, repetitive motions, excessive muscular force, or computer use?
- Could any work processes, tools, or equipment cause health hazards (such as back injuries, soft tissue injuries, whole body vibration, hearing loss, infections, and so forth)?
- Could departures from safe work practices cause illnesses?
- Can any potential health hazards be detected with the senses (smell, taste, touch, hearing, and sight)?
- Is there a presence of harmful stress in the workplace?
- Are there any complaints from workers about workplace-related health problems?

10.1.6 HEALTH HAZARDS ASSESSMENT

Once a health hazard is identified, the risk it poses to workers must be assessed. The safety and health staff or other responsible parties can help the employer do this by using monitoring equipment to assess exposure levels and by determining the probability and severity of any potential exposure.

There are many different monitors for detecting and assessing health hazards. Some, such as air monitors, sample the work environment at specific places for specific chemical hazards. Others measure the levels of noise, vibration, and so on. Safety and health personnel or others assigned to monitor the health aspects of the workplace can get advice on how to use monitors from the supplier. Advice on how to interpret monitoring results can be obtained from consultants and agencies.

10.2 CHEMICAL HAZARDS

If possible, use monitoring equipment to determine exactly what the exposure levels for health hazards are in the workplace and at workstations. Different hazards require different monitoring techniques and equipment. The employer may decide to bring in experts to do the monitoring. Supervisors and workers should be kept informed of the reasons for the monitoring (see Figure 10.1).

Once an exposure level is determined, compare it with standards set by the organization, industry, legislation, and so on. Review MSDSs as well as industry and product literature for advice. More detailed coverage of chemical hazards is found in Chapter 12.

10.3 BIOLOGICAL HAZARDS

Some biological hazards can be detected by monitoring. However, the risk of catching an illness can usually be assessed by applying knowledge of the disease involved, including how it spreads and infects people (see Figure 10.2). Biological safety data sheets provide useful information such as survival characteristics of the microorganism outside of the body, how it is transmitted, and how likely workers are to contract the disease. Biological hazards are covered in Chapter 11.

FIGURE 10.1   Application of pesticides can lead to both acute and chronic health effects.

10.4   PHYSICAL HEALTH HAZARDS

Physical health hazards are sources of energy strong enough to cause harm. They include noise, vibration, heat or cold, and radiation.

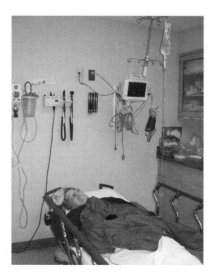

FIGURE 10.2   Diseases can be contracted from others who are infected.

10.4.1 NOISE

Common noise sources include equipment, work processes, compressors, ventilation systems, and power tools. Generally, if ordinary conversation cannot be understood at normal distances, noise levels are too loud. Hazard identification techniques, such as inspections, monitoring, and conversations with workers will usually detect noise concerns. High levels of noise can result in hearing loss as well as other health issues such as increased blood pressure.

10.4.2 VIBRATION

Vibration is a rapid back and forth or up and down motion that may affect all or part of the body. It can gradually damage nerves and circulation systems in limbs and affect internal organs. Standard hazard identification techniques can detect jobs and workers that are exposed to vibration. Monitoring and assessing vibration usually requires technical specialists.

10.4.3 HEAT AND COLD

The health effects of too much heat include heat cramps, heat exhaustion, and heat stroke. Cold can produce frostbite and hypothermia. As well as causing serious health problems, heat and cold stress disorders can reduce performance and increase the risk of accidents.

Employers should maintain thermal conditions that are reasonable and appropriate for the work performed. If it is not reasonably practicable to adequately control indoor conditions, or where work is done outdoors, the employer must take effective measures to protect workers from heat and cold stress disorders.

Workers can dress in layers, keep dry, and take warm up breaks to prevent cold-related problems. Heat problems can come about even when workers are acclimatized to the heat. The uses of a regulated work/rest cycle tied to the temperature and the types of work often helps as well as prevents dehydration, but remember that dehydration can occur in cold environments also.

The employer must provide suitable monitoring equipment if workers are concerned about thermal conditions. The assessment must consider factors such as temperature, humidity, airflow, wind, and work levels.

10.4.4 RADIATION

Radiation is made up of moving particles or waves of energy. It is divided into two groups: (1) Ionizing radiation; and (2) nonionizing radiation, both of which are covered in Chapters 17 and 19.

Ionizing radiation is given off by decaying radioactive elements, such as uranium. Specialized monitoring equipment is used to measure and assess radiation exposures. Radiation workers are also required to wear badges that measure the radiation dose they receive. Ionizing radiation can result in acute and chronic health issue such as burns, organ destruction, and lung and other cancers.

Nonionizing radiation includes

- Ultraviolet radiation given off by sun lamps and welding equipment. Ultraviolet radiation can burn the skin and cause eye damage.
- Infrared radiation (radiated heat) used in cooking and warming equipment in food processing and industrial packaging.
- Lasers producing concentrated beams of light used in a variety of commercial, medical, and industrial purposes. Care must be taken to ensure that lasers are set up properly, adequately shielded, and cannot damage the eyes or skin of workers.
- Microwave and high radio frequency radiation used in cooking equipment, radar, and in high-energy radio transmission and communications equipment. If not properly shielded, some equipment may injure the skin, eyes, and other organs.
- Long wave radiation used in radio and other communications equipment. Some equipment can heat the entire body.

Exposure to nonionizing radiation can result in burns of skin, eye damage, and damage of some organs damage.

10.5   ERGONOMIC HAZARDS

Hazards can exist in the design of the workplace, the workstation, tools and equipment, and the workflow. Ergonomics is concerned with identifying and controlling these hazards by reducing the physical, environmental, and mental stresses associated with a job. It does this by trying to balance the capabilities of the worker with the demands of the job. Ideally, the job should fit the person's mental, physical, and psychological characteristics. Chapter 13 provides more specific information on musculoskeletal disorders.

Common problems caused by work design hazards include repetitive strain injuries (RSIs), cumulative trauma disorders (CTDs), and musculoskeletal injuries (MSIs), including back injuries. Ergonomic-related injuries are the fastest growing occupational health problem.

In determining the potential hazards that exist and could be considered ergonomic risk factors, examine these factors when assessing the risk of ergonomic hazards:

- The posture a worker must use to do the job (stooping, bending, and crouching). For example, a potential hazard is when a worker is in a static posture, such as when sitting or standing without a break, the muscles are held in a fixed position without movement. Over time, work requiring a static posture can cause health problems. Complaints of back, shoulder, and neck pain can indicate static posture problems.
- The muscular force (exertion) required (lifting, pulling, pushing, and twisting). Muscular force describes the amount of force required to do the

work. Consider the weight of the loads or tools involved; the fit of hand-grips to the worker; the force required; the muscles used; and the adequacy of work gloves provided.

- The number of repetitive motions needed (frequency, speed, duration, and position). Doing the same job rapidly over and over again can cause injury. Jobs that must be repeated in less than 30 s, such as data entry, are classed as highly repetitive.
- The physical condition of the person doing the job.
- Vibration of all or part of the body such as when using jackhammers and chainsaws or when operating mobile equipment.
- Work organization factors such as where, when, and how the work is done and at what pace.
- Poorly designed tasks can force workers to do too much too fast. This can increase stress and reduce work efficiency, increasing the risk of accidents.
- Work environment problems including vibration, heat, cold, and contaminants in the atmosphere.

Remember that these factors can interact, worsening the situation. A good rule of thumb is, the more awkward or static the posture required by a job, the more excessive the force needed to do the work; and the more repetitive the tasks, then the greater will be the risk of injury. Factors contributing to ergonomic problems can include

- Problems in the work environment (light, heat, cold, vibration, and so forth) as well as the health of the worker can promote ergonomic health problems.
- Lack of work variation during shifts can prevent workers from resting their muscles adequately.
- Poorly shaped, heavy, or vibrating hand tools can encourage workers to grip the tool too hard, reducing blood flow to muscles and increasing fatigue. Bulky or clumsy gloves can do the same thing.

10.5.1 Stress Hazards

Anything that affects the health of workers and is part of the overall work environment is considered by most professionals to be ergonomically related and puts the worker at risk for accidents and stress related health problems. This type of stress may be from job expectation, extremes of pressure from supervisor and peers, bullying and harassment, as well as shift work or excess overtime. Stress can seriously harm the health and wellbeing of workers. It can also interfere with efficiency and productivity.

Shift workers have irregular patterns of eating, sleeping, working, and socializing, which may lead to health and social problems. Shift work can also reduce performance and attentiveness. In turn, this may increase the risk of accidents and injuries. Statistics suggest that certain shift workers (such as employees in convenience stores and other workplaces that are open 24 h a day) are more likely to encounter violent situations when working alone.

10.6 SUMMARY

The employer and workers can work together to identify and control health hazards by assessing the risks to the worker's health and safety posed by the work and informing the worker about the nature and extent of the risks and how to eliminate or reduce them.

11 Biological Safety

BIOHAZARD

Recognized symbol for biohazards. (Courtesy of Occupational Safety and Health Administration.)

The handling and dealing with biological organisms of all types pose specific hazards. Some viruses, bacteria, molds, and fungi pose limited problems when they exist in the workplace while others could be potentially deadly if misused or handled. In this chapter, guidelines will be provided for safe handling of these organisms as well as a discussion regarding bloodborne pathogens, which are regulated by Occupational Safety and Health Administration (OSHA).

11.1 BIOSAFETY LEVELS

11.1.1 BIOSAFETY LEVEL 1

Biosafety Level 1 (BSL-1) is suitable for work involving agents of no known or minimal potential hazard to workplace personnel and the environment. The work area may be an integral part to general traffic patterns in the building. Work may be conducted on open bench tops. Special containment equipment is neither required nor generally used. Personnel shall have specific training in procedures conducted in the facility.

11.1.2 BIOSAFETY LEVEL 2

Biosafety Level 2 (BSL-2) is similar to Level 1 and is suitable for work involving agents of moderate potential hazard to personnel and the environment. It differs in that (1) personnel are specifically trained to handle pathogenic agents and are directed by supervisors who are experienced in working with these agents, (2) access to the actual workplace is limited when work is conducted, (3) extreme precautions are taken with contaminated sharp items, and (4) certain procedures that may result in the creation of infectious aerosols or splashes are conducted in biological safety cabinets or other physical containment equipment.

11.1.3 BIOSAFETY LEVEL 3

All work to be conducted with agents assigned to Biosafety Level 3 (BSL-3) must be approved in advance by the Chemical and Biological Safety Committee. BSL-3 work must be conducted in accordance with the facility safeguards, standard micro-biological practices, special practices, and safety equipment described in Centers for Disease Control and Prevention's book entitled, *Biosafety in Microbiological and Biomedical Laboratories.*

BSL-3 is applicable to clinical, diagnostic, teaching, research, or production facilities in which work is done with indigenous or exotic agents that may cause serious or potentially lethal disease as a result of exposure by inhalation. Workplace personnel have specific training in handling pathogenic and potentially lethal agents and are supervised by experienced supervisors in working with these agents.

All procedures involving the manipulation of infectious materials are conducted within biological safety cabinets or other physical containment devices, or by personnel wearing appropriate personal protective clothing and equipment. The worksite should have special engineering and design features such as access zones, sealed penetrations, and directional airflow.

Many workplaces may not have all the facility safeguards recommended for BSL-3. In these circumstances, acceptable safety may be achieved for routine or repetitive operations (e.g., diagnostic procedures involving the propagation of an agent for identification, typing, and susceptibility testing) in BSL-2 facilities. However, the recommended standard microbiological practices, special practices, and safety equipment for BSL-3 must be rigorously followed.

11.1.4 BIOSAFETY LEVEL 4

Biological agents requiring Biosafety Level 4 (BSL-4) containment practices are an unusual occurrence. Working at BSL-4 will require special planning and approval processes since these are the most hazardous organisms that are to be handled and will at times require specially designed handling procedures, specific policies and procedures, and safety plans.

11.2   REQUIREMENTS OF BIOSAFETY LEVEL 1

11.2.1   Standard Microbiological Practices for BSL-1

The following are some general guidelines for facilities working with BSL-1 organisms:

- Limited access or restricted access to the worksite area while work is in progress.
- Post a biohazard sign at the entrance to the work area whenever infectious agents are present. The sign must include the name of the agent(s) in use and the name and phone number of the responsible individual.
- Work surfaces are to be decontaminated once a day and after any spill of viable material.
- All contaminated liquid or solid wastes must be decontaminated before disposal. Contaminated materials that are to be decontaminated at a site outside the work area shall be placed in a durable, leak proof, closed container before they are removed from the facility.
- The work areas shall have an established policy for the safe handling of sharps.
- Mechanical pipetting devices shall be used; mouth pipetting is prohibited.
- Eating, drinking, smoking, and applying cosmetics are not permitted in the work area. Food may be stored in cabinets and refrigerators designated and used for this purpose only. Food storage cabinets and refrigerators shall be located outside the work area.
- All personnel shall wash their hands after they handle viable materials and animals and before leaving the work area (see Figure 11.1).
- All procedures shall be performed carefully to minimize the creation of aerosols.
- An insect and rodent control program is to be in effect. At northwestern facilities, vermin control should be employed for the control of pests and should be contacted if insects or rodents are seen.

11.2.2   Safety Equipment for BSL-1

Adequate safety equipment should be available and in use:

- Special containment equipment is generally not required for manipulation of agents assigned to BSL-1.
- It is recommended that laboratory coats, gowns, or uniforms be worn to prevent contamination or soiling of street clothes.
- Gloves should be worn if skin is broken or afflicted by a rash.

11.2.3   Facilities for BSL-1

A work area should be constructed to provide the best conditions for the safe handling of BSL-1 organisms and should include the following:

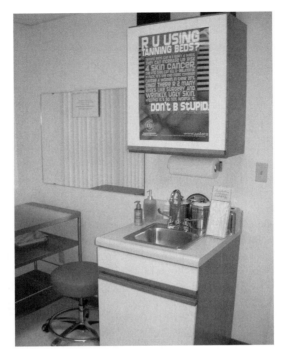

FIGURE 11.1   Hand washing is always an appropriate procedure to control the spread of potential harmful organisms.

- The laboratory shall be designed so that it can be easily cleaned.
- Bench tops shall be impervious to water and resistant to acids, alkalis, organic solvents, and moderate heat.
- Furniture shall be sturdy. Spaces between benches, cabinets, and equipment shall be accessible for cleaning.
- Each work area shall contain a sink for hand washing.
- If the facility has windows that open, they shall be fitted with fly screens.

11.3   BIOSAFETY LEVEL 2 REQUIREMENTS

The following standard and special practices, safety equipment, and facilities apply to agents assigned to BSL-2.

11.3.1   Standard Microbiological Practices for BSL-2

The standard practices for handling BSL-2 are the same as for BSL-1 standard microbiological practices.

11.3.2 SPECIAL PRACTICES FOR BSL-2

Because of the risk involved in handling more hazardous biological organisms than BSL-1, special procedures are to be employed as follows:

- Access to the laboratory is limited or restricted by the principal supervisor when work with infectious agents is in progress. In general, persons at increased risk of contracting infection or for whom infection may be unusually hazardous are not allowed in the work area or animal rooms. Persons who are immunocompromised or immunosuppressed may be at unusual risk of acquiring infections.
- The principal supervisor establishes policies and procedures whereby only persons who have been advised of the potential hazard and meet specific requirements (e.g., immunization) enter the work area or animal rooms.
- When an infectious agent requires special provisions (e.g., immunization) for entering a work area where it is in use, a hazard warning sign incorporating the universal biohazard symbol is posted on the access door to the laboratory work area. The hazard warning sign identifies the infectious agent, lists the name and telephone number of the principal supervisor or other responsible person, and indicates the special requirements for entering the area.
- Personnel are to receive appropriate immunizations for the agents handled or potentially present in the work area.
- When appropriate, baseline serum samples for work area and other at-risk personnel are collected and stored. Additional specimens may be collected periodically.
- A biosafety manual shall be developed for the specific work area. Personnel are advised of special hazards and are required to read and follow instructions on practices and procedures.
- Personnel are to receive appropriate training on the potential hazards associated with the work involved, the necessary precautions to prevent exposures, and the exposure evaluation procedures. Personnel must also receive annual retraining and receive additional training when procedures or policies change.
- A high degree of precaution must always be taken with any contaminated sharp items, including needles and syringes, slides, pipettes, capillary tubes, and scalpels. Needles and syringes should be used in the work area only when there is no alternative, such as when parenteral injection, phlebotomy, or aspiration of fluids from animals and diaphragm bottles are conducted. Plastic ware should be substituted for glassware whenever possible.
- Only needle-locking syringes or disposable syringe-needle units (i.e., the needle is integral to the syringe) are used for injection or aspiration of infectious materials. Used disposable needles must not be bent, sheared, broken, recapped, removed from disposable syringes, or otherwise manipulated by hand before disposal; rather they must be carefully placed in conveniently located puncture-resistant containers used for sharps disposal.

Nondisposable sharps must be placed in a hard-walled container for transport to a processing area for decontamination, preferably by autoclaving.

- Syringes that resheathe the needle, needleless systems, and other safe devices should be used when appropriate.
- Broken glassware must not be handled directly by hand but must be removed by mechanical means such as a brush and dustpan, tongs, or forceps. Containers of contaminated needles, sharp equipment, and broken glass are decontaminated before disposal.
- Cultures, tissues, and specimens of body fluids are placed in a container that prevents leakage during collection, handling, processing, storage, transport, or shipping.
- Equipment and work surfaces should be decontaminated with an appropriate disinfectant on a routine basis as well as after work with infectious material is finished and, especially, after overt spills, splashes, or other contamination by infectious materials. Contaminated equipment must be decontaminated before it is sent for repair or maintenance or packaged for transport.
- Spills or accidents that result in overt exposures to infectious materials are immediately reported to the principal or responsible supervisor. Medical evaluation, surveillance, and treatment are provided as appropriate at no cost to employees, and written records are maintained.
- Animals not involved in the work performed are not permitted in the work area.

11.3.3 Safety Equipment for BSL-2

Properly maintained biological safety cabinets, preferably Class II, or other appropriate personal protective equipment (PPE) or physical containment devices are used in the following cases:

- Procedures with a potential for creating infectious aerosols or splashes are conducted. These may include centrifuging, grinding, blending, vigorous shaking or mixing, sonic disruption, opening containers of infectious materials in which internal pressure may differ from ambient pressure, inoculating animals intranasally, and harvesting infected tissues from animals or eggs.
- High concentrations or large volumes of infectious agents are used. Such materials may be centrifuged in the open work area if sealed rotor heads or centrifuge safety cups are used and if these rotors or safety cups are opened only in a biological safety cabinet.
- Face protection (goggles, mask, face shield, or other splatter guards) is used for anticipated splashes or sprays of infectious or other hazardous materials to the face when the microorganisms must be manipulated outside the biological safety cabinet.
- Protective laboratory coats, gowns, smocks, or uniforms designated for lab use are worn in the work area. This protective clothing is removed and left

in the work area before personnel leave for other areas (e.g., cafeteria, library, or offices) in the facility. All protective clothing is either disposed of in the work area or sent to the laundry service (only after being decontaminated). They are never taken home.

- Personnel are to wear gloves when handling infected animals and when hands may come in contact with infectious materials or contaminated surfaces or equipment. Wearing two pairs of gloves may be appropriate; if a spill or splatter occurs, the hand will be protected after the contaminated glove is removed. Gloves are disposed of when contaminated, removed when work with infectious materials is completed, and not worn outside the work area. Disposable gloves are not washed or reused.

11.3.4 Facilities (Secondary Barriers) for BSL-2

Facilities for BSL-2 organisms must have special design and construction features such as the following:

- Facilities that house restricted agents must be provided with lockable doors.
- Each work area must contain a sink for handwashing.
- Work area is designed so that it can be easily cleaned. Rugs are not appropriate in areas where biological organisms are present.
- Bench tops are to be impervious to water and resistant to acids, alkalis, organic solvents, and moderate heat.
- Furniture is to be sturdy, and spaces between benches, cabinets, and equipment accessible for cleaning.
- If the work area has windows that open, they must be fitted with fly screens.
- Eyewash facility is to be readily available.
- Biological safety cabinets shall be installed in such a manner that fluctuations of the room supply and exhaust air do not cause them to operate outside their parameters for containment. Biological safety cabinets shall be located away from doors, from windows that can be opened, from heavily traveled work areas, and from other potentially disruptive equipment so as to maintain the biological safety cabinets' airflow parameters for containment.
- Illumination shall be adequate for all activities, avoiding reflections and glare, which could impede vision.

11.4 BIOLOGICAL SPILLS

A biological spill shall be followed by prompt action to contain and clean it up. When a spill occurs, warn everyone in the area and call for assistance as needed. The degree of risk involved in the spill depends on the volume of material spilled, the potential concentration of organisms in the material spilled, the hazard of the organisms involved, the route of infection of the organisms, and the diseases caused by the organisms.

Spills of biological agents can contaminate areas and lead to infection of laboratory workers. Prevention of exposure is the primary goal in spill containment and cleanup, exactly as in chemical spills. In evaluating the risks of spill response, generation of aerosols or droplets is a major consideration.

If an accident generates droplets or aerosols in the workplace atmosphere, especially if the agent involved requires containment at BSL-2 or higher, the room shall be evacuated immediately.

Doors shall be closed and clothing decontaminated. In general, a 30 min wait is sufficient for the droplets to settle and aerosols to be reduced by air changes. Longer waiting periods may be imposed depending on the situation. Supervisory personnel and other responsible parties must exercise judgment as to the need for outside emergency help in evacuation.

If a spill of a biological agent requiring containment at BSL-2 or higher occurs in a public area, evacuation of the area shall be immediate. The areas supervisor shall be responsible for designating the extent of evacuation until other emergency personnel arrive. Prevention of exposure to hazardous aerosols is of primary importance.

Anyone cleaning a spill shall wear PPE (for example, laboratory coat, shoe covers, gloves, and possible respiratory protection) to prevent exposure to organisms. An air-purifying negative-pressure respirator with P-100 filter cartridges is generally adequate protection against inhalation of most biological agents. However, there may be exceptions.

An appropriate chemical disinfectant (see Figure 11.2), which is effective against the organisms involved in the spill, should be chosen such as

- Sterilizer or sterilant: An agent intended to destroy all microorganisms and their spores on inanimate surfaces.
- Disinfectant: An agent intended to destroy or irreversibly inactivate specific viruses, bacteria, or pathogenic fungi, but not necessarily their spores, on inanimate surfaces. Most disinfectants are not effective sterilizers.
- Hospital disinfectant: An agent shown to be effective against specific organisms such as *Staphylococcus aureus*, *Salmonella choleraesuis*, and *Pseudomonas aeruginosa*. It may also be effective against other organisms and some viruses.
- Antiseptic: A chemical germicide formulated for use on skin or tissue. Antiseptics should not be used as disinfectants.
- Decontamination: A procedure that eliminates or reduces microbial contamination to a safe level with respect to the transmission of infection. Sterilization and disinfection procedures are often used for decontamination.
- Alcohols: They are effective for killing hepatitis B virus (HBV) but are not recommended for this purpose because of their rapid evaporation and the consequent difficulty of maintaining proper contact times.
- Chlorine compounds: They are probably the most widely used disinfectants in the laboratory. You can easily prepare an inexpensive, broad-spectrum disinfectant by diluting common household bleach.
- Formaldehyde: An OSHA-regulated chemical that is a suspect carcinogen, so its use as a disinfectant is not recommended.

FIGURE 11.2   Use of the proper chemical for the cleanup and disinfection of areas that have or had the potential for contamination of biological organisms is critical.

- Iodophors: Those registered with the EPA may be effective hard-surface decontaminants when used according to manufacturer's instructions, but iodophors formulated as antiseptics are not suitable for use as disinfectants.
- Peracetic (peroxyacetic) acid and hydrogen peroxide mixtures: They minimize the negative effects of corrosiveness sometimes seen with chlorine compounds and high concentrations of peracetic acid alone.
- Quaternary ammonium compounds: They are low-level disinfectants and are not recommended for spills of human blood, blood products, or other potentially infectious materials.

11.4.1   DECONTAMINATION OF SPILLS

The following procedure is recommended for decontaminating spills of agents used at BSL-2:

- Wear gloves and a laboratory coat or gown. Heavyweight, puncture-resistant utility gloves, such as those used for housecleaning and dishwashing, are recommended.
- Do not handle sharps with the hands. Clean up broken glass or other sharp objects with sheets of cardboard or other rigid, disposable material. If a broom and dustpan are used, they must be decontaminated later.

- Avoid generating aerosols by sweeping.
- Absorb the spill. Most disinfectants are less effective in the presence of high concentrations of protein, so absorb the bulk of the liquid before applying disinfectants. Use disposable absorbent material such as paper towels. After absorption of the liquid, dispose of all contaminated materials as waste.
- Clean the spill site of all visible spilled material using an aqueous detergent solution (e.g., any household detergent). Absorb the bulk of the liquid to prevent dilution of the disinfectant.
- Disinfect the spill site using an appropriate disinfectant, such as a household bleach solution. Flood the spill site or wipe it down with disposable towels soaked in the disinfectant.
- Absorb the disinfectant or allow it to dry.
- Rinse the spill site with water.
- Dispose of all contaminated materials properly. Place them in a biohazard bag or other leak proof, labeled biohazard container for sterilization.

11.4.2 BIOLOGICAL SPILL ON A PERSON

If a biological material is spilled on a person, emergency response is based on the hazard of the biological agent spilled, the amount of material spilled, and whether significant aerosols were generated. If aerosol formation is believed to have been associated with the spill, a contaminated person shall leave the contaminated area immediately. If possible, he or she should go to another laboratory area so that hallways and other public areas do not become contaminated.

Contaminated clothing is removed and placed in red or orange biohazard bags for disinfecting. Contaminated skin shall be flushed with water and thoroughly washed with a disinfectant soap. Showering may be appropriate, depending on the extent of the spill.

11.5 INFECTIOUS WASTE MANAGEMENT

Infectious waste materials shall be treated properly to eliminate the potential hazard that these wastes pose to human health and the environment. Treatment commonly involves steam sterilization or incineration.

11.5.1 SEPARATION AND PACKAGING OF INFECTIOUS WASTE

Infectious wastes shall be separated from general, noninfectious waste materials and from wastes containing radioactive, carcinogenic, or toxic materials. Some wastes may contain multiple hazards. These shall be handled such that priority is given to the greatest hazard present.

Disposable infectious materials shall be placed in red or orange plastic bags. The bags shall be seamless, tear-resistant, and autoclavable. Single bags shall have a minimum thickness of 3.0 mils and double bags, 1.5–2.0 mils. Bags shall be closed by folding or tying when full, at the end of the day, or before transporting. To minimize formation of aerosols, infectious wastes shall not be compacted before decontamination.

11.5.2   Storage and Transport of Infectious Waste

Infectious wastes that are removed from a work area or stored temporarily shall be closed and double-bagged or placed inside a covered, unbreakable outer container.

11.5.3   Infectious Waste Treatment

Infectious wastes are generally rendered noninfectious by autoclaving, although incineration is occasionally used. After sterilization, previously infectious wastes are disposed of as noninfectious. They may be placed in noninfectious trash collection containers and sent to a sanitary landfill. Treated wastes in red or orange bags shall be overpacked into opaque plastic bags of another color (not yellow) for noninfectious disposal. The custodial staff has been instructed not to touch or remove red, orange, or yellow bags. Sterilized liquid wastes may be discarded to the sewer.

As described under procedures for BSL-2 and -3 and in the Bloodborne Pathogens Program, syringes and needles shall be handled with extreme caution to avoid autoinoculation and the generation of aerosols. Needles shall not be bent, sheared, replaced in the sheath or guard, or removed from the syringe following use. The needle and syringe shall be promptly placed in a puncture-resistant container.

All human blood, blood products, nonfixed human tissues, and other potentially infectious materials are considered infectious and shall be disinfected, steam sterilized, or incinerated. Sterilized blood-related waste materials are discarded as nonhazardous.

Infectious wastes, including cultures and stocks of etiologic agents, shall be made noninfectious by steam sterilization. Sterilized wastes are disposed of as nonhazardous. Pathological wastes are generally incinerated. Steam sterilization is acceptable, although after sterilization, pathological wastes shall be either incinerated or ground and flushed to a sewer. Animal carcasses, bedding, and wastes are generally incinerated.

11.5.4   Steam Sterilization

Most infectious wastes are sterilized, based on the type of waste, load volume, packaging material, and load configuration. The frequency of monitoring depends on the hazard of the organism used and the frequency of waste sterilization. Infectious wastes that also contain volatile chemicals should be autoclaved only if a chemical (hydrophobic) filter is on line.

11.5.5   Incineration

Incineration is preferred for pathological and animal wastes.

11.5.6   Chemical Disinfection

Chemical treatment is usually a disinfection rather than sterilization. Thus, it is usually intended as a temporary measure to control infectious wastes until sterilization can treat the hazard. Disinfection may be used as final treatment on a case-by-case basis.

11.6 BLOODBORNE PATHOGENS

OSHA has promulgated a regulation to minimize the serious health risks faced by workers exposed to blood and other potentially infectious materials. Among the risks are human immunodeficiency virus (HIV), hepatitis B, and hepatitis C.

The emphasis on engineering controls is based on ways to better protect workers from contaminated needles or other sharp objects. Many safety medical devices are already available and effective in controlling these hazards and wider use of such devices would reduce thousands of injuries each year. OSHA issued a final regulation (29 CFR 1910.1030) on occupational exposure to bloodborne pathogens in 1991 to protect nearly 6 million workers in health care and related occupations at risk of exposure to bloodborne diseases. The occupational exposure to blood and other potentially infectious materials that may contain bloodborne pathogens—or microorganisms—that cause bloodborne diseases is of concern to those who have exposures.

An annual review of exposure control plan must be conducted. Employers must ensure that their plans reflect consideration and use of commercially available safer medical devices. An emphasis should be placed on the use of effective engineering controls, to include safer medical devices, work practices, administrative controls, and PPE. Employers should rely on relevant evidence in addition to FDA approval to ensure effectiveness of devices designed to prevent exposure to bloodborne pathogens.

Multiemployer worksites such as employment agencies, personnel services, home health services, independent contractors, and physicians in independent practice should receive special attention.

The purpose of the regulation is to limit occupational exposure to blood and other potentially infectious materials since any exposure could result in transmission of bloodborne pathogens, which could lead to disease or death. It covers all employees who could be reasonably anticipated as the result of performing their job duties to face contact with blood and other potentially infectious materials. OSHA has not attempted to list all occupations where exposures could occur. Good Samaritan acts such as assisting a co-worker with nosebleed would not be considered occupational exposure.

Infectious materials include semen, vaginal secretions, cerebrospinal fluid, synovial fluid, pleural fluid, pericardial fluid, peritoneal fluid, amniotic fluid, saliva in dental procedures, any body fluid visibly contaminated with blood, and all body fluids in situations where it is difficult or impossible to differentiate between body fluids. They also include any unfixed tissue or organ other than intact skin from a human (living or dead) and HIV—containing cell or tissue cultures and organ cultures and HIV or hepatitis (HBV)—containing culture medium or other solutions as well as blood, organs, or other tissues from experimental animals infected with HIV or HBV.

This regulation requires employers to develop an exposure control plan to identify, in writing tasks and procedures as well as job classifications, where occupational exposure to blood occurs—without regard to personal protective clothing and equipment. It must also set forth the schedule for implementing other provisions of the standard and specify the procedure for evaluating circumstances surrounding exposure incidents. The plan must be accessible to employees and available to OSHA. Employers must review and update it at least annually—more often if necessary to accommodate workplace changes.

The regulation mandates universal precautions, (treating body fluids/materials as if infectious) emphasizing engineering and work practice controls. The standard stresses handwashing and requires employers to provide facilities and ensure that employees use them following exposure to blood. It sets forth procedures to minimize needlesticks, minimize splashing and spraying of blood, ensure appropriate packaging of specimens and regulated wastes, and decontaminate equipment or label it as contaminated before shipping to servicing facilities.

Employers must provide, at no cost, and require employees to use appropriate PPE such as gloves, gowns, masks, mouthpieces, and resuscitation bags and must clean, repair, and replace these when necessary. Gloves are not necessarily required for routine phlebotomies in voluntary blood donation centers but must be made available to employees who want them.

The standard requires a written schedule for cleaning and identifying the method of decontamination to be used in addition to cleaning following contact with blood or other potentially infectious materials. It specifies methods for disposing of contaminated sharps and sets forth standards for containers for these items and other regulated waste. Further, the standard includes provisions for handling contaminated laundry to minimize exposures (see Figure 11.3).

HIV and HBV research laboratories and production facilities must follow standard microbiological practices and specify additional practices intended to minimize exposures of employees working with concentrated viruses and reduce the risk of accidental exposure for other employees at the facility. These facilities must include required containment equipment and an autoclave for decontamination of regulated waste and must be constructed to limit risks and enable easy cleanup. Additional training and experience requirements apply to workers in these facilities.

OSHA also requires vaccinations (i.e., hepatitis B) to be made available to all employees who have occupational exposure to blood within 10 working days of assignment, at no cost, at a reasonable time and place, under the supervision of licensed physician/licensed healthcare professional, and according to the latest recommendations of the U.S. Public Health Service (USPHS). Prescreening may not be required as a condition of receiving the vaccine. Employees must sign a declination form if they choose not to be vaccinated, but may later opt to receive the vaccine at no cost to them. Should booster doses later be recommended by the USPHS, employees must be offered the recommended booster vaccine.

Specific procedures for postexposure and follow-up must be made available to all employees who have had an exposure incident and any laboratory tests must be conducted by an accredited laboratory at no cost to the employee. Follow-ups must

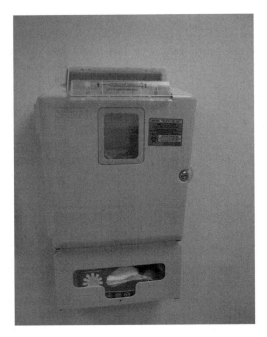

FIGURE 11.3 Container for the proper disposal of sharps helps prevent needlestick contamination.

include a confidential medical evaluation documenting the circumstances of exposure, identifying and testing the source individual if feasible, testing the exposed employee's blood if he/she consents, postexposure prophylaxis, counseling, and evaluation of reported illnesses. Healthcare professionals must be provided specific information to facilitate the evaluation and their written opinion on the need for hepatitis B vaccination following the exposure. Information such as the employee's ability to receive the hepatitis B vaccine must be supplied to the employer. All diagnoses must remain confidential.

Warning labels are required including the orange or orange-red biohazard symbol affixed to containers of regulated waste, refrigerators and freezers, and other containers that are used to store or transport blood or other potentially infectious materials (see Figure 11.1). Red bags or containers may be used instead of labeling. When a facility uses universal precautions in its handling of all specimens, labeling is not required within the facility. Likewise, when all laundry is handled with universal precautions, the laundry need not be labeled. Blood that has been tested and found free of HIV or HBV and released for clinical use, and regulated waste that has been decontaminated, need not be labeled. Signs must be used to identify restricted areas in HIV and HBV research laboratories and production facilities.

There is a mandate for training employees initially upon assignment and annually— those who have received appropriate training within the past year need only receive additional training in items not previously covered. Training must include making accessible a copy of the regulatory text of the standard and explanation of its contents, general discussion on bloodborne diseases and their transmission, exposure control plan, engineering and work practice controls, PPE, hepatitis B vaccine, response to emergencies involving blood, how to handle exposure incidents, the postexposure evaluation and follow-up program, and signs/labels/color-coding. There must be opportunity for questions and answers, and the trainer must be knowledgeable in the subject matter. Laboratory and production facility workers must receive additional specialized initial training.

The maintenance of medical records for each employee with occupational exposure for the duration of employment plus 30 years must be confidential and must include name and social security number; hepatitis B vaccination status (including dates); results of any examinations, medical testing and follow-up procedures; a copy of the healthcare professional's written opinion; and a copy of information provided to the healthcare professional. Training records must be maintained for 3 years and must include dates, contents of the training program or a summary, trainer's name and qualifications, and names and job titles of all persons who attended the sessions. Medical records must be made available to the subject employee, anyone with written consent of the employee, OSHA, and NIOSH—they are not available to the employer. Disposal of records must be in accordance with OSHA's standard covering access to records.

As can be seen, the whole intent of this bloodborne pathogen regulation is based on the prevention of the diseases that can result from exposure in the workplace to hazardous organisms.

12 Chemical Safety

Chemicals are found in almost all workplaces and some may be hazardous.

The information in this section will help workers to identify the hazards associated with chemicals that are encountered by those working within the people-oriented services of the service industry. Not only are chemicals hazardous, but at time the waste generated is also a hazard to contend with. Some of the factors that must be considered regarding hazardous chemicals are as follows:

- The substance exhibits any of the characteristics of hazardous materials or waste.
- The substance has been found to be fatal to humans in low dose or, in the absence of data on human toxicity, it has been shown in studies to have an oral LD_{50} toxicity (rat) of less than 50 mg/kg, an inhalation LC_{50} toxicity (rat) of less than 2 mg/cu m or a dermal LD_{50} toxicity (rabbit) of less than 200 mg/kg, or is otherwise capable of causing or significantly contributing to an increase in serious irreversible or incapacitating reversible illness.
- The chemical contains any number of toxic constituents.

12.1 HAZARDOUS CHEMICALS

Chemicals that have any of the following four characteristics are considered to be hazardous.

12.1.1 IGNITABILITY

A liquid, other than an aqueous solution, containing less than 24% alcohol by volume, which has a flash point of less than 60°C, is considered ignitable. This category includes almost all organic solvents. Some examples are acetone, methanol, ethanol, toluene, benzene, pentane, hexane, and xylene.

12.1.2 CORROSIVITY

An aqueous solution that has a pH of less than or equal to 2.0, or greater than or equal to 12.5, is considered corrosive. Corrosive materials also include substances such as thionyl chloride, solid sodium hydroxide, and some other nonaqueous acids or bases.

12.1.3 REACTIVITY

A chemical exhibits the characteristics of reactivity if a representative sample of the waste has any of the following properties:

- It is normally unstable and readily undergoes violent changes without detonating
- It reacts violently with water
- It forms potentially explosive mixtures with water
- When mixed with water, it generates toxic gases, vapors, or fumes in a quantity sufficient to present a danger to public health or the environment
- It is a cyanide- or sulfide-bearing waste that, when exposed to pH conditions between 2.0 and 12.5, can generate toxic gases, vapors, or fumes in a quantity sufficient to present a danger to public health or the environment.
- It is capable of detonation or explosive reaction if it is subjected to a strong initiating source or if heated under confinement
- It is readily capable of detonation, explosive decomposition, or reaction at standard temperatures and pressure
- It is a forbidden explosive as defined in 49 CFR 173.51 or a Class A explosive as defined in 40 CFR 173.53, or a Class B explosive as defined in 49 CFR 173.88

Toxicity

- Has an acute oral LD50 less than 2,500 mg/kg
- Has an acute dermal LD50 less than 4,300 mg/kg
- Has an acute inhalation LC50 less than 10,000 ppm as a gas or vapor
- Has an acute aquatic 96-hour LC50 less than 500 mg/l
- Has been shown through experience or testing to pose a hazard to human health or environment because of its carcinogenicity (carcinogen, mutagen, tetratogen), acute toxicity, chronic toxicity, bioaccumulative properties, or persistence in the environment

12.2 DEFINITION OF A HAZARDOUS CHEMICAL

A hazardous chemical is defined by Occupational Safety and Health Administration (OSHA) as any chemical that is a health hazard or a physical hazard.

12.2.1 HEALTH HAZARD

Occupational Safety and Health Administration defines a health hazard as a chemical for which there is statistically significant evidence, based on at least one study conducted in accordance with established scientific principles, that acute or chronic health effects may occur in exposed employees. Chemicals covered by this definition include carcinogens, toxic or highly toxic agents, reproductive toxins, irritants, corrosives, sensitizers, hepatotoxins, nephrotoxins, neurotoxins, agents that act on the hematopoietic system, and agents that damage the lungs, skin, eyes, or mucous membranes.

12.2.2 PHYSICAL HAZARD

Occupational Safety and Health Administration defines a physical hazard as a chemical for which there is scientifically valid evidence that it is a combustible liquid, a compressed gas, explosive, flammable, an organic peroxide, an oxidizer, pyrophoric, is unstable (reactive), or water-reactive.

12.2.3 ADDITIONAL HAZARDOUS CHEMICALS

The broad definition OSHA uses to define hazardous chemicals includes not only generic chemicals but also paints, cleaning compounds, inks, dyes, and many other common substances. Chemical manufacturers and importers are required to determine if the chemicals they produce or repackage meet the definition of a hazardous chemical. A chemical mixture may be considered as a whole or by its ingredients to determine its hazards. It may be considered as a whole if it has been tested as a whole and a material safety data sheet (MSDS) has been issued accordingly. Otherwise, the mixture must be evaluated by its components. If the mixture contains 1.0% or more of a hazardous chemical or 0.1% of an ingredient listed as a carcinogen or suspected carcinogen, the whole mixture is assumed to have the same health or carcinogenic hazards as its components.

12.3 CHEMICAL HAZARDS

The term toxin is defined as poison for practical reasons. It does not matter what the substance is. In the right doses or concentration, it can be toxic or poisonous. Not long ago a woman died when she drank too much water. Therefore, the term toxin is used interchangeably with poison.

A thorough discussion of toxicity is beyond the scope of any single publication. Individuals who handle chemicals should supplement the information on a chemical with specific details applicable to their workplace situation. Such information is available in MSDSs and other reference materials. The complex relationship between a material and its biological effect in humans involves considerations of dose, duration and frequency of the exposure, route of exposure, and many other factors, including sex, allergic factors, age, previous sensitization, and lifestyle.

Chemicals enter the body through the following routes:

- Inhalation—Absorption through the respiratory tract by inhalation. This is probably the easiest way for chemicals to enter the body.
- Ingestion—Absorption through the digestive tract by eating or smoking with contaminated hands or in contaminated work areas. Depending on particle or droplet size, aerosols may also be ingested.
- Skin or eye contact—Absorption through the skin or eyes. Skin contact is the most common cause of the widespread occupational disease dermatitis. Eyes are very porous and can easily absorb toxic vapors that cause permanent eye damage.
- Injection—Percutaneous injection through the skin. This can occur through misuse of sharp items, especially hypodermic needles.

12.4 TOXIC EFFECTS

Toxic effects can be immediate or delayed, reversible or irreversible, local or systemic. Toxicity is the measure of a poisonous material's adverse effect on the human body or its ability to damage or interfere with the metabolism of living tissue. Generally, toxicity is divided into two types, acute and chronic (see Figure 12.1). Many chemicals may cause both types of toxicity, depending on the pattern of use. These two types are

FIGURE 12.1   This is a symbol for toxic chemicals. (Courtesy of U.S. Department of Transportation.)

- Acute toxicity is an adverse effect with symptoms of high severity coming quickly to a crisis. Acute effects are normally the result of short-term exposures and are of short duration. Examples of acutely toxic chemicals are hydrogen cyanide and ammonia.
- Chronic toxicity is an adverse effect with symptoms that develop slowly over a long period of time as a result of frequent exposure. The dose during each exposure period may frequently be small enough that no effects are noticed at the time of exposure. Chronic effects are the result of long-term exposure and are of long duration. Carcinogens as well as many metals and their derivatives exhibit chronic toxicity.

Cumulative poisons are chemicals that tend to build up in the body as a result of numerous chronic exposures, leading to chronic toxicity. The effects are not seen until a critical body burden is reached. Examples of cumulative poisons are lead and mercury.

With substances in combination, such as exposure to two or more hazardous materials at the same time, the resulting effect can be greater than the combined effect of the individual substances. This is called a synergistic or potentiating effect. One example is concurrent exposure to alcohol and chlorinated solvents.

The published toxicity information for a given substance is general (human data may not be available) and the actual effects can vary greatly from one person to another. Do not underestimate the risk of toxicity. All substances of unknown toxicity should be handled as if they are toxic, with the understanding that any mixture may be more toxic than its most toxic component.

12.4.1 OTHER TYPES OF TOXINS

A carcinogen is a chemical that causes malignant (cancerous) tumors. Some individual carcinogens are currently regulated by OSHA. Other carcinogens are recognized and suspected carcinogens identified by other agencies. There is no safety level of exposure for carcinogens.

Reproductive toxins can affect both adult male and female reproductive systems. Chemicals may also affect a developing fertilized ovum, embryo, or fetus through exposure to the mother (teratogenic effects). Reproductive hazards affect people in a number of ways, including mental disorders, loss of sexual drive, impotence, infertility, sterility, mutagenic effects on cells, teratogenic effects on the fetus, and transplacental carcinogenesis. Consult the MSDS for information on possible reproductive hazards.

12.4.2 CARCINOGENS

Carcinogens are any substance or agent that has the potential to cause cancer. Whether these chemicals or agents have been shown to cause cancer only in animals makes little difference to workers. Workers should consider these as cancer causing on a precautionary basis since not all is known regarding their effects upon humans long term. Since most scientist say that there is no known safe level of a carcinogen, then no exposure should be the goal of workplace health and safety. Do not let the label "suspect" carcinogen or agent put your mind at ease. This chemical or agent can cause cancer. The OSHA has identified thirteen chemicals as carcinogens. They are as follows:

- 4-Nitrobiphenyl, Chemical Abstracts Service Register Number (CAS No.) 92933
- Alpha-naphthylamine, CAS No. 134327
- Methyl chloromethyl ether, CAS No. 107302
- 3,3'-Dichlorobenzidine (and its salts) CAS No. 91941
- Bis-chloromethyl ether, CAS No. 542881
- Beta-naphthylamine, CAS No. 91598
- Benzidine, CAS No. 92875
- 4-Aminodiphenyl, CAS No. 92671
- Ethyleneimine, CAS No. 151564
- Beta-propiolactone, CAS No. 57578
- 2-Acetylaminofluorene, CAS No. 53963
- 4-Dimethylaminoazo-benezene, CAS No. 60117
- N-Nitrosodimethylamine, CAS No. 62759

Many other chemicals probably should be identified as carcinogens, but have not stood the scrutiny of the regulatory process to make them such. This is probably true in many cases due to special interest by manufacturers and other groups.

The OSHA regulation 29 CFR 1910.1003 pertains to solid or liquid mixtures containing less than 0.1% by weight or volume of 4-nitrobiphenyl; methyl chloromethyl ether; bis-chloromethyl ether; beta-naphthylamine; benzidine or 4-aminodiphenyl; and solid or liquid mixtures containing less than 1.0% by weight or volume of alpha-naphthylamine; 3,3'-dichlorobenzidine (and its salts); ethyleneimine; beta-propiolactone; 2-acetylaminofluorene; 4-dimethylaminoazo-benzene, or N-nitrosodimethylamine.

Any equipment, material, or other item taken into or removed from a regulated area shall be done so in a manner that does not cause contamination in non-regulated areas or the external environment. Decontamination procedures shall be established and implemented to remove carcinogens addressed by this section from the surfaces of materials, equipment, and the decontamination facility.

Dry sweeping and dry mopping are prohibited for 4-nitrobiphenyl, alpha-naphthylamine, 3,3'-dichlorobenzidine (and its salts), beta-naphthylamine, benzidine, 4-aminodiphenyl, 2-acetylaminofluorene, 4-dimethylaminoazo-benzene and N-nitrosodimethylamine.

Entrances to regulated areas shall be posted with signs bearing the legend: "Cancer-Suspect Agent Authorized Personnel Only." Entrances to regulated areas containing operations covered by OSHA's regulations are to be posted with signs bearing the legend: "Cancer-Suspect Agent Exposed in this Area."

These are requirements for areas containing a carcinogen. A regulated area shall be established by an employer where a carcinogen is manufactured, processed, used, repackaged, released, handled, or stored. All such areas shall be controlled in accordance with the requirements for the following category or categories describing the operation involved:

- Isolated systems—Employees working with a carcinogen addressed by this section within an isolated system such as a glove box shall wash their hands and arms upon completion of the assigned task and before engaging in other activities not associated with the isolated system.
- Closed system operation—Within regulated areas where the carcinogens are stored in sealed containers, or contained in a closed system, including piping systems, with any sample ports or openings closed while the carcinogens are contained within, access shall be restricted to authorized employees only.
- Open-vessel system operations—Open-vessel system operations are prohibited.

Employees exposed to 4-nitrobiphenyl; alpha-naphthylamine; 3,3'-dichlorobenzidine (and its salts); beta-naphthylamine; benzidine; 4-aminodiphenyl; 2-acetylaminofluorene; 4-dimethylaminoazo-benzene; and N-nitrosodimethylamine shall be required to wash hands, forearms, face, and neck upon each exit from the regulated areas, close to the point of exit, and before engaging in other activities.

Transfer from a closed system, charging or discharging point operations, or otherwise opening a closed system is prohibited. In operations involving laboratory-type hoods, or in locations where the carcinogens are contained in an otherwise closed system, but is transferred, charged, or discharged into other normally closed containers, the provisions that apply are

- Access shall be restricted to authorized employees only.
- Each operation shall be provided with continuous local exhaust ventilation so that air movement is always from ordinary work areas to the operation. Exhaust air shall not be discharged to regulated areas, nonregulated areas, or the external environment unless decontaminated. Clean makeup air shall be introduced in sufficient volume to maintain the correct operation of the local exhaust system.
- Employees shall be provided with, and required to wear, clean, full body protective clothing (smocks, coveralls, or long-sleeved shirt and pants), shoe covers, and gloves before entering the regulated area.
- Employees engaged in handling operations involving the carcinogens addressed by this section must be provided with, and required to wear and use, a half-face filter-type respirator with filters for dusts, mists, and fumes, or air-purifying canisters or cartridges. A respirator affording higher levels of protection than this respirator may be substituted.
- Prior to each exit from a regulated area, employees shall be required to remove and leave protective clothing and equipment at the point of exit and at the last exit of the day and to place used clothing and equipment in impervious containers at the point of exit for purposes of decontamination or disposal. The contents of such impervious containers shall be identified.
- Drinking fountains are prohibited in the regulated area.

- Employees shall be required to wash hands, forearms, face, and neck on each exit from the regulated area, close to the point of exit, and before engaging in other activities and employees exposed to 4-nitrobiphenyl, alpha-naphthylamine, 3,3'-dichlorobenzidine (and its salts), beta-naphthylamine, benzidine, 4-aminodiphenyl, 2-acetylaminofluorene, 4-dimethylaminoazo-benzene, and N-nitrosodimethylamine shall be required to shower after the last exit of the day.

Maintenance and decontamination activities that occur during the cleanup of leaks or spills, maintenance, or repair operations on contaminated systems or equipment, or any operations involving work in an area where direct contact with a carcinogen addressed by this section could result, each authorized employee entering that area shall

- Be provided with and required to wear clean, impervious garments, including gloves, boots, and continuous-air supplied hood
- Be decontaminated before removing the protective garments and hood
- Be required to shower upon removing the protective garments and hood

The requirements for general regulated areas are

- The employer must implement a respiratory protection program in accordance with 29 CFR 1910.134.
- In an emergency, immediate measures including, but not limited to, the following requirements shall be implemented.
- The potentially affected area shall be evacuated as soon as the emergency has been determined.
- Hazardous conditions created by the emergency shall be eliminated and the potentially affected area shall be decontaminated prior to the resumption of normal operations.
- Special medical surveillance by a physician shall be instituted within 24 h for employees present in the potentially affected area at the time of the emergency. A report of the medical surveillance and any treatment shall be included in the incident report.
- Where an employee has a known contact with a carcinogen addressed by this section, such employee shall be required to shower as soon as possible, unless contraindicated by physical injuries.
- An incident report on the emergency shall be created.
- Emergency deluge showers and eyewash fountains supplied with running potable water shall be located near, within sight of, and on the same level with locations where a direct exposure to ethyleneimine or beta-propiolactone only would be most likely as a result of equipment failure or improper work practice.

Storage or consumption of food, storage or use of containers of beverages, storage or application of cosmetics, smoking, storage of smoking materials, tobacco products or other products for chewing, or the chewing of such products are prohibited in regulated areas. Where employees are required to wash, washing facilities and shower facilities shall be provided in accordance with 29 CFR 1910.141. Where employees wear protective clothing and equipment, clean change rooms shall be provided for the number of such employees required to change clothes, in accordance with 29 CFR 1910.141. Where toilets are in regulated areas, such toilets shall be in a separate room.

Except for outdoor systems, regulated areas shall be maintained under pressure negative with respect to nonregulated areas. Local exhaust ventilation may be used to satisfy this requirement. Clean makeup air in equal volume shall replace the air removed.

Appropriate signs and instructions shall be posted at the entrance to, and exit from, regulated areas, informing employees of the procedures that must be followed in entering and leaving a regulated area.

Containers of a carcinogen are to be accessible only to and handled only by authorized employees or by other employees trained. Many containers may have their content identification limited to a generic or proprietary name or other proprietary identification of the carcinogen and percent, and identification that includes the full chemical name in the Chemical Abstracts Service Registry. Containers shall have the warning words "Cancer-Suspect Agent" displayed immediately under or adjacent to the content identification. Containers whose contents are carcinogens addressed by this section with corrosive or irritating properties shall have label statements warning of such hazards noting, if appropriate, particularly sensitive or affected portions of the body. Lettering on signs and instructions required should be a minimum letter height of 2 in. Labels on containers required must not be less than one-half the size of the largest lettering on the package and not less than eight point type in any instance. No such required lettering need be more than 1 in. in height. No statement that contradicts or detracts from the effect of any required warning, information, or instruction shall appear on or near any required sign, label, or instruction.

Before being authorized to enter a regulated area, each employee is to receive a training and indoctrination program including, but not necessarily limited to, the nature of the carcinogenic hazards of a carcinogen addressed by this section, local and systemic toxicity, the specific nature of the operation involving a carcinogen addressed by this section that could result in exposure, the purpose and application of the medical surveillance program, including, as appropriate, methods of self-examination, the purpose and application of decontamination practices and purposes, the purpose and significance of emergency practices and procedures, the employee's specific role in emergency procedures, specific information to aid the employee in recognition and evaluation of conditions and situations that may result in the release of a carcinogen addressed by this section, the purpose and application of specific first-aid procedures and practices, and a review of this section at the employee's first training and indoctrination program and annually thereafter.

Specific emergency procedures shall be prescribed and posted. Employees must be familiarized with their terms, and rehearsed in their application. All materials relating to the program are to be provided upon request to authorized representatives of the assistant secretary and the director.

Any changes in such operations or other information must be reported in writing within 15 calendar days of such change and include

- Brief description and in-plant location of the area(s) regulated and the address of each regulated area
- Name(s) and other identifying information as to the presence of a carcinogen addressed by this section in each regulated area
- Number of employees in each regulated area during normal operations and including maintenance activities
- The manner in which carcinogens addressed by this section are present in each regulated area; for example, whether it is manufactured, processed, used, repackaged, released, stored, or otherwise handled

Incidents that result in the release of a carcinogen into any area where employees may be potentially exposed are to be reported. A report of the occurrence of the incident and the facts obtainable at that time including a report on any medical treatment of affected employees shall be made within 24 h to the nearest OSHA area director. A written report shall be filed with the nearest OSHA area director within 15 calendar days thereafter and shall include

- A specification of the amount of material released, the amount of time involved, and an explanation of the procedure used in determining this figure
- A description of the area involved, and the extent of known and possible employee exposure and area contamination
- A report of any medical treatment of affected employees and any medical surveillance program implemented
- An analysis of the circumstances of the incident and measures taken or to be taken, with specific completion dates, to avoid further similar releases

At no cost to the employee, a program of medical surveillance must be established and implemented for employees considered for assignment to enter regulated areas and for authorized employees. Before an employee is assigned to enter a regulated area, a pre-assignment physical examination by a physician is to be provided. The examination must include the personal history of the employee, family, and occupational background, including genetic and environmental factors. Authorized employees shall be provided periodic physical examinations, not less often than annually, following the pre-assignment examination.

In all physical examinations, the examining physician should consider whether there exist conditions of increased risk, including reduced immunological competence, those undergoing treatment with steroids or cytotoxic agents, pregnancy, and cigarette smoking.

Employers of employees examined must maintain complete and accurate records of all such medical examinations. Records shall be maintained for the duration of the employee's employment. Upon termination of the employee's employment, including retirement or death, or in the event that the employer ceases business without a successor, records, or notarized true copies thereof, shall be forwarded by registered mail to the Director of NIOSH. Records shall be provided upon request to employees, designated representatives, and the assistant secretary of OSHA. Any physician who conducts a medical examination is to furnish to the employer a statement of the employee's suitability for employment in the specific exposure.

The specificity of the previous requirements is also an indicator as to the danger presented by the exposure to or working with carcinogens that are regulated by OSHA. There are other carcinogens that OSHA regulates but are not part of the original thirteen. These carcinogens are

- Vinyl chloride (1910.1017)
- Inorganic arsenic (1910.1018)
- Cadmium (1910.1027 and 1926.1127)
- Benzene (1910.1028)
- Coke oven emissions (1910.1029)
- 1,2-Dibromo-3-chloropropane (1910.1044)
- Acrylonitrile (1910.1045)
- Ethylene oxide (1910.1047)
- Formaldehyde (1910.1048)
- Methylenedianiline (1910.1050)
- 1,3-Butadiene (1910.1051)
- Methylene chloride (1910.1052)

Recently, OSHA has reduced the permissible exposure limit (PEL) for methylene chloride from 400 to 25 ppm. This is a huge reduction in the PEL, which equates to a 15 times decrease in what a worker can be exposed to. This reduction is an indicator of the potential of the chemical to cause cancer in workers. This should raise the flag that chemicals that are regarded to be possibly causing cancer are not to be taken lightly since information and research are continuously evolving and provide new insight into the dangers of these chemical and agents.

12.5   WORKING WITH TOXINS

The chemical-handling guidelines described in this chapter are based on several basic principles:

- Substitute less hazardous chemicals whenever possible
- Minimize chemical exposures
- Avoid underestimating risks
- Provide adequate ventilation

Since most chemicals are hazardous to some degree, it is prudent to minimize exposure to chemicals as a general rule, rather than implementing safety protocols only for specific compounds. Avoid skin contact with chemicals as much as possible. Assume that mixtures are more toxic than their components and that all substances of unknown toxicity are toxic. Do not work with a volatile or aerosolizing material without adequate ventilation from chemical fume hoods or other protective devices. Remember: Prepare yourself, and then protect yourself.

FIGURE 12.2 Depending on the carcinogen or toxin, appropriate personal protective equipment must be selected and used (self-contained breathing apparatus (SCBA), full facepiece).

12.5.1 GUIDELINES FOR USING TOXINS

The following general guidelines are applicable to nearly all uses of chemicals. They apply to most hazardous chemicals, such as acids, bases, and flammable liquids (see Figure 12.2). They are also applicable to chemicals that display low carcinogenic potency in animals and are not considered carcinogens. The general guidelines are not, by themselves, adequate for chemicals with high acute toxicity or high chronic toxicity such as heavy metals, chemical carcinogens, or reproductive toxins. However, these guidelines will go a long way in protecting workers from hazardous chemicals when the potential for exposure or contact exists:

- Wear eye protection at all times where chemicals are used or stored.
- Wear a lab coat or other protective clothing (e.g., aprons).
- Wear gloves selected on the basis of the hazard. Inspect them before use. Wash reusable gloves before removal. Turn disposable gloves inside out carefully when removing to avoid contaminating hands.
- Wash hands immediately after removing gloves, after handling chemical agents, and before leaving the workplace, even though you wore gloves. Protective clothing and gloves are worn only in the work area. They are not taken outside the workplace to lunchrooms or offices nor are they worn outdoors. All protective clothing shall be cleaned frequently.
- Confine long hair and loose clothing.
- Wear sturdy shoes that cover feet completely.
- Do not store or prepare food, eat, drink, chew gum, apply lip balm or cosmetics, or handle contact lenses in areas where hazardous chemicals are

present. Check with your supervisor regarding contact lens policy in the workplace. If wearing them is acceptable, take appropriate precautions such as informing fellow workers and have a suction-type removal device in the first-aid kit.

- Store food in cabinets or refrigerators designated for such use only.
- Label all chemical containers.
- Store chemicals by hazard class or type. Chemicals are not stored merely by alphabetical order.
- Never smell or taste chemicals. Again, label containers properly to avoid confusion about contents.
- Keep work areas clean and uncluttered.
- Keep personal belongings away from chemicals.
- Obtain an MSDS for each chemical, and consult the MSDS before using a chemical.
- Know the emergency procedures of the particular workplace and the chemicals used.
- Vent into local exhaust devices any apparatus that may discharge toxic vapors, fumes, mists, dusts, or gases. Never release toxic chemicals into cold rooms or warm rooms that have recirculating atmospheres.
- Use chemical fume hoods or other engineering controls to minimize exposure to airborne contaminants.
- Properly handle, collect, and dispose of surplus and waste chemicals.

12.5.2 GUIDELINES FOR USING ACUTE TOXINS

Chemicals that exhibit acute toxicity are defined by OSHA as those that cause rapid effects as a result of a short-term exposure—generally sudden and severe, as in the case of a leak from equipment. Acute toxic effects include irritation, corrosion, sensitization, and narcosis.

To illustrate, hydrofluoric acid (HF) is a chemical of high acute toxicity because of its destructive effect on skin and bone tissue. Inhalation of high concentrations of carbon monoxide can cause immediate poisoning and death, as the gas directly interferes with oxygen transport in the body by preferentially binding to hemoglobin. Hydrogen cyanide inhalation inhibits enzyme systems vital to cellular uptake of oxygen.

When working with significant quantities of such chemicals, the aim is to minimize exposure to the material. Special care should be taken in the selection of protective clothing to ensure that it is appropriate for the hazard. Personal hygiene and work practices should also be carefully evaluated to minimize exposure. The following guidelines should be practiced in addition to the general guidelines for handling chemicals:

- When performing procedures that may result in the release of airborne contaminants, use a chemical fume hood.
- Trap or treat effluents to remove gases, fumes, vapors, and particulates before discharging them to facility's exhaust.
- Restrict access to the work area.

- Establish and label a designated area for work with acutely toxic chemicals. Keep materials within the designated area.
- Use plastic-backed paper or trays under work areas. Replace the paper when contaminated.
- Develop and know special emergency procedures. Keep emergency supplies at hand for immediate use.

12.5.3 Guidelines for Using Chronic Chemicals, Carcinogens, and Reproductive Toxins

In addition to the general guidelines for handling chemicals, use the following guidelines for handling chemicals with chronic toxins, which include most heavy metals, chemicals displaying moderate to high carcinogenic potency in animals, and reproductive toxins.

For carcinogens, determine whether the chemical is regulated by OSHA in a substance-specific standard. Also, follow the guidelines provided below:

- Designate a work and storage area for carcinogens, chemicals with chronic toxicity, and reproductive toxins. Materials are to be kept within the designated area to the extent possible.
- Label designated work and storage areas for chemical carcinogens, including chemical fume hoods and refrigerators that are labeled "Chemical Carcinogen." The outer door to the laboratory shall also be labeled chemical carcinogen.
- Label designated work and storage areas used for chemicals with chronic toxicity or reproductive toxins toxic chemical or toxic substance.
- Use specific and secure access procedures if work involves moderate or greater amounts of carcinogens or moderate to lengthy procedures. These procedures may include
  - Closed doors
  - Restricted access—only authorized personnel permitted
  - Written access procedures posted on the outer door
- Cover all surfaces, including chemical fume hood surfaces, with plastic-backed paper or protective trays. Inspect work surfaces following procedures, and remove the paper if contamination is present. Dispose of the used paper as hazardous waste.
- Dispose of disposable gloves as hazardous waste. Wash reusable gloves before removing them. Contact the responsible party for chemical waste management prior to washing to determine if the wash water must be collected for disposal as a hazardous waste.
- Transport highly toxic or carcinogenic materials through public areas, such as hallways, in closed containers within unbreakable outer containers. Sealed plastic bags may be used as secondary containment in many cases.
- To avoid potential inhalation hazards, handle powdered carcinogens and toxins in a chemical fume hood, even during weighing procedures. Inside the chemical fume hood, measure the powder with a spatula into a

pre-weighed vessel, then seal or cover the vessel, remove it from the chemical fume hood, and take it to the balance to be weighed. If more or less material is needed, return the container to the chemical fume hood for addition or subtraction of material. Close the container again and reweigh it. Repeat these steps until the desired amount is obtained. This procedure eliminates contamination of the air, the workbench, and the scale. Procedures generating either solid or liquid airborne contaminants or involving volatile chemicals are always to be performed in a chemical fume hood.

- Protect vacuum pumps against contamination (e.g., traps and filters in lines) and vent into direct exhaust ventilation. Pumps and other equipment and glassware shall be decontaminated before they are removed from the designated area. The designated area shall be decontaminated before other normal work is conducted. Vacuum pump oil shall be collected as a contaminated waste and disposed using the chemical waste management procedures.
- Water vacuum lines shall be equipped with traps to prevent vapors from entering the wastewater stream.
- Floors shall be wet-mopped or cleaned with a high-efficiency particulate air filter (HEPA) vacuum cleaner if powdered materials are used.

12.6 STORING HAZARDOUS CHEMICALS

With the potential for the presence of so many chemicals in workplaces, only a limited discussion of the safe storage of these types of chemicals can be undertaken in this chapter (see Figure 12.3). All chemicals should be dated on receipt.

Chemicals shall be stored only with other compatible chemicals. Do not store them alphabetically, except within a grouping of compatible chemicals. Chemical groupings are listed below:

- Highly toxic (poisons) and habit-forming organic chemicals
- Inorganic bases and inorganic reducers/salts
- Flammable organic chemicals and organic acids
- Organic bases and other organic compounds
- Inorganic (mineral) acids and inorganic oxidizers (some additional separation may be required because of the reactivity of these materials)
- Carcinogens

12.6.1 STORAGE FACILITIES

Highly toxic chemicals (such as cyanide), shock-sensitive chemicals (such as picric acid), and habit-forming chemicals (amyl nitrite) shall be stored in locked cabinets to prevent theft.

Peroxide-forming chemicals and those that may become shock-sensitive with long-term storage shall be stored separately and shall be labeled and dated. Peroxide-forming chemicals shall be stored in a cool, dark, dry place.

FIGURE 12.3   Chemical should be stored in appropriate cabinets. Proper labeling and neatness assures the best safety practices.

Flammable liquids should be stored in flammable liquid cabinets if a total of 10 gal or more is present, including flammable liquid wastes.

Volatile or highly odorous chemicals shall be stored in a well-ventilated area; a ventilated cabinet is preferable. Chemical fume hoods shall not be used for storage, as containers block proper airflow in the hood and take up workspace.

Storage areas for carcinogens shall be labeled chemical carcinogen. This requirement for cancer-warning labels applies even to chemicals that exhibit more than one hazard (e.g., carcinogenic and flammable).

12.6.2   INSPECTION OF STORED CHEMICALS

Chemical storage areas shall be inspected at least annually and any unwanted or expired chemicals shall be removed. During this inspection, the list of chemicals present should be updated or verified and the date and name recorded.

Although the deterioration in storage of a specific compound cannot be predicted in detail, generalizations can often be made about the reaction characteristics of groups of compounds. Some general conclusions about the stability of classes of chemicals can be reached, and corresponding storage time spans can be identified. Visual inspection of stored chemicals is important in the disposal decision. Chemicals showing any of the indications listed below shall be recommended for safe disposal:

- Slightly cloudy liquids
- Darkening or change in color
- Spotting on solids
- Caking of anhydrous materials
- Existence of solids in liquids or liquids in solids
- Pressure buildup in containers
- Evidence of reaction with water
- Corrosion or damage to the container
- Missing or damaged (i.e., illegible) labels

12.6.3 REFRIGERATOR STORAGE

Flammable liquids shall not be stored in ordinary domestic refrigerators. Refrigerator temperatures are almost universally higher than the flash points of flammable liquids, and ignition sources are readily available inside the storage compartment. Furthermore, the compressor and its circuits are typically located at the bottom of the units, where vapors (from flammable liquid spills or leaks, for example) may easily accumulate. Specially designed explosive proof refrigerators are to be used.

12.7 TRANSPORTATION OF HAZARDOUS CHEMICALS

12.7.1 USE SECONDARY CONTAINERS

The container-within-a-container concept will protect the primary containers from shock during any sudden change of movement. Secondary containment is especially important when chemicals are moved in public areas, such as hallways or elevators, where the effects of a spill would be more severe. Other rules to follow during transportation of hazardous chemicals are

- A sturdy cart should always be used, and it must be ensured that the cart has a low center of gravity. Carts with large wheels are best for negotiating irregularities in floors and at elevator doors.
- Freight elevators shall be used for moving chemicals and biological materials. Passenger elevators shall not be used for this purpose.
- Incompatible chemicals shall not be transported together on the same cart.
- All chemical containers transported shall have labels identifying the contents.
- Large containers of corrosives shall be transported in a chemical-resistant bucket or other containers designed for this purpose.
- Sudden backing up or changes in direction from others should be anticipated. If you stumble or fall while carrying glassware or chemicals, try to project them away from yourself and others.

12.8 CHEMICAL WASTE MANAGEMENT

Proper handling of reaction by-products, surplus and waste chemicals, and contaminated materials is an important part of chemical safety procedures. Each worker is responsible for ensuring that wastes are handled in a manner that minimizes personal exposure and the potential for environmental contamination.

The first steps in managing chemical wastes are selecting the least hazardous chemicals for the task and ordering chemicals only in quantities really needed. Chemicals should not be kept in if they will not be needed, especially if they are peroxide-forming chemicals such as ethyl ether or dioxane, polynitro compounds such as picric acid or dinitrophenyl hydrazine, or chemicals that are air- or water-reactive.

U.S. Environmental Protection Agency (EPA) regulations for disposal of hazardous chemical wastes should be followed by transporting shipments of waste to disposal sites.

It is a federal offense to dispose of chemicals improperly. It is a violation of both safety and environmental regulations to pour chemicals down the drain unless they are treated or neutralized and local regulation allows them in the sanitary sewer system. Nothing except water or dilute aqueous solutions of low-toxicity material shall be disposed of in the sink.

12.8.1 STORAGE

Storage of waste chemicals shall include separation of incompatible materials, as in chemical storage. Separate organic and inorganic chemicals. Keep caps loose to allow vapors to escape, and keep wastes in well-ventilated areas. Waste containers shall be capped at all times and uncapped only for addition of more waste.

12.8.2 WASTE MINIMIZATION

EPA's policy for hazardous waste management places the highest priority on waste minimization. Under current environmental laws, employers must certify that they have a waste minimization program in place and must annually report to the government on efforts they have made to reduce hazardous wastes.

Waste minimization as defined by the EPA means a reduction in both the volume and physical hazard or toxicity of the material. The benefits of waste minimization include reduced disposal costs, decreased liability, improved working conditions, and less impact on the environment at the time of disposal.

12.9 HAZARDOUS CHEMICAL EMERGENCY PROCEDURES

12.9.1 PROCEDURES FOR SPILLS OF VOLATILE, TOXIC, OR FLAMMABLE MATERIALS

A specific set of procedures should be developed for spills of hazardous chemicals and should have at least the minimum set of rules as follows:

- Warn all persons nearby
- Turn off any ignition sources such as burners, motors, and other spark-producing equipment
- Leave the work area and close the door if possible
- Call 911 to report a life-threatening hazardous material spill

Small spills can be absorbed with paper towels or other absorbents. However, these materials can increase the surface area and evaporation rate, increasing the potential fire hazard if the material is flammable and the airborne concentration reaches the flammability level.

12.9.2 INCIDENTAL SPILLS

Incidental spills procedure for small amounts of low-toxicity chemicals should be prepared for such events. Keep appropriate spill-containment material on hand for emergencies. Workers must have received training to distinguish between the types of spills they can handle on their own and those spills that are classified as "Major." Major spills dictate the need for outside help.

Workers are to be trained and qualified to clean up spills that are incidental. OSHA defines an incidental spill as a spill that does not pose a significant safety or health hazard to employees in the immediate vicinity nor does it have the potential to become an emergency within a short time frame. The period that constitutes a short time is not defined.

The procedures for addressing incidental spills are as follows:

- Alert persons in the area that a spill has occurred.
- Evaluate the toxicity, flammability, and other hazardous properties of the chemical as well as the size and location of the spill (e.g., chemical fume hood or elevator) to determine whether evacuation or additional assistance is necessary. Large or toxic spills are beyond the scope of this procedure.
- Contain any volatile material within a room by keeping doors closed. Increase exhaust efficiency by minimizing sash height of the chemical fume hood or activating the emergency purge, if available.
- Consult your MSDS, the laboratory emergency plan, workplace emergency plan, or written procedures provided by the company.
- Obtain cleaning equipment and protective gear, if needed.
- Wear protective equipment such as goggles, apron, laboratory coat, gloves, shoe covers, or respirator. Base the selection of the equipment on the hazard.
- First, cordon off the spill area to prevent inadvertently spreading the contamination over a much larger area.
- Absorb liquid spills using paper towels, spill pillows, vermiculite, or sand. Place the spill pillow over the spill and draw the free liquid into the pillow. Sprinkle vermiculite or sand over the surface of the free liquid.
- Place the used pillows or absorbent materials in plastic bags for disposal along with contaminated disposable gear, such as gloves.
- Neutralize spills of corrosives and absorb, if appropriate. Sweep up waste and place in plastic bags for disposal.
- Complete a chemical waste collection form to trigger vendor pickup of the wastes.
- Complete an incident report describing the spill and send a copy to the appropriate department. A copy may be kept by the department head, if required.

Workers can handle incidental spills because they are expected to be familiar with the hazards of the chemicals they routinely handle during an average workday. If the spill exceeds the scope of the workers' experience, training, or willingness to respond, the workers must be able to determine that the spill cannot be dealt with internally.

Emergency assistance is provided by in house responders or an outside agency. Spills requiring the involvement of individuals outside the regular workforce are those exceeding the exposure one would expect during the normal course of work. Spills in this category are those that have truly become emergency situations in that workers are overwhelmed beyond their level of training. Their response capability is compromised by the magnitude of the incident.

12.9.3 Major Spills

Some of the factors that would indicate that the workers need to call for help and that a major spill has occurred are

- The need to evacuate employees in the area
- The need for response from outside the immediate release area
- The release poses, or has potential to pose, conditions that are immediately dangerous to life and health
- The release poses a serious threat of fire and explosion
- The release requires immediate attention due to imminent danger
- The release may cause high levels of exposure to toxic substances
- There is uncertainty that the worker can handle the severity of the hazard with the PPE and equipment that have been provided, and the exposure limit could be easily exceeded
- The situation is unclear or data are lacking regarding important factors

At times special spill procedures are needed for specific chemicals such as mercury. Consult company chemical hazard manual for chemicals that require special handling procedures.

12.10 WORKER CONTAMINATION

12.10.1 Chemical Spill on a Worker

At times, a worker might have a chemical spilled on him or her. If this happens a set of steps should be in place and followed such as

- Know where the nearest eyewash and safety shower are located.
- For small spills on the skin, flush immediately under running water for at least fifteen minutes, removing any jewelry that might contain residue. If there is no sign of a burn, wash the area with soap under warm running water.
- When assisting a victim of chemical contamination, use appropriate PPE.
- For a chemical splash in the eyes, immediately flush the eyes under running potable water for fifteen minutes, holding the eyes open and rotating the

eyeballs. This is preferably done at an eyewash fountain with tepid water and properly controlled flow. Hold the eyelids open and move the eye up, down, and sideways to ensure complete coverage. Use an irrigator loop to thoroughly flush the conjunctiva under the upper eyelid, if available in your first-aid kit. If no eyewash fountain is available, put the victim on his or her back and gently pour water into the eyes for fifteen minutes or until medical personnel arrive.

- For spills on clothing, immediately remove contaminated clothing, including shoes and jewelry, while standing under running water or the safety shower. When removing shirts or pullover sweaters, be careful not to contaminate the eyes. Cutting off such clothing will help prevent spreading the contamination. To prepare for emergencies, shears (rounded-tip scissors) should be available in the first-aid kit to allow safe cutting of contaminated clothing.

- Consult the MSDS to see if any delayed effects should be expected, and keep the MSDS with the victim. Call for emergency transportation to have the victim taken to the emergency room for medical attention. Be sure to inform emergency personnel of the decontamination procedures used before their arrival (e.g., flushing for 15 min with water). Be certain that emergency room personnel are told exactly what the victim was contaminated with so that they can treat the victim accordingly.

12.10.2 PROCEDURE FOR CRYOGENIC LIQUID SPILL ON A PERSON

Contact with cryogenic liquids may cause crystals to form in tissues under the spill area, either superficially or more deeply in the fluids and underlying soft tissues. The first-aid procedure for contact with cryogenic liquids is identical to that for frostbite. Re-warm the affected area as quickly as possible by immersing it in warm, but not hot, water (between 102°F and 105°F). Do not rub the affected tissues. Do not apply heat lamps or hot water and do not break blisters. Cover the affected area with a sterile covering and seek assistance as you would for burns.

12.11 CHEMICALS AND WORKER HEALTH

12.11.1 MEDICAL SURVEILLANCE

Medical surveillance is needed when workers show any signs and symptoms. Whenever an employee develops signs or symptoms associated with a hazardous chemical exposure, that person shall be provided an opportunity to receive an appropriate medical examination (see Figure 12.4).

If exposure monitoring reveals that the airborne concentration of a chemical is above the action level or the PEL (if no action level is set) for a chemical regulated by OSHA, medical surveillance shall be implemented for affected persons as prescribed in the OSHA standard for the material.

If a spill, leak, explosion, or other occurrence results in the likelihood of a hazardous chemical exposure, affected employees shall be provided an opportunity for a medical consultation. The consultation will determine whether there is a need for a medical examination.

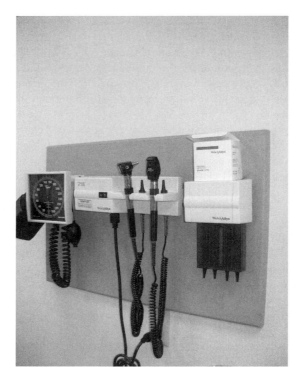

FIGURE 12.4   Medical examinations should be provided to those workers who have been exposed to hazardous chemicals when signs and symptoms exist.

12.11.2   MEDICAL CONSULTATION AND EVALUATION

Medical consultation and evaluation shall be performed under the direct supervision of a licensed physician without cost to the employee, without loss of pay, and at a reasonable time and place.

The supervisor or company shall ensure that the following information is provided to the physician: the identity of the chemical involved in the exposure, a description of conditions relating to the exposure, any quantitative data available regarding the exposure, and a description of signs and symptoms experienced by the affected person. The physician should provide the following information to the company in writing:

- Recommendation for medical follow-up
- Results of the medical examination and associated tests
- Any medical condition revealed in the course of the examination that may place the affected person at increased risk as a result of the exposure
- A statement that the physician has informed the affected person of the results of the consultation or examination and any medical condition that may require further treatment

The physician shall not reveal specific findings or diagnoses unrelated to the chemical exposure. All medical records shall be kept as part of an employee's medical records.

Medical surveillance for chemicals of chronic toxicity may require routine medical surveillance for individuals working with chemicals of chronic toxicity, including carcinogens.

Although no restriction of hiring can be made, candidates for work with carcinogens shall be informed of the possibility of increased risk associated with these conditions:

- Strong family history of cancer, comprising at least two first-generation relatives from maternal and paternal ancestry or a specific pattern of cancer incidence that can be recognized as a genetic trait
- Precancerous condition or past history of cancer
- History of exposure to therapeutic doses of radiation
- History of treatment with cytotoxic drugs
- History of impaired immunity or current use of therapeutic doses of steroids or other immunosuppressive drugs
- Concurrent pregnancy or likelihood of pregnancy during employment

13 Preventing Musculoskeletal Disorders

A well-designed medical office reduces chances for musculoskeletal disorders (MSDs).

13.1 MUSCULOSKELETAL DISORDERS

This chapter is intended to help employers reduce the number and severity of work-related musculoskeletal disorders (MSDs) in their facilities. MSDs include conditions such as low back pain, sciatica, rotator cuff injuries, epicondylitis, and carpal tunnel syndrome. More remains to be learned about the relationship between workplace activities and the development of MSDs.

However, these tasks can entail high physical demands due to the large amount of weight involved, awkward postures that may result from leaning over an object or working in a confined area, shifting of weight that may occur if someone loses their balance or strength while moving, and many other factors. The risks that workers face in the workplace include

- Force
- Repetition
- Awkward postures
- Contact stress

Excessive exposure to these risk factors can result in a variety of disorders in affected workers. These conditions are collectively referred to as musculoskeletal disorders. MSDs include conditions such as low back pain, sciatica, rotator cuff injuries, epicondylitis, and carpal tunnel syndrome. Early indications of MSDs can include persistent pain, restriction of joint movement, or soft tissue swelling.

While some MSDs develop gradually over time, others may result from instantaneous events such as a single heavy lift. Activities outside of the workplace that involve substantial physical demands may also cause or contribute to MSDs. In addition, the development of MSDs may be related to genetic causes, gender, age, and other factors. Finally, there is evidence that reports of MSDs may be linked to certain psychosocial factors such as job dissatisfaction, monotonous work, and limited job control. This chapter addresses only physical factors in the workplace that are related to the development of MSDs.

Over the years, the science of ergonomics has found application in addressing the prevention of MSDs. Ergonomics is a science that directs its efforts toward fitting the workplace to the worker and not the worker to the workplace. By applying the principles of ergonomics, much has been accomplished in making the workplace a more comfortable environment and in alleviating risks that can result in the development of MSDs.

In the people-oriented service sectors of the service industry, these risk factors can lead to injury and illness. In this chapter, the term MSDs refers to a variety of injuries and illnesses, including

- Carpal tunnel syndrome
- Tendonitis (tendon inflammation)
- Rotator cuff injuries (a shoulder problem)
- Epicondylitis (sometimes called tennis or golfers' elbow)
- Trigger finger
- Muscle strains and back injuries that occur from repeated use or overexertion

Employers should consider an MSD to be work-related if an event or exposure in the work environment either caused or contributed to the MSD or significantly aggravated a preexisting MSD. For example, when an employee develops carpal tunnel syndrome, the employer needs to look at the hand activity required for the job and the amount of time spent doing the activity. If an employee develops carpal tunnel syndrome, and his or her job requires frequent hand activity, or forceful or sustained awkward hand motions, then the problem may be work-related. If the job requires very little hand activity, then the disorder may not be work-related.

13.2 ORGANIZED APPROACH

13.2.1 Provide Management Support

Strong support by management creates the best opportunity for success. It is recommended that employers develop clear goals, assign responsibilities to designated staff members to achieve those goals, provide necessary resources, and ensure that assigned responsibilities are fulfilled. Providing a safe and healthful workplace requires a sustained effort, allocation of resources, and frequent follow-up, which can only be achieved through the active support of management.

Company management personnel should consider the general steps discussed below when establishing and implementing an ergonomics process. It should be noted, however, that each workplace will have different needs and limitations that should be considered when identifying and correcting workplace problems. Different businesses may implement different types of programs and activities and may assign different staff to accomplish the goals of the ergonomics process.

One key to successful ergonomic efforts is management support to lead the efforts. Management support improves the company's ability to maintain a sustained effort, allocate needed resources, and follow up on program implementation.

13.2.2 INVOLVE EMPLOYEES

Employees are a vital source of information about hazards in their workplace. Their involvement adds problem-solving capabilities and hazard identification assistance, enhances worker motivation and job satisfaction, and leads to greater acceptance when changes are made in the workplace. Employees can

- Submit suggestions or concerns
- Discuss the workplace and work methods
- Participate in the design of work, equipment, procedures, and training
- Evaluate equipment
- Respond to employee surveys
- Participate in task groups with responsibility for ergonomics
- Participate in developing the nursing home's ergonomics process

13.2.3 IDENTIFY PROBLEMS

Identifying where and how a job may exceed the physical capabilities of workers (worksite analysis) is the first step in addressing ergonomic concerns. Worksite analysis involves looking at specific tasks and conditions to identify whether ergonomic risk factors are present and pose a risk of injury to workers. A number of ways are available to look for the ergonomic risk factors that may be present in a job, and to identify the risk factors that may cause a risk of injury, such as

- Analyzing OSHA 300 and 301 injury and illness information, workers' compensation records, or employee reports of problems to identify the types of injuries that may have occurred. A detailed analysis is needed to determine the cause of these injuries and the solutions that may be appropriate to prevent future injuries.
- Talking with employees who work in that job.
- Walking through the workplace to observe employees performing the job.
- Evaluating what various studies have suggested are risk factors for MSDs.

FIGURE 13.1   Maintenance worker performing tasks overhead and above the shoulders.

The ergonomic risk factors that should be looked for include

- Force—the amount of physical effort required to perform a task (such as heavy lifting) or to maintain control of equipment or tools
- Repetition—performing the same motion or series of motions continually or frequently for an extended period of time
- Awkward or static postures—include repeated or prolonged reaching, twisting, bending, kneeling, squatting, or working overhead, or holding fixed positions (see Figure 13.1)
- Contact stress—pressing the body or part of the body against a hard or sharp edge, or using the hand as a hammer

When there are several risk factors in a job, there can be a greater risk of injury. However, the presence of risk factors in a job does not necessarily mean that the job poses a risk of injury. Whether certain work activities put an employee at risk of injury depends on the duration (how long), frequency (how often), and magnitude (how intense) of the employee's exposure to the risk factors in the activity. A checklist of risk-related items that might be observed during an ergonomic assessment is found in Figure 13.2.

13.2.4 IMPLEMENT SOLUTIONS

Examples of potential solutions for various concerns are located in this section. To implement an effective ergonomics program, employers may need to modify work-stations, purchase equipment, and change work practices. Simple, low-cost solutions are often available to solve problems. For example, carts or anti-fatigue mats can be

---

Checklist for identifying potential ergonomics concerns by workplace activity

If the answer to any of the following questions is yes, the activity may be a potential source of ergonomic concern, depending on the duration, frequency, and magnitude of the activity. For example, occasionally lifting items into overhead storage areas may not present a problem while doing so frequently may present a problem.

Force in lifting

Yes ☐ No ☐ Does the lift require pinching to hold the object?
Yes ☐ No ☐ Is the lift made with one hand?
Yes ☐ No ☐ Are very heavy items lifted without the assistance of a mechanical device?
Yes ☐ No ☐ Are heavy items lifted by bending over or reaching above shoulder height?

Force in pushing, pulling, carrying

Yes ☐ No ☐ Are dollies, pallet jacks, or other carts difficult to get started?
Yes ☐ No ☐ Are there cracks in the floor, debris (e.g., broken pallets), or uneven surfaces (e.g., dock plates) that catch the wheels while pushing?
Yes ☐ No ☐ Is pulling routinely used to move an object?
Yes ☐ No ☐ Are heavy objects carried manually for a long distance?

Force to use tools

Yes ☐ No ☐ Are tool handles too narrow or too wide for the employee's hand?
Yes ☐ No ☐ Do tools require the use of a pinch grip or trigger finger to operate?

Repetitive tasks

Yes ☐ No ☐ Are multiple motions needed?
Yes ☐ No ☐ Is a quick wrist motion used in the task?
Yes ☐ No ☐ Are most items lifted rather than slid on constant basis?
Yes ☐ No ☐ Do repetitive motions last for several hours without a break?
Yes ☐ No ☐ Is a quick wrist motion used while performing the task?
Yes ☐ No ☐ Does the job require repeated finger force (e.g., kneading, squeezing, or using trigger operated tool)?

Awkward and static postures

Yes ☐ No ☐ Is the back twisted while lifting or holding heavy items?
Yes ☐ No ☐ Are objects lifted from or into a cramped space?

FIGURE 13.2 Checklist for identifying potential ergonomics concerns by workplace activity.

(*continued*)

Yes ☐	No ☐	Do routine tasks involve leaning, bending over, kneeling, or squatting?
Yes ☐	No ☐	Do routine tasks involve working with the wrists in a bent or twisted position?
Yes ☐	No ☐	Are routine tas ks done with the hands below the waist or above the shoulders?
Yes ☐	No ☐	Are routine tasks done behind (e.g., pushing items) or to the sides of the body?
Yes ☐	No ☐	Are routine tasks performed too far in front of the body?
Yes ☐	No ☐	Does the job require standing for most of the time at work?
Yes ☐	No ☐	Do employees work with their arms or hands in the same position for long periods of time without changing positions or resting?

Contact stress

Yes ☐	No ☐	Are there sharp edges the worker may come into contact with?
Yes ☐	No ☐	Do employees use their hands as a hammer (e.g., closing containers)?

FIGURE 13.2 (continued)

used to reduce risk factors. Employers should consider ergonomic issues when designing new workplaces or redesigning existing workplaces. At that time, major changes are easier to implement, and ergonomic design elements can be incorporated at little or no additional cost.

Some of the factors to consider when designing and implementing solutions to ergonomic issues are

- Existence of prolonged repetitive flexion and extension of the wrist
- Prolonged or repetitive bending at the waist (see Figure 13.3)
- Prolonged standing or sitting without shifting of the workers' body
- Suspending an outstretched arm for extended periods of time
- Holding or turning the head consistently to one side
- Any unnatural posture that is held repeatedly or for a prolonged time
- Repeated motion without periods of rest
- Repeated motion with little or no variation
- Repeated motion with great force
- Resting or compressing a body part on or against a surface
- Lifting heavy objects far away from the body
- Frequent reaching or working above the shoulders

Some factors that can contribute to MSDs are

- Furniture or work area arrangement that produces bad posture
- Physically demanding work that the worker is not used to doing
- Home or recreational activities that produce unaccustomed stress on the body
- Being out of shape
- Diminished muscle strength or joint flexibility
- Suffering from arthritis

FIGURE 13.3 Loading a hand truck in close quarters requires twisting and turning.

Implementing change will help prevent reoccurrence of ergonomic issues. There is no substitute for ergonomically sound work environments and work practices. In order to make a difference something needs to be changed. These changes may be medical intervention to treat MSDs, work restrictions to decrease or eliminate stress on the body, educating employees and supervisors on the hazards and simple preventive techniques, and the use of long-term solutions.

The solution may be to recognize the MSD early and take medical action by using anti-inflammatories, ice and heat, splints, exercise, physical therapy and as a last resort surgery.

Recognition has to come from knowledge and this is usually in the form of training. With knowledge both supervisors and workers can begin to make improvements in workstations, change work practices, react to signs and symptoms, take reports of injury or disorder as serious, start and maintain a fitness regime, take time for recovery by breaks or performing other unrelated tasks, plan ahead to reduce stressors that have the potential to cause MSDs, and look for ergonomic solutions to solve problems. Do all those things that can be done to

**Checklist for identifying potential ergonomics concerns
at job-specific workstations**

If the answer to any of the following questions is no, the activity may be a potential source of ergonomic concern, depending on the duration, frequency, and magnitude of the activity. For example, infrequently carrying light items without a cart may not present a problem while frequently carrying heavy items without a cart may present a problem.

Yes ☐   No ☐   Are items within easy reach?

Yes ☐   No ☐   Are keyboard supports adjustable?

Yes ☐   No ☐   Can the work be done with items at about elbow height?

Yes ☐   No ☐   Can the display be read without twisting?

Yes ☐   No ☐   Are all edges smoothed or rounded so the cashier does not come into contact with sharp edges?

Yes ☐   No ☐   Do the work benches or tables appear to be the right height?

Yes ☐   No ☐   Does the worker standing have an antifatigue mat and/or footrest?

Yes ☐   No ☐   Can workers adjust the height of the stand?

Yes ☐   No ☐   Do loads have handles?

Yes ☐   No ☐   Can materials be placed without leaning over the some object or twisting the back?

Yes ☐   No ☐   Are step stools available to reach high shelves?

Yes ☐   No ☐   Do totes and boxes have handles?

Yes ☐   No ☐   Are cutter blades sharp?

Yes ☐   No ☐   Are carts available to move heavy items?

Yes ☐   No ☐   Are carts or pallet jacks used to keep lifts at waist height?

Yes ☐   No ☐   Are lightweight pallets used?

Yes ☐   No ☐   Are box weights within the lifting ability of employees?

Yes ☐   No ☐   Are counter heights and widths appropriate for employees?

Yes ☐   No ☐   Are knives kept sharp?

Yes ☐   No ☐   Are counter heights and widths appropriate for employees?

Yes ☐   No ☐   Are work tables, etc. positioned so that the work can be performed at elbow height?

FIGURE 13.4 Checklist for identifying potential ergonomics concerns at job-specific workstations.

improve the situation. Evaluate all workstations and use a checklist similar to the one found in Figure 13.4.

13.3   BACK AND BACK INJURIES

Back disorders can develop gradually as a result of micro-trauma brought about by repetitive activity over time or can be the product of a single traumatic event. Because of the slow and progressive onset of this internal injury, the condition is often ignored until the symptoms become acute, often resulting in disabling injury. Acute back injuries can be the immediate result of improper lifting techniques or lifting loads that are too heavy for the back to support. While the acute injury may seem to be caused by a single well-defined incident, the real cause is often a combined interaction of the observed stressor coupled with years of weakening of the musculoskeletal support mechanism by repetitive micro-trauma. Injuries can arise in muscles, ligaments, vertebrae, and discs, either singly or in combination.

Although back injuries account for no work-related deaths, they do account for a significant amount of human suffering, loss of productivity, and economic burden on compensation systems. Back disorders are one of the leading causes of disability for people in their working years and afflict over 600,000 employees each year with a cost of about $50 billion annually (data in 1991) according to National Institute for Occupational Safety and Health (NIOSH). The frequency and economic impact of back injuries and disorders on the work force are expected to increase over the next several decades as the average age of the work force increases and medical costs go up.

13.3.1 FACTORS ASSOCIATED WITH BACK DISORDERS

Statistics on injuries resulting from improper lifting and carrying usually indicate that manual handling of materials accounts for approximately 25% of all occupational injuries. In some industries, 90% of those employees suffering from back pain attribute it to work-related accidents—as opposed to only 30%–50% in the general population. Low back pain is second only to head colds as the leading cause of absenteeism. Right now, 7 million employees are treated for chronic back problems and new cases occur at the rate of about 2 million a year (Figure 13.5).

FIGURE 13.5 Worker attempting to make a safe lift based on training and supervised practice.

Back disorders result from exceeding the capability of the muscles, tendons, discs, or the cumulative effect of several contributors:

- Reaching while lifting
- Poor posture—how one sits or stands
- Stressful living and working activities—staying in one position for too long
- Bad body mechanics—how one lifts, pushes, pulls, or carries objects
- Poor physical condition—losing the strength and endurance to perform physical tasks without strain
- Poor design of job or workstation
- Repetitive lifting of awkward items, equipment, or patients
- Twisting while lifting
- Bending while lifting
- Maintaining bent postures
- Heavy lifting
- Fatigue
- Poor footing such as slippery floors or constrained posture
- Lifting with forceful movement
- Vibration, such as that with lift truck drivers, delivery drivers, etc.

Other marginal contributing factors that have seemed to be linked to back disorders are

- Congenital defects of the spine
- Increase in static standing or sitting tasks
- An aging work force
- Decreases in physical conditioning and exercise
- Increased awareness of workplace hazards
- Job dissatisfaction

Signs and symptoms include pain when attempting to assume normal posture, decreased mobility, and pain when standing or rising from a seated position. Manual material handling is the principal source of compensable injuries in the American work force, and four out of five of these injuries affect the lower back.

13.3.2 TYPES OF INJURIES FROM LIFTING TASKS

The types and kinds of injuries that can result from improper lifting and carrying besides back injuries are

- Injuries to the hands and fingers—cuts, splinters, pinched fingers, etc.
- Injuries to the feet, toes, and legs—losing balance or dropping a heavy object
- Injuries to the eyes, head, and trunk—opening a wire bound box or bale, handling cable or metal strapping, etc.
- Hernias and back injuries—most common of all

Most of the strains, sprains, fractures, and bruises that result from handling materials on the job are caused by unsafe practices such as

- Improper lifting
- Carrying too heavy a load
- Incorrect gripping
- Failing to observe proper foot or hand clearances
- Failing to use or wear proper equipment

Some ways to avoid these injuries are

- Eliminate the need to lift materials manually by using mechanical lifting devices whenever possible
- Select employees who are physically qualified for the job; preemployment physical exams are used to determine the loads an employee can safely lift
- Train employees to lift properly

13.3.3   INHERENT HAZARDS

As a walkthrough of the workplace is conducted, the following should be accomplished:

- Ask employees about their opinion on the difficulty of the task as well as personal experiences of back pain
- Observe worker postures and lifting
- Determine weight of objects lifted
- Determine the frequency and duration of lifting tasks
- Measure the dimensions of the workplace and lift
- Repetitive material handling increases the likelihood of an injury

Principal variables in evaluating manual lifting tasks to determine how heavy a load can be lifted are the horizontal distance from the load to the employee's spine, the vertical distance through which the load is handled, the amount of trunk twisting the employee used during the lifting, the ability of the hand to grasp the load, and the frequency with which the load is handled. Additional variables include floor and shoe traction, space constraints, two-handed lifts, and size and stability of the load.

13.4   PREVENTION AND CONTROL

13.4.1   ENGINEERING CONTROLS

Alter the task to eliminate the hazardous motion or change the position of the object in relation to the employee's body—such as adjusting the height of a pallet or shelf. Engineering controls are always preferred. Other engineering controls that can be utilized are

- Material-handling tasks should be designed to minimize the weight, range of motion, and frequency of the activity.
- Work methods and stations should be designed to minimize the distance between the person and the object that is handled.
- Platforms and conveyors should be built at about waist height to minimize awkward postures. Conveyors or carts should be used for horizontal motion whenever possible. Reduce the size or weight of the object(s) lifted.
- High-strength push–pull requirements are undesirable, but pushing is better than pulling. Material-handling equipment should be easy to move, with handles that can be easily grasped in an upright posture.
- Workbench or workstation configurations can force people to bend over.
- Corrections should emphasize adjustments necessary for the employee to remain in a relaxed upright stance or fully supported seated posture.
- Bending the upper body and spine to reach into a bin or container is highly undesirable.
- The bins should be elevated, tilted, or equipped with collapsible sides to improve access.
- Repetitive or sustained twisting, stretching, or leaning to one side are undesirable.
- Corrections could include repositioning bins and moving employees closer to parts and conveyors.
- Heavy objects should be stored at waist level.
- Lift-assist devices and lift tables must be provided.

13.4.2 WORK PRACTICES

Worker training and education should include general principles of ergonomics, recognition of hazards and injuries, procedures for reporting hazardous conditions, and methods and procedures for early reporting of injuries. Additionally, job-specific training should be given on safe work practices, hazards, and controls. Some of the work practices that can be employed are

- Strength and fitness training, which can reduce compensation costs
- Rotating of employees, providing short breaks every hour, or using a two-person lift may be helpful
- Rotation is not simply a different job, but must be a job that utilizes a completely different muscle group from the ones that have been overexerted

13.4.3 ADJUSTMENT TO WORK

There are often simple yet effective solutions to back injury mitigation issues such as

- Decrease standing for extended periods because it places excessive stress on the back and legs
- Provide a footrest or rail, resilient floor mats, height-adjustable chairs or stools, and opportunities for the employee to change position

- Ensure that employees who are seated have chairs or stools that have been properly chosen
- Provide proper adjustable lumbar support for workers
- Avoid static seated postures with bending or reaching

13.4.4 OTHER SOLUTIONS

Workers need to be trained to lift and carry safely. It should not be assumed that workers inherently know how to lift safely. Before workers lift anything, they should follow this procedure.

- Inspect materials for slivers, jagged edges, burrs, rough or slippery surfaces, protruding nails, etc.
- Make sure your hands are free of oil and grease
- Wipe off any wet, greasy, slippery, or dirty objects before trying to handle them
- Get a firm grip on the object
- Keep your fingers away from pinch points
- Wear the appropriate protective clothing (safety shoes, hand leathers or gloves, etc.)

Assuming the correct posture is an important part of achieving a safe lift. Some recommendations for this are

- Keep your feet parted—one alongside, one behind the object
- Keep your back straight, but not necessarily vertical
- Tuck your chin in
- Grip the object with the whole hand
- Tuck your elbows and arms in
- Keep your body weight directly over your feet

The following steps should be followed for lifting, carrying, and setting the load down:

- Size up the load, estimating weight, size, and shape.
- If the load is too much to handle, get help.
- Stand close to the object, keeping your feet 8–12 in. apart for good balance.
- Bend the knees to a comfortable position and get a good handhold.
- Using both leg and back muscles, lift the load straight up. Move smoothly and easily, pushing with the legs and keeping the load close to your body.
- Lift the object to carrying position. Avoid twisting and turning about until the lift is completed.
- To turn the body, change foot positions and check to see that your path of travel is clear before moving.
- To set the load down, bend the knees using leg and back muscles.
- When load is securely positioned, release it.

Each lift should be considered unique since it may require handling of specific or unusual weights and shapes of boxes or cartons, bags or sacks, barrels and drums, machinery, equipment, long objects (ladders, lumber, pipe, etc.), and flat material (glass, sheet metal, plywood, etc.). Such lifts may require a team lifting and carrying, which needs to be carefully orchestrated by making sure that the load is evenly distributed. The team's movements are coordinated with those helping with the lift both at start and at finish of the lifting action and result in the same timing of the lift and turning together. Whenever possible manual lifts should not be done; it is better to use accessories for lifting and carrying such as the following:

- Hooks and handles
- Rollers and cradles
- Jacks
- Hand trucks, dollies, wheelbarrows
- Cranes
- Hoists
- Forklifts

If manual lifts must be performed, it is important that workers be physically fit and capable of making any lifts. A lifting safety checklist is included to help make each lift a safe one while protecting the back and preventing back injuries (see Figure 13.6).

Lifting safety checklist

Yes ☐ No ☐ Has all workers been trained on proper lifting techniques?
Yes ☐ No ☐ Was the object to be lifted inspected to decide the best way to grasp it?
Yes ☐ No ☐ Has the load been sized up to insure it can be lifted?
Yes ☐ No ☐ Are load kept small to prevent heavy lifts?
Yes ☐ No ☐ Is the feet placed close to the object?
Yes ☐ No ☐ Can a good grip be made on the load?
Yes ☐ No ☐ Are the knees bent while keeping the back straight?
Yes ☐ No ☐ Is the load held close to the workers body?
Yes ☐ No ☐ Can the worker see past the load?
Yes ☐ No ☐ Did the worker get help for large or heavy objects?
Yes ☐ No ☐ Is the lift more appropriate for a team lift?
Yes ☐ No ☐ Are harsh jerking movements when pushing, pulling, or lifting a load avoided?
Yes ☐ No ☐ Are materials stack so they are positioned between the knees and waist?
Yes ☐ No ☐ Are gloves worn when handling sharp or rough objects?
Yes ☐ No ☐ Is power lifting equipment used when possible to prevent injuries instead of manually?
Yes ☐ No ☐ Is care taken not to drop materials that might hit someone?
Yes ☐ No ☐ Is equipment, carts, and/or table kept at a proper height to help prevent back injuries?

FIGURE 13.6 Lifting safety checklist.

13.5 SUMMARY

Training is critical for employers and employees to safely use and develop solutions to ergonomic problems. Training should include

- Knowledge of the work tasks that may lead to pain or injury
- Understanding of the proper work practices for tasks that employees will be performing
- Ability to recognize MSDs and their early indications
- Advantage of addressing early indications of MSDs before serious injury has developed
- Awareness of reporting work-related injuries and illnesses as required by OSHA's injury and illness recording and reporting regulation (29 CFR 1904)

Occupational Safety and Health Administration recommends that staff members who coordinate and direct ergonomics efforts also receive training to give them the knowledge to effectively carry out their responsibilities. These designated staff members will benefit from information and training that will allow them to

- Appropriately use checklists and other tools to analyze tasks in the workplace
- Address problems by selecting proper equipment and work practices
- Identify the potential benefits of specific workplace changes
- Help other workers implement solutions
- Assess the effectiveness of ergonomics efforts

Management and supervisory personnel may benefit from training that focuses on general information describing ergonomics, its purpose, common risk factors in the workplace, common ergonomic solutions, and how to respond to injury reports. Managers will also be interested in potential benefits, including increased efficiency, reduced workers' compensation claims, and improved customer service.

Occupational Safety and Health Administration recommends that employers evaluate the effectiveness of their ergonomic efforts and follow up on unresolved problems. Evaluation and follow-up help sustain continuous improvement in reducing injuries and illnesses, track the effectiveness of specific ergonomic solutions, identify new problems, and show areas where further attention is needed.

How often an employer evaluates the program will vary by the size and complexity of the facility. Management should revise the program in response to identified deficiencies and communicate the results of the program evaluation and any program revisions to employees.

13.5.1 LIFTING

Most workplace jobs involve some lifting. It is important that employers provide employees with help to lift heavy or bulky items (see Figure 13.7). Whether a

FIGURE 13.7   Servers have to carry heavy, awkward, and unbalanced loads.

particular lift will require assistance depends on several factors, including the weight of the object, how frequently the object is lifted, how close the object is to the ground, how high it must be lifted, and how far it must be carried. Assistance can include a dolly or cart, or help from a coworker. For lighter items, the employer should ensure that employees use good lifting techniques:

- Use good lifting techniques when holding, lifting or carrying items
- Before lifting boxes and cases, check to see if the weight is given so that you can prepare to lift properly
- Keep the item close to your body
- Turn with the feet, not the torso
- Keep your back straight
- Use your legs to do the lifting
- Get close to where you want to set the item down

Prevent ergonomic injuries and illnesses as well as back disorders by applying the principles of good ergonomics and safe lifting practices to each workplace.

14 Electrical Safety

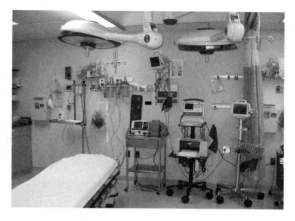

Just look at the amount of electricity-driven devices in this hospital examination room.

14.1   ELECTRICAL DANGERS

All workers have some exposure to electricity everyday. The risk of electrocution (death caused by contact with electricity) and injury vary with their workplace and their job tasks.

Whenever a worker uses power tools or works on electrical circuits, there is a risk of electrical hazards, especially electrical shock. Anyone can be exposed to these hazards at home or at work. Workers are exposed to more hazards because job sites can be cluttered with tools, equipment, and materials, fast-paced, and open to the weather at times. Risk is also higher at work because many jobs involve electric power tools or work on electrical circuits or machinery.

Electrical workers must pay special attention to electrical hazards because they work on electrical circuits. Coming in contact with an electrical voltage can cause current to flow through the body, resulting in electrical shock and burns. Serious injury or even death may occur. As a source of energy, electricity is used without much thought about the hazards it can cause. Because electricity is a familiar part of our lives, it is often not treated with enough caution. As a result, an average of one worker is electrocuted on the job every day of every year. Electrocution is the third leading cause of work-related deaths among 16- and 17-year-olds, after motor vehicle deaths and workplace homicide. Electrocution is the cause of as high as 12% of all workplace deaths among young workers. Electrocution results in between 5% and 6% of all occupationally related deaths.

The electrical current in regular businesses and homes has enough power to cause death by electrocution. Even changing a light bulb without unplugging the lamp can be hazardous because coming in contact with the hot or live part of the socket could kill a person.

14.2   ELECTRICAL PATHWAYS

An electrical shock is received when electrical current passes through the body. Current will pass through the body in a variety of situations. Whenever two wires are of different voltages, current will pass between them if they are connected. Your body can connect the wires if you touch both of them at the same time. Current will pass through your body.

In most household wiring, the black wires and the red wires are at 120 V. The white wires are at 0 V because they are connected to ground. The connection to ground is often through a conducting ground rod driven into the earth. The connection can also be made through a buried metal water pipe. If a worker comes in contact with an energized black wire—and they are also in contact with the neutral white wire—current will pass through his/her body. He/she will receive an electrical shock.

There are four main types of injuries resulting from contact with an electrical circuit: electrocution (fatal), electric shock, burns, and falls. These injuries can happen in various ways:

- Direct contact with electrical energy
- When the electricity arcs (jumps) through a gas (such as air) to a person who is grounded (that would provide an alternative route to the ground for the electricity)
- Thermal burns including flash burns from heat generated by an electric arc, and flame burns from materials that catch fire from heating or ignition by electrical currents
- High-voltage contact burns can burn internal tissues while leaving only very small injuries on the outside of the skin
- Muscle contractions, or a startled reaction, can cause a person to fall from a ladder, scaffold, or aerial bucket. The fall can cause serious injuries

14.3   RESULT OF ELECTRICAL CONTACT

The severity of injury from electrical shock depends on the amount of electrical current and the length of time the current passes through the body. For example, 1/10 of an ampere (A) of electricity going through the body for just 2 s is enough to cause death. The amount of internal current a person can withstand and still be able to control the muscles of the arm and hand can be less than 10 mA. Currents above 10 mA (0.010 A) can paralyze or freeze muscles. When this freezing happens, a person is no longer able to release a tool, wire, or other object. In fact, the electrified object may be held even more tightly, resulting in longer exposure to the shocking current. For this reason, handheld tools that give a shock can be very

TABLE 14.1
Estimated Effects of 60 Hz AC Currents

1 mA	Barely perceptible
16 mA	Maximum current an average man can grasp and let go
20 mA	Paralysis of respiratory muscles
100 mA	Ventricular fibrillation threshold
2 A	Cardiac standstill and internal organ damage
15/20 A	Common fuse or breaker opens circuit[a]

[a] Contact with 20 mA of current can be fatal. As a frame of reference, a common household circuit breaker may be rated at 5, 10, 15, 20, and 30 A.

dangerous. If a worker cannot let go of the tool, current continues through their body for a longer time, which can lead to respiratory paralysis (the muscles that control breathing cannot move). Breathing is stopped for a period. People have stopped breathing when shocked with currents from voltages as low as 49 V. Usually, it takes about 30 mA of current to cause respiratory paralysis.

Currents greater than 75 mA may cause ventricular fibrillation (very rapid, ineffective heartbeat) to occur. This condition will cause death within a few minutes unless a special device called a defibrillator is used to save the victim. Heart paralysis occurs at 4 A, which means the heart does not pump at all. Tissue is burned with currents greater than 5 A.

Table 14.1 shows what usually happens for a range of currents (lasting 1 s) at typical household voltages. Longer exposure times increase the danger to the shock victim. For example, a current of 100 mA applied for 3 s is as dangerous as a current of 900 mA applied for a fraction of a second (0.03 s). The muscle structure of the person also makes a difference. People with less muscle tissue are typically affected at lower current levels. Even low voltages can be extremely dangerous because the degree of injury depends not only on the amount of current but also on the length of time the body is in contact with the circuit.

14.4 VOLTAGE

Sometimes high voltages usually above 600 V lead to additional injuries. High voltages can cause violent muscular contractions. You may lose your balance and fall, which can cause injury or even death if you fall into machinery that can crush you. High voltages can also cause severe burns.

At 600 V, the current through the body may be as great as 4 A, causing damage to internal organs such as the heart. High voltages also produce burns. In addition, internal blood vessels may clot. Nerves in the area of the contact point may be damaged. Muscle contractions may cause bone fractures either from the contractions themselves or from falls.

A severe shock can cause much more damage to the body than is visible. A person may suffer internal bleeding and destruction of tissues, nerves, and muscles.

TABLE 14.2
Number of High-/Low-Voltage Electrocutions

Voltages (V)	Number of Deaths
110–120	26
220–240	15
270–277	12
440–480	16
600	6
601–7,199	15
7,200–7,620	80
7,621–12,999	13
13,000–13,800	18
>20,000	21

Sometimes the hidden injuries caused by electrical shock result in delayed death. Shock is often only the beginning of a chain of events. Even if the electrical current is too small to cause injury, your reaction to the shock may cause you to fall, resulting in bruises, broken bones, or even death.

The length of time of the shock greatly affects the amount of injury. If the shock is short in duration, it may only be painful. A longer shock (lasting a few seconds) could be fatal if the level of current is high enough to cause the heart to go into ventricular fibrillation. This is not much current when you realize that a small power drill uses 30 times as much current as what will kill. At relatively high currents, death is certain if the shock is long enough. However, if the shock is short and the heart has not been damaged, a normal heartbeat may resume if contact with the electrical current is eliminated. (This type of recovery is rare.)

NIOSH data for 1982 through 1994 indicates that of 224 electrocutions, the following voltages resulted in these deaths (see Table 14.2). The indication is that both low and high voltages have the capacity to cause death from contact with electrical sources.

14.5 CURRENT

The amount of current passing through the body also affects the severity of an electrical shock. Greater voltages produce greater currents. Therefore, there is greater danger from higher voltages. Resistance hinders current. The lower the resistance (or impedance in AC circuits), the greater the current will be. Dry skin may have a resistance of 100,000 Ω or more. Wet skin may have a resistance of only 1000 Ω. Wet working conditions or broken skin will drastically reduce resistance. The low resistance of wet skin allows current to pass into the body more easily and give a greater shock. When more force is applied to the contact point or when the contact area is larger, the resistance is lower, causing stronger shocks (see Figure 14.1).

FIGURE 14.1   Warning signs should be used when dangerous currents exist. (Courtesy of the Department of Energy.)

The path of the electrical current through the body affects the severity of the shock. Currents through the heart or nervous system are most dangerous. If contact is made with a live wire with the head, the nervous system will be damaged. Contacting a live electrical part with one hand while you are grounded at the other side of the body will cause electrical current to pass across your chest, possibly injuring your heart and lungs.

There have been cases where an arm or leg is severely burned by high-voltage electrical current to the point of coming off, and the victim is not electrocuted. In these cases, the current passes through only a part of the limb before it goes out of the body and into another conductor. Therefore, the current does not go through the chest area and may not cause death, even though the victim is severely disfigured. If the current does go through the chest, the person will almost surely be electrocuted. A large number of serious electrical injuries involve current passing from the hands to the feet. Such a path involves both the heart and lungs. This type of shock is often fatal. The danger from electrical shock depends on

- Amount of the shocking current through the body
- Duration of the shocking current through the body
- Path of the shocking current through the body

14.6   ELECTRICAL BURNS

14.6.1   Voltage Burns

The most common shock-related, nonfatal injury is a burn. Burns caused by electricity may be of three types: electrical burns, arc burns, and thermal contact burns. Electrical burns can result when a person touches electrical wiring or equipment that is used or maintained improperly. Typically, such burns occur on the hands. Electrical burns are one of the most serious injuries you can receive. They need to be given immediate attention. Additionally, clothing may catch fire and a thermal burn may result from the heat of the fire.

14.6.2 Arcing Burns

Arc-blasts occur when powerful, high-amperage currents arc through the air. Arcing is the luminous electrical discharge that occurs when high voltages exist across a gap between conductors and current travels through the air. This situation is often caused by equipment failure due to abuse or fatigue. Temperatures as high as 35,000°F have been reached in arc-blasts.

14.6.3 Thermal Burns

Electricity is one of the most common causes of fires and thermal burns in homes and workplaces. Defective or misused electrical equipment is a major cause of electrical fires. If there is a small electrical fire, be sure to use only a Class C or multipurpose (ABC) fire extinguisher, or you might make the problem worse.

14.7 HANDLING ELECTRICAL HAZARDS

If you do not recognize, evaluate, and control hazards, you may be injured or killed by the electricity itself, electrical fires, or falls. If you use the safety model to recognize, evaluate, and control hazards, you are much safer.

14.7.1 Identify Hazards

The first part of the safety model is identifying the hazards around you. Only then can one avoid or control the hazards. It is best to discuss and plan hazard recognition tasks with co-workers. Sometimes workers take risks, but when they are responsible for others, they are more careful. Sometimes others see hazards that we overlook. Of course, it is possible to be talked out of our concerns by someone who is reckless or dangerous. Do not take a chance. Careful planning of safety procedures reduces the risk of injury. Decisions to lockout and tagout circuits and equipment need to be made during this part of the safety model. Plans for action must be made now.

14.7.2 Evaluate Hazards

When evaluating hazards, it is best to identify all possible hazards first and then evaluate the risk of injury from each hazard. Do not assume the risk is low until you evaluate the hazard. It is dangerous to overlook hazards. Job sites are especially dangerous because they are always changing. Many people are working at different tasks. Job sites are frequently exposed to bad weather. A reasonable place to work on a bright, sunny day might be very hazardous in the rain. The risks in the work environment need to be evaluated all the time. Then, whatever hazards are present need to be controlled.

14.7.3 Control Hazards

Once electrical hazards have been recognized and evaluated, they must be controlled. Controlling of electrical hazards is in two main ways: (1) create a safe work environment and (2) use safe work practices. Controlling electrical hazards (as well as other hazards) reduces the risk of injury or death.

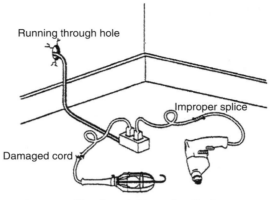

Running through hole

Improper splice

Damaged cord

Unsafe portable cord methods

FIGURE 14.2 Damaged or improper wiring should not be used. (Courtesy of the Department of Energy.)

14.8 IDENTIFYING HAZARDS

The first step toward protecting oneself is recognizing the many hazards faced on the job. To do this, workers must have knowledge of situations that can place them in danger. Knowing where to look helps in recognizing hazards such as the following:

- Inadequate wiring is dangerous (see Figure 14.2)
- Exposed electrical parts are dangerous
- Overhead power lines are dangerous
- Wires with bad insulation can give you a shock
- Electrical systems and tools that are not grounded or double-insulated are dangerous (see Figure 14.3)

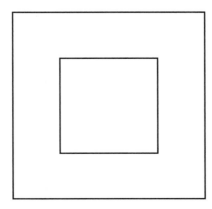

FIGURE 14.3 Example of the symbol used on electrical tools and devices to indicate double insulation. (Courtesy of the Department of Energy.)

- Overloaded circuits are dangerous
- Damaged power tools and equipment are electrical hazards
- Using the wrong PPE is dangerous
- Using the wrong tool is dangerous
- Some on-site chemicals are harmful
- Defective ladders and scaffolding are dangerous
- Ladders that conduct electricity are dangerous
- Electrical hazards can be made worse if the worker, location, or equipment is wet

14.9   SPECIFIC HAZARDS

14.9.1   INADEQUATE WIRING HAZARDS

An electrical hazard exists when the wire is too small a gauge for the current it carries. Normally, the circuit breaker in a circuit is matched to the wire size. However, in older wiring, branch lines to permanent ceiling light fixtures could be wired with a smaller gauge than the supply cable. Let us say a light fixture is replaced with another device that uses more current. The current capacity (ampacity) of the branch wire could be exceeded. When a wire is too small for the current it is supposed to carry, the wire will heat up. The heated wire could cause a fire.

When you use an extension cord, the size of the wire placed into the circuit may be too small for the equipment. The circuit breaker could be the right size for the circuit but not right for the smaller-gauge extension cord. A tool plugged into the extension cord may use more current than the cord can handle without tripping the circuit breaker. The wire will overheat and could cause a fire.

The kind of metal used as a conductor can cause an electrical hazard. Special care needs to be taken with aluminum wire. Since it is more brittle than copper, aluminum wire can crack and break more easily. Connections with aluminum wire can become loose and oxidize if not made properly, creating heat or arcing.

14.9.2   EXPOSED ELECTRICAL PARTS HAZARDS

Electrical hazards exist when wires or other electrical parts are exposed. Wires and parts can be exposed if a cover is removed from a wiring or breaker box. The overhead wires coming into a home may be exposed. Electrical terminals in motors, appliances, and electronic equipment may be exposed. Older equipment may have exposed electrical parts. If a worker contacts exposed live electrical parts, they will be shocked.

14.9.3   OVERHEAD POWER LINE HAZARDS

Most people do not realize that overhead power lines are usually not insulated. More than half of all electrocutions are caused by direct shocks and electrocutions occur where physical barriers are not in place to prevent contact with the wires. When dump trucks, cranes, work platforms, or other conductive materials (such as pipes and ladders) contact overhead wires, the equipment operator or other workers can be killed. If an adequate distance or the required clearance distances from power lines are not maintained, workers can be shocked and killed. (The minimum distance for voltages up to 50 kV is 10 ft. For voltages over 50 kV, the minimum distance is 10 ft plus 4 in. for every 10 kV over 50 kV.) Never store materials and equipment under or near overhead power lines.

14.9.4   DEFECTIVE INSULATION HAZARDS

Insulation that is defective or inadequate is an electrical hazard. Usually, a plastic or rubber covering insulates wires. Insulation prevents conductors from coming in contact with each other. Insulation also prevents conductors from coming in contact with people.

Extension cords may have damaged insulation. Sometimes the insulation inside an electrical tool or appliance is damaged. When insulation is damaged, exposed metal parts may become energized if a live wire inside touches them. Electric hand tools that are old, damaged, or misused may have damaged insulation inside. If damaged power tools or other equipment are touched, a shock will be received. Workers are more likely to receive a shock if the tool is not grounded or double-insulated. (Double-insulated tools have two insulation barriers and no exposed metal parts.)

14.9.5   IMPROPER GROUNDING HAZARDS

When an electrical system is not grounded properly, a hazard exists. The most common Occupational Safety and Health Administration (OSHA) electrical violation is improper grounding of equipment and circuitry. The metal parts of an electrical wiring system that we touch (switch plates, ceiling light fixtures, conduit, etc.) should be grounded and at 0 V. If the system is not grounded properly, these parts may become energized. Metal parts of motors, appliances, or electronics that are plugged into improperly grounded circuits may be energized. When a circuit is not grounded properly, a hazard exists because unwanted voltage cannot be safely eliminated. If there is no safe path to ground for fault currents, exposed metal parts in damaged appliances can become energized.

Extension cords may not provide a continuous path to ground because of a broken ground wire or plug. If a person comes into contact with a defective electrical device that is not grounded (or grounded improperly), a shock will be received.

Electrical systems are often grounded to metal water pipes that serve as a continuous path to ground. If plumbing is used as a path to ground for fault current, all pipes must be made of conductive material (a type of metal). Many electrocutions and fires occur because (during renovation or repair) parts of metal plumbing are replaced with plastic pipe, which does not conduct electricity. In these cases, the path to ground is interrupted by nonconductive material.

14.9.6 Ground Fault Circuit Interrupter

A ground fault circuit interrupter (GFCI) works by detecting any loss of electrical current in a circuit. When a loss is detected, the GFCI turns the electricity off before severe injuries or electrocution can occur. GFCIs are set to kick off at 5 mA. A painful shock may occur during the time that it takes for the GFCI to cut off the electricity so it is important to use the GFCI as an extra protective measure rather than a replacement for safe work practices.

GFCI wall outlets can be installed in place of standard outlets to protect against electrocution for just that outlet, or a series of outlets in the same branch. A GFCI circuit breaker can be installed on some circuit breaker electrical panels to protect an entire branch circuit. Plug-in GFCIs can be plugged into wall outlets where appliances will be used (see Figure 14.4).

Test the GFCI monthly. First plug a night-light or lamp into the GFCI-protected wall outlet (the light should be turned on), then press the "test" button on the GFCI. If the GFCI is working properly, the light should go out. If not, have the GFCI repaired or replaced. Reset the GFCI to restore power. If the "reset" button pops out but the light does not go out, the GFCI has been improperly wired and does not offer shock protection at that wall outlet. Contact a qualified electrician to correct any wiring errors.

14.9.7 Overload Hazards

Overloads in an electrical system are hazardous because they can produce heat or arcing. Wires and other components in an electrical system or circuit have a maximum amount of current they can carry safely. If too many devices are plugged

FIGURE 14.4   A plug in GFCI attached to an electric drill.

into a circuit, the electrical current will heat the wires to a very high temperature. If any one tool uses too much current, the wires will heat up.

The temperature of the wires can be high enough to cause a fire. If their insulation melts, arcing may occur. Arcing can cause a fire in the area where the overload exists, even inside a wall.

In order to prevent too much current in a circuit, a circuit breaker or fuse is placed in the circuit. If there is too much current in the circuit, the breaker trips and opens like a switch. If an overloaded circuit is equipped with a fuse, an internal part of the fuse melts, opening the circuit. Both breakers and fuses do the same thing: open the circuit to shut off the electrical current.

If the breakers or fuses are too big for the wires they are supposed to protect, an overload in the circuit will not be detected and the current will not be shut off. Overloading leads to overheating of circuit components (including wires) and may cause a fire.

Overcurrent protection devices are built into the wiring of some electric motors, tools, and electronic devices. For example, if a tool draws too much current or if it overheats, the current will be shut off from within the device itself. Damaged tools can overheat and cause a fire.

14.9.8 WET CONDITIONS HAZARDS

Working in wet conditions is hazardous because a worker may become an easy path for electrical current. If workers touch a live wire or other electrical component and the workers are well grounded because they are standing in even a small puddle of water, they will receive a shock.

Damaged insulation, equipment, or tools can expose you to live electrical parts. A damaged tool may not be grounded properly, so the housing of the tool may be energized, causing you to receive a shock. Improperly grounded metal switch plates and ceiling lights are especially hazardous in wet conditions. If you touch a live electrical component with an uninsulated hand tool, you are more likely to receive a shock when standing in water.

Remember a worker does not have to be standing in water to be electrocuted. Wet clothing, high humidity, and perspiration also increase your chances of being electrocuted.

14.10 SAFE USE OF ELECTRICAL POWER TOOLS

When using electrical power tools the following rules should be followed:

- Disconnect power supply before making adjustments.
- Ensure tools are properly grounded or double-insulated. The grounded tool must have an approved 3 wire cord with a 3 prong plug. This plug should be plugged in a properly grounded 3 pole outlet.
- Test all tools for effective grounding with a continuity tester or a GFCI before use.

- Do not bypass the switch and operate the tools by connecting and disconnecting the power cord.
- Do not use electrical tools in wet conditions or damp locations unless tool is connected to a GFCI.
- Do not clean tools with flammable or toxic solvents.
- Do not operate tools in an area containing explosive vapors or gases.

14.11 SAFE USE OF POWER CORDS

Since power cords are often used, it is important that they be used in a proper manner as follows:

- Keep power cords clear of tools during use.
- Suspend power cords over aisles or work areas to eliminate stumbling or tripping hazards.
- Replace open front plugs with dead front plugs. Dead front plugs are sealed and present less danger of shock or short circuit.
- Do not use light-duty power cords.
- Do not carry electrical tools by the power cord.
- Do not tie power cords in tight knots. Knots can cause short circuits and shocks. Loop the cords or use a twist lock plug.

14.12 APPLICABLE OSHA REGULATIONS

14.12.1 ELECTRICAL (29 CFR 1910.303, .304, .305, .331, AND .333)

Electricity is accepted as a source of power without much thought to the hazards encountered. Some employees work with electricity directly, as is the case with engineers, electricians, or people who do wiring, such as overhead lines, cable harnesses, or circuit assemblies. Others, such as office workers and salespeople, work with it indirectly. Approximately 700 workers are electrocuted each year with many workers suffering injuries such as burns, cuts, etc.

OSHA's electrical standards address the government's concern that electricity has long been recognized as a serious workplace hazard, exposing employees to such dangers as electric shock, electrocution, fires, and explosions. The objective of the standards is to minimize such potential hazards by specifying design characteristics of safety in the use of electrical equipment and systems.

Electrical equipment must be free from recognized hazards that are likely to cause death or serious physical harm to employees. Flexible cords and cables (extension cords) should be protected from accidental damage. Unless specifically permitted, flexible cords and cable may not be used as a substitute for the fixed wiring of a structure, where attached to building surfaces, where concealed or where they run through holes in walls, ceilings, or floors, or where they run through doorways, windows, or similar openings. Flexible cords are to be connected to devices and fittings so that strain relief is provided, which will prevent pulls from being directly transmitted to joints or terminal screws.

A grounding electrode conductor is to be used for a grounding system to connect both the equipment grounding conductor and the grounded circuit conductor to the grounding electrode. Both the equipment grounding conductor and the grounding electrode conductor are to be connected to the ground circuit conductor on the supply side of the service disconnecting means or on the supply side of the system disconnecting means or overcurrent devices if the system is separately derived. For ungrounded service-supplied systems, the equipment grounding conductor shall be connected to the grounding electrode conductor at the service equipment. The path to ground from circuits, equipment, and enclosures shall be permanent and continuous.

Electrical equipment shall be free from recognized hazards that are likely to cause death or serious physical harm. Each disconnecting means shall be legibly marked to indicate its purpose, unless it is located such that the purpose is evident. Listed or labeled equipment is to be used or installed in accordance with any instructions included in the listing or labeling. Unused openings in cabinets, boxes, and fittings must be effectively closed.

Safety-related work practices are to be employed to prevent electric shock or other related injuries resulting from either direct or indirect electrical contacts, when work is performed near or on equipment of circuits that are or may be energized. Electrical safety-related work practices cover both qualified persons (those who have training in avoiding the electrical hazards of working on or near exposed energized parts) and unqualified persons (those with little or no such training).

There must be written lockout or tagout procedures. Overhead power lines must be de-energized and grounded by the owner or operator of the power lines, or other protective measures must be provided before work is started. Protective measures, such as guarding or insulating the lines, must be designed to prevent employees from contacting the lines.

Unqualified employees and mechanical equipment must be at least 10 ft away from overhead power lines. If the voltage exceeds 50,000 V, the clearance distance should be increased 4 in. for each 10,000 V.

OSHA requires portable ladders to have nonconductive side rails if used by employees who would be working where they might contact exposed energized circuit parts.

Conductors are to be spliced or joined with devices identified for such use or by brazing, welding, or soldering with a fusible alloy or metal. All splices, joints, and free ends of conductors shall be covered with an insulation equivalent to that of the conductor or with an insulating device suitable for the purpose.

All employees should be required to report as soon as practical any obvious hazard to life or property observed in connection with electrical equipment or lines. Employees need to be instructed to make preliminary inspections or appropriate tests to determine the conditions that exist before starting work on electrical equipment or lines.

All portable electrical tools and equipment are to be grounded or of double-insulated type. Electrical appliances such as vacuum cleaners, polishers, and vending machines must be grounded. Extension cords that are used are to have a grounding conductor and multiple plug adapters are prohibited.

GFCIs should be installed on each temporary 15 or 20 A, 120 V AC circuit at locations where construction, demolition, modifications, alterations, or excavations are performed. All temporary circuits are to be protected by suitable disconnecting switches or plug connectors at the junction with permanent wiring.

If electrical installations in hazardous dust or vapor areas exist, they need to meet the National Electrical Code (NEC) for hazardous locations. In wet or damp locations, the electrical tools and equipment must be appropriate for this use or all locations or otherwise protected. The location of electrical power lines and cables (overhead, underground, under the floor, other side of walls) is to be determined before digging, drilling, or similar work is begun.

All energized parts of electrical circuits and equipment are to be guarded against accidental contact by approved cabinets or enclosures, and sufficient access and working space must be provided and maintained about all electrical equipment to permit ready and safe operations and maintenance.

Low-voltage protection is to be provided in the control device of motors driving machines or equipment that could cause probable injury from inadvertent starting. Each motor disconnecting switch or circuit breaker shall be located within sight of the motor control device and each motor shall be located within sight of its controller or the controller disconnecting means capable of being locked in the open position, or is a separate disconnecting means installed in the circuit within sight of the motor.

Employees who regularly work on or around energized electrical equipment or lines should be instructed in cardiopulmonary resuscitation (CPR) methods.

14.13   SUMMARY

Some general tips for working with electricity are as follows:

- Inspect tools, power cords, and electrical fittings for damage or wear prior to each use. Repair or replace damaged equipment immediately.
- Always tape cords to walls or floors when necessary. Nails and staples can damage cords causing fire and shock hazards.
- Use cords or equipment that is rated for the level of amperage or wattage that you are using.
- Always use the correct size fuse. Replacing a fuse with one of a larger size can cause excessive currents in the wiring and possibly start a fire.
- Be aware that unusually warm or hot outlets may be a sign that unsafe wiring conditions exist. Unplug any cords to these outlets and do not use until a qualified electrician has checked the wiring.
- Always use ladders made of wood or other nonconductive materials when working with or near electricity or power lines.
- Place halogen lights away from combustible materials such as clothes or curtains. Halogen lamps can become very hot and may be a fire hazard.
- Risk of electric shock is greater in areas that are wet or damp. Install GFCIs as they will interrupt the electrical circuit before a current sufficient to cause death or serious injury occurs.

- Make sure that exposed receptacle boxes are made of nonconductive materials.
- Know where the breakers and boxes are located in case of an emergency. Label all circuit breakers and fuse boxes clearly. Each switch should be positively identified as to which outlet or appliance it is for.
- Do not use outlets or cords that have exposed wiring.
- Do not use power tools with the guards removed.
- Do not block access to circuit breakers or fuse boxes.
- Do not touch a person or electrical apparatus in the event of an electrical accident. Always disconnect the current first.

Some the basic electrical checks before working with or in electrical circuitry are

- Inspect cords and plugs.
- Check power cords and plugs daily. Discard if worn or damaged. Have any cord that feels more than comfortably warm checked by an electrician.
- Eliminate octopus connections.
- Do not plug several power cords into one outlet.
- Pull the plug, not the cord.
- Do not disconnect power supply by pulling or jerking the cord from the outlet. Pulling the cord causes wear and may cause a shock.
- Never break off the third prong on a plug.
- Replace broken 3 prong plugs and make sure the third prong is properly grounded.
- Never use extension cords as permanent wiring.
- Use extension cords only to temporarily supply power to an area that does not have a power outlet.
- Keep power cords away from heat, water, and oil. They can damage the insulation and cause a shock.
- Do not allow vehicles to pass over unprotected power cords. Cords should be put in conduit or protected by placing planks alongside them.

15 Emergency and Fire Safety

At amusement parks, serious accidents and emergencies that require rescue of riders have transpired.

Emergencies and fires come in all shapes and sizes and no industrial sector is immune to them, even though it may seem that an employer could roll the dice and take a chance on not planning for these events. The result could be the end of their business and the problem of legal issues from the loss of life and human suffering. It surely seems that it is worth the time and expense to formulate and implement a plan as a part of good business practices.

Emergencies come in all types and have increased over the years with the addition of acts of terrorism. Some of the events that need to be planned for depending upon the risk faced by a business are

- Fires
- Explosions
- Bomb threats
- Smoke (without fire)
- Toxic vapors
- Airborne chemicals or biological agents
- Nuclear radiation exposure
- Storms (tornados, hurricanes, etc.)

- Flash floods
- Terrorism
- Political unrest and radicals
- Mentally ill persons
- Domestic or workplace violence
- Chemical spills
- Biological hazards
- Suspicious mail (bombs and biological pathogens)
- Power outages
- Utility failures
- Natural gas leaks

It would be highly unusual if a business did not face at least one of the preceding emergencies and in most cases multiple risks of occurrence exist. Even though Occupational Safety and Health Administration (OSHA) may not mandate an emergency action plan (EAP), it is certainly called for.

15.1   EMERGENCY ACTION PLANS

Not every employer is required to have an EAP. The standards that require such plans include the following:

- Process safety management of highly hazardous chemicals, 1910.119
- Fixed extinguishing systems, general, 1910.160
- Fire detection systems, 1910.164
- Grain handling, 1910.272
- Ethylene oxide, 1910.1047
- Methylenedianiline, 1910.1050
- 1,3-Butadiene, 1910.1051

If the employer has 10 or fewer employees, then he/she can communicate the plan orally instead of the required written plan that is to be maintained in the workplace and available to all workers. When required, employers must develop emergency action plans that

- Describe the routes for workers to use and procedures to follow
- Account for all evacuated employees
- Remain available for employee review
- Include procedures for evacuating disabled employees
- Address evacuation of employees who stay behind to shut down critical plant equipment
- Include preferred means of alerting employees to a fire emergency
- Provide for an employee alarm system throughout the workplace
- Require an alarm system that includes voice communication or sound signals such as bells, whistles, or horns
- Make the evacuation signal known to employees

- Ensure emergency training
- Require employer review of the plan with new employees and with all employees whenever the plan is changed

In addition, employers must designate and train employees to assist in safe and orderly evacuation of other employees. Employers must also review the EAP with each employee covered when the following occur:

- Plan is developed or an employee is assigned initially to a job
- Employee's responsibilities under the plan change
- Plan is changed

Before the development of an EAP it is necessary to perform a hazard assessment to determine potentially toxic materials and unsafe conditions. For information on chemicals, the manufacturer or supplier can be contacted to secure Material Safety Data Sheets (MSDSs).

15.1.1 CHAIN OF COMMAND

The employer should designate an emergency response coordinator and backup coordinator. The coordinator is responsible for the facility-wide operations, public information, and ensuring that outside aid is called. Having a backup coordinator ensures that a trained person is always available. Employees should know who the designate coordinator is. Duties of the coordinator include

- Determining what emergencies may occur and ensuring that emergency procedures are developed to address each situation
- Directing all emergency activities, including evacuation of personnel
- Ensuring that outside emergency services are notified when necessary
- Directing shutdown of operations when necessary

15.1.2 EMERGENCY RESPONSE TEAMS

Emergency response team members should be thoroughly trained for potential crises and physically capable of carrying out their duties. The team members need to know about toxic hazards in the workplace and be able to judge when to evacuate personnel or when to rely on outside help (e.g., when a fire is too large to handle). One or more teams must be trained in

- Use of various types of fire extinguishers (see Figure 15.1)
- First aid, including cardiopulmonary resuscitation (CPR) and self-contained breathing apparatus (SCBA)
- Requirements of the OSHA bloodborne pathogen standard
- Shutdown procedures
- Search and emergency rescue procedures
- Hazardous material emergency response

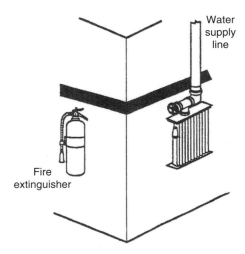

FIGURE 15.1   Adequate firefighting equipment must be provided at the worksite. (Courtesy of the Department of Energy.)

15.1.3 RESPONSE ACTIVITIES

Effective emergency communication is vital. An alternative area for a communications center other than management offices should be established in the plans and the emergency response coordinator should operate from this center. Management should provide emergency alarms and ensure that the employees know how to report emergencies. An updated list of key personnel and off-duty telephone numbers should be maintained.

Accounting for personnel following evacuation is critical. A person in the control center should notify police and emergency response team members of persons believed missing.

Effective security procedures can prevent unauthorized access and protect vital records and equipment. Duplicate records of essential accounting files, legal documents, and list of employees' relatives—to be notified in case of emergency—can be kept at off-site locations.

15.1.4 TRAINING

Every employee needs to know details of the emergency action plan, including evacuation plans, alarm systems, reporting procedures for personnel, shutdown procedures, and types of potential emergencies. Any special hazards such as flammable materials, toxic chemicals, radioactive sources, or water-reactive substances should be discussed with employees. Drills should be held at random intervals, at least annually, and should include outside police and fire authorities.

Training must be conducted at least annually and when employees are hired or when their job changes. Additional training is needed when new equipment, materials, or processes are introduced, when the layout or design of the facility changes, when procedures have been updated or revised or when exercises show that employee performance is inadequate.

15.1.5 PERSONAL PROTECTION

Employees exposed to or near accidental chemical splashes, falling objects, flying particles, unknown atmospheres with inadequate oxygen or toxic gases, fires, live electrical wiring, or similar emergencies need and should be supplied appropriate protective equipment.

15.1.6 MEDICAL ASSISTANCE

First aid must be available within 3–4 min of an emergency. Worksites more than 4 to 5 min from an infirmary, clinic, or hospital should have at least one person onsite trained in first aid (available during all work times), have medical personnel readily available for advice and consultation, and develop written emergency medical procedures.

It is essential that first-aid supplies are available to the trained first-aid providers, that emergency phone numbers are placed in conspicuous places near or on telephones, and prearranged ambulance services are available for any emergency. It may help to coordinate an EAP with outsider responders such as the fire department, hospital emergency rooms, emergency medical services (EMS) providers, and local HAZMAT teams.

15.2 EXITS AND EXIT ROUTES

Every workplace must have enough exits suitably located to enable everyone to get out of the facility quickly. Considerations include the type of structure, the number of persons exposed, the fire protection available, the type of industry involved, and the height and type of construction of the building or structure. In addition, fire doors must not be blocked or locked when employees are inside. Delayed opening of fire doors, however, is permitted when an approved alarm system is integrated into the fire door design. Exit routes from buildings must be free of obstructions and properly marked with exit signs. See 29 CFR Part 1910.36 for details about all requirements.

An exit route is a continuous and unobstructed path of exit travel from any point within a workplace to a place of safety. An exit route consists of three parts:

1. Exit access is the portion of the route that leads to an exit
2. Exit is the portion of an exit route that is generally separated from other areas to provide a protected way of travel to the exit discharge
3. Exit discharge is the part of the exit rout that leads directly outside or to a street, walkway, refuge area, public way, or open space with access to the outside

Normally, a workplace must have at least two exit routes to permit prompt evacuation of employees and other facility occupants during an emergency. More than two exits are required if the number of employees, size of the facility, or arrangement of the workplace will not allow a safe evacuation. Exit routes must be located as far away as practical from each other in case one is blocked by fire or smoke.

15.2.1   Requirements for Exits

The requirements for exits are as follows:

- Exits must be separated from the workplace by fire-resistant materials, that is, a 1 h fire-resistance rating if the exit connects three or fewer stories and a 2 h fire-resistance rating if the exit connects more than three floors.
- Exits can have only those openings necessary to allow access to the exit from occupied areas of the workplace or to the exit discharge. Openings must be protected by a self-closing, approved fire door, which remains closed or automatically closes in an emergency.
- The line of sight to exit signs must be kept clearly visible always.
- Exit signs must be installed using plainly legible letters.

15.2.2   Safety Features for Exit Routes

Exit routes should have special safety features such as the following:

- Keep exit routes free of explosives or highly flammable furnishings and other decorations.
- Arrange exit routes such that employees will not have to travel toward a high-hazard area unless the path of travel is effectively shielded from the high-hazard area.
- Ensure that exit routes are free and unobstructed by materials, equipment, locked doors, or dead-end corridors (see Figure 15.2).

FIGURE 15.2   Exits should be marked and free of obstructions. (Courtesy of the Department of Energy.)

- Provide lighting for exit routes adequate for employees with normal vision.
- Keep exit route doors free of decorations or signs that obscure their visibility.
- Post signs along the exit access indicating the direction of travel to the nearest exit and exit discharge if that direction is not immediately apparent.
- Mark doors or passages along an exit access that could be mistaken for an exit "Not an Exit" or with a sign identifying its use (such as closet).
- Renew fire-retardant paints or solutions when needed.
- Maintain exit routes during construction, repairs, or alterations.

15.2.3 DESIGN AND CONSTRUCTION REQUIREMENTS

There are specific design and construction requirements for exits and exit routes, which include the following:

- Exit routes must be permanent parts of the workplace.
- Exit discharges must lead directly outside or to a street, walkway, refuge area, public way, or open space with access to the outside.
- Exit discharge areas must be large enough to accommodate people likely to use the exit route.
- Exit route doors must unlock from the inside. They must be free of devices or alarms that could restrict use of the exit route if the device or alarm fails.
- Exit routes can be connected to rooms only by side-hinged doors, which must swing out in the direction of travel if the room may be occupied by more than 50 people.
- Exit routes must support the maximum permitted occupant load for each floor served, and the capacity of an exit route may not decrease in the direction of exit route travel to the exit discharge.
- Exit routes must have ceilings at least 7 ft, 6 in. high.
- An exit access must be at least 28 in. wide at all points. Objects that project into the exit must not reduce its width.

15.3 FIRE PREVENTION PLANS

Employers should train workers about fire hazards in the workplace and about what to do in a fire emergency. If employers want workers to evacuate, they should train them on how to escape. If workers are expected to use firefighting equipment, employers should give them appropriate equipment and train them to use the equipment safely. (See Title 29 of the Code of Federal Regulations Part 1910 Subpart L). It is recommended that employers develop a fire prevention plan as part of good business practices. OSHA standards that require fire prevention plans include the following:

- Ethylene oxide, 1910.1047
- Methylenedianiline, 1910.1050
- 1,3 Butadiene, 1910.1051

Employers covered by these standards must implement plans to minimize the frequency of evacuations. All fire prevention plans must

- Be available for employee review
- Include housekeeping procedures for storage and cleanup of flammable materials and flammable waste
- Address handling and packaging of flammable waste (recycling of flammable waste such as paper is encouraged)
- Cover procedures for controlling workplace ignition sources such as smoking, welding, and burning
- Provide for proper cleaning and maintenance of heat-producing equipment such as burners, heat exchangers, boilers, ovens, stoves, and fryers and require storage of flammables away from this equipment
- Inform workers of the potential fire hazards of their jobs and plan procedures
- Require plan review with all new employees and with all employees whenever the plan is changed

15.4　PORTABLE FIRE SUPPRESSION EQUIPMENT

The guidelines from OSHA fire protection (Subpart L) describes the requirements for the prevention of fires.

15.4.1　SCOPE, APPLICATION, AND DEFINITIONS—1910.155

This subpart contains requirements for fire brigades and all portable and fixed fire suppression equipment, fire detection systems, and fire and employee alarm systems installed to meet the fire protection requirements of 29 CFR 1910. It applies to employment other than maritime, construction, and agriculture. This discussion is limited to fire brigades and portable fire suppression equipment.

There are many important definitions included in Subpart L. Some of these are

- Class A fire—A fire involving ordinary combustible materials such as paper, wood, cloth, and some rubber and plastic materials.
- Class B fire—A fire involving flammable or combustible liquids, flammable gases, greases and similar materials, and some rubber and plastic materials.
- Class C fire—A fire involving energized electrical equipment where safety to the employee requires the use of electrically nonconductive extinguishing media.
- Class D fire—A fire involving combustible metals such as magnesium, titanium, zirconium, sodium, lithium, and potassium.
- Dry chemical—An extinguishing agent primarily composed of very small particles of chemicals, for example, sodium bicarbonate, potassium bicarbonate, monoammonium phosphate.
- Dry powder—A compound used to extinguish or control Class D fires.

- Extinguisher rating—The numerical rating given to an extinguisher, which indicates the extinguishing potential of the unit based on standardized tests developed by Underwriters' Laboratories (UL).
- Fire brigade—An organized group of employees who are knowledgeable, trained, and skilled in at least basic firefighting operations.
- Halon 1211—A colorless, faintly sweet smelling, electrically nonconductive liquefied gas ($CBrClF_2$), which is a medium for extinguishing fires by inhibiting the chemical chain reaction of fuel and oxygen. It is also known as bromocholorodifluoridemethane. Halon 1211 is no longer manufactured.
- Halon 1301—Colorless, odorless, electrically nonconductive gas ($CBrF_3$), which is a medium for extinguishing fire by inhibiting the chemical chain reaction of fuel and oxygen. It is also known as bromotrifluoromethane. Halon 1301 is no longer manufactured.
- Incipient stage fire—A fire that is in the initial or beginning stage and can be controlled or extinguished by portable fire extinguishers, Class II standpipe, or small hose system with the need for protective clothing or breathing apparatus.
- Interior structural firefighting—The physical activity of fire suppression, rescue, or both of buildings or enclosed structures that are involved in a fire situation beyond the incipient stage.
- Multipurpose dry chemical—A dry chemical that is approved for use on Class A, Class B and Class C fires.
- Standpipe systems
  - Class I system means a 2½ in. hose connection for use by fire departments and those trained in handling heavy fire streams.
  - Class II system means a 1½ in. hose system, which provides a means for the control or extinguishment of incipient stage fires.
  - Class III system means a combined system of hose, which is for the use of employees trained in hose operations and which is capable of furnishing effective water discharge during the more advanced stages of fire (beyond the incipient stage) in the interior of workplaces. Hose outlets are available for both 1½ and 2½ in. hose.
  - Small hose system means a system of hose (5/8 to 1½ in. diameter) that is for the use of employees for the control or extinguishment of incipient stage fires.

15.4.2 FIRE BRIGADES—1910.156

This section contains requirements for the organization, training, and personal protective equipment (PPE) of fire brigades whenever they are established by an employer. It should be noted that this regulation does not require an employer to establish fire brigades. If they are established, however, the requirements of this section must be met.

The requirements of this section apply to fire brigades, industrial fire departments, and private or contractual fire departments. This section does not apply to airport crash rescue or forest firefighting operations.

The employer shall prepare and maintain a written policy statement, which

- Establishes the fire brigade and its organizational structure
- Defines the functions to be performed
- States training program requirements

The employer must ensure that employees who are expected to carry out interior structural firefighting are physically capable of performing duties that may be assigned to them during emergencies.

Training shall be conducted prior to assignment and at least annually for all fire brigade members. Quarterly training or education sessions are required for fire brigade members expected to perform interior structural firefighting. Some sources of qualified training instructors are

- Local Fire Department
- State Fire Marshal's Office
- State University Extension Service
- International Society of Fire Service Instructors
- Community College Fire Science Programs

Firefighting equipment—The employer shall maintain and inspect, at least annually, firefighting equipment to ensure safe operational condition of the equipment. Portable fire extinguishers and respirators shall be inspected at least monthly.

Protective clothing—This requirement applies to employees who perform interior structural firefighting. The requirements do not apply to employees who use fire extinguishers or standpipe systems to control or extinguish fires only in the incipient stage.

Requirements for protective clothing are specified for the following components: foot and leg protection, body protection, hand protection, and head, eye, and face protection.

The employer shall ensure that the respiratory protection devices worn by fire brigade members meet the requirements of 1910.134 and the requirements contained in this section. These respirators must also be certified under 42 CFR Part 84.

15.4.3   PORTABLE FIRE EXTINGUISHERS—1910.157

The requirements of this section apply to the placement, use, maintenance, and testing of portable fire extinguishers provided for the use of employees. The selection and distribution requirements of this section do not apply to extinguishers provided for employee use on the outside of workplace buildings or structures.

Where extinguishers are provided but are not intended for employee use and the employer has an EAP and a fire prevention plan that meets the requirements of 1910.38, then only the requirements of this section dealing with inspection, maintenance, and testing apply.

The standard does not require the employees to use extinguishers. Where the employer has a total evacuation policy, an EAP, and a fire prevention plan that meets the requirements of 1910.38, and extinguishers are not available in the workplace, the employer is exempt from all requirements of this section unless a specific standard in Part 1910 requires that a portable extinguisher be provided.

Where the employer has an EAP meeting the requirements of 1910.38, which establishes fire brigades and requires all other employees to evacuate, the employer is exempt from the distribution requirements of this section.

General requirements for portable fire extinguishers include

- Mount, locate, and identify extinguishers so that they are readily accessible to employees
- Use only approved extinguishers
- Do not use carbon tetrachloride or chlorobromomethane extinguishing agents as they are prohibited
- Maintain extinguishers in a fully charged and operable condition and keep them in their designated places at all times except during use
- Permanently remove from service soldered or riveted shell inverting-type extinguishers

Extinguishers shall be provided for employee use and selected and distributed based on the classes of anticipated workplace fires and on the size and degree of hazards that would affect their use (see Figure 15.3). Extinguishers shall be distributed so that the following maximum travel distances apply:

- Class A—75 ft
- Class B—50 ft
- Class C—Based on appropriate pattern for existing Class A or B hazards
- Class D—75 ft

Extinguishers shall be visually inspected monthly, maintained annually, and hydro-statically tested periodically as per Table L-l of this standard.

Employees shall be educated in the use of extinguishers and associated hazards upon initial employment and at least annually thereafter. Employees designated to use firefighting equipment shall be trained.

15.4.4   Standpipe and Hose Systems—1910.158

This section applies to all small hose, Class II, and Class III standpipe systems required by other OSHA standards. It does not apply to Class I standpipe systems. Standpipes shall be located or otherwise protected against mechanical damage (see Figure 15.4). Damaged standpipes shall be repaired promptly.

Where reels or cabinets are provided to contain fire hoses, the employer shall ensure that they are designed to facilitate prompt use at the time of an emergency. Hose outlets and connections must be located high enough above the floor to avoid being obstructed and to be accessible to employees. Each required hose outlet shall be equipped with the hose connected and ready for use. Where hose may be damaged by extreme cold, it may be kept in a protected location as long as it is readily

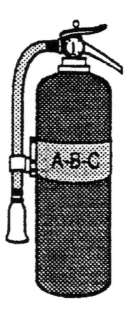

FIGURE 15.3 Appropriate fire extinguisher should be available for the type of fire and close enough to access. (Courtesy of the Department of Energy.)

available to be connected for use. The employer shall ensure that standpipe hose is equipped with shut-off type nozzles. There are two basic nozzle types:

1. Straight stream
2. Fog (also referred to as variable stream, spray, or combination)

While fog is generally preferred, straight stream is acceptable. The minimum water supply for standpipe and hose systems, which are provided for the use of employees, shall be sufficient to provide 100 gal/min for at least 30 min. Tests and maintenance are critical to ensure systems are in working order. Piping and hose of Class II and Class III systems shall be hydrostatically tested before placing in service. The following maintenance items are required for standpipe and hose systems:

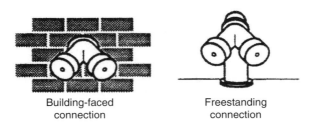

Building-faced
connection

Freestanding
connection

FIGURE 15.4 Example of standpipes. (Courtesy of the Department of Energy.)

- Water supply tanks are to be kept filled except during repairs.
- Valves in the main piping connections to the automatic sources of water supply must always be kept fully open except during repairs.
- Hose systems must be inspected at least annually and after each use.
- Any unserviceable portion of the system must be removed immediately and replaced with equivalent protection during repair.
- Hemp or linen hoses shall be unracked, inspected for deterioration, and reracked using a different fold pattern at least annually. Defective hose shall be replaced.
- Trained persons shall be designated to conduct all these required inspections.

15.4.5 FIXED FIRE SUPPRESSION EQUIPMENT—1910.159

This section applies to all automatic sprinkler systems installed to meet a particular OSHA standard. Systems installed solely for property protection are not covered.

All automatic sprinkler designs must provide the necessary discharge patterns, densities, and water flow characteristics for complete coverage. Only approved equipment and devices shall be used.

Systems shall be properly maintained. A main drain flow test must be performed on each system annually. The inspector's test valve shall be opened at least every 2 years to ensure proper operation of the system. New systems shall have proper acceptance tests conducted including

- Flushing of underground connections
- Hydrostatic tests of system piping
- Air-tests in dry-pipe systems
- Tests of drainage facilities

Every automatic sprinkler system must be provided with at least one automatic water supply capable of providing the designed water flow for at least 30 min. In order to provide a maximum protection area per sprinkler and a minimum of interference to the discharge pattern, the vertical clearance between sprinklers and material below shall be at least 18 in.

15.4.6 FIXED EXTINGUISHING SYSTEMS, GENERAL—1910.160

This section applies to all fixed extinguishing systems installed to meet a particular OSHA standard except for automatic sprinkler systems covered by 1910.159. Certain paragraphs of this section also apply to fixed systems not installed to meet a particular OSHA standard, but which, by their operation, may expose employees to possible injury, death, or adverse health consequences caused by the extinguishing agent. Specific fixed extinguishing systems using dry chemical, gaseous agents, water spray and foam are regulated by 1910.161 through 1910.163.

Fixed extinguishing system components and agents must be designed and approved for use on the specific fire hazards they are expected to control. If the system becomes inoperable, the employer shall notify employees and take the necessary temporary precautions to ensure their safety until the system is restored to operating order.

Except where discharge is immediately recognizable, a distinctive alarm or signaling system that complies with 1910.165 and is capable of being perceived above ambient noise or light levels shall be provided on all extinguishing systems in areas covered by the system.

Effective safeguards shall be provided to warn employees against entry into discharge areas where the atmosphere remains hazardous to employee safety or health. Hazard warning or caution signs shall be posted at the entrance to, and inside of, areas protected by systems that use agents in hazardous concentrations.

Fixed systems shall be inspected annually by a person knowledgeable in the design and function of the system. The weight and pressure of refillable containers and the weight of non-refillable containers shall be checked at least semiannually.

Total flooding systems with potential health and safety hazards to employees: the employer shall provide an EAP according to 1910.38 for each area protected by a total flooding system that provides agent concentrations exceeding the maximum safe levels specified in 1910.162(b)(5) and (b)(6).

All systems must have a predischarge alarm, which complies with 1910.165 and is capable of being perceived above ambient light or noise levels, which will give the employees time to safely exit from the discharge area before discharge. Automatic actuation of the system shall be provided by an approved fire detection device interconnected with the predischarge employee alarm system.

15.4.7   FIXED EXTINGUISHING SYSTEMS, DRY CHEMICAL—1910.161

This section applies to all fixed systems using dry chemical as the extinguishing agent, installed to meet a particular OSHA standard. These systems must also comply with 1910.160. Specific requirements are that dry chemical agents must be compatible with any foams or wetting agents with which they are used.

When dry chemical discharge may obscure vision, a predischarge employee alarm is required, which complies with 1910.165 and which will give employees time to safely exit from the discharge area before system discharge.

The rate of application of dry chemicals must be such that the designed concentration of the system will be reached within 30 s of initial discharge.

15.4.8   FIXED EXTINGUISHING SYSTEMS, GASEOUS AGENT—1910.162

This section applies to all fixed extinguishing systems, using a gas as the extinguishing agent, installed to meet a particular OSHA standard. These systems shall also comply with 1910.160.

Specific requirements for total flooding systems are designed for the extinguishing concentration to be reached within 30 s of initial discharge except for Halon systems, which must achieve design concentration within 10 s.

For total flooding systems, a predischarge alarm is required on Halon 1211 and carbon dioxide systems with a design concentration of 4% or greater and for Halon

1301 systems with a design concentration of 10% or greater. The alarm must provide employees time to safely exit the discharge area prior to system discharge.

For total flooding systems using Halon 1301 where egress from an area takes more than 1 min, agent concentrations of more than 7% shall not be used. Where egress takes longer than 30 s, but less than 1 min, concentrations are limited to 10%. Concentrations greater than 10% are permitted only in areas not normally occupied, provided any employee in the area can escape within 30 s.

15.4.9 Fixed Extinguishing Systems, Water Spray and Foam—1910.163

This section applies to all fixed extinguishing systems, using water or foam solution as the extinguishing agent, installed to meet a particular OSHA standard. These systems must also comply with 1910.160. This section does not apply to automatic sprinkler systems, which are covered by 1910.159.

The foam and water spray systems must be designed to be effective in at least controlling fire in the protected area or on protected equipment. Drainage of water spray systems must be directed away from areas where employees are working and no emergency egress is permitted through the drainage path.

15.4.10 Fire Detection Systems—1910.164

This section applies to all automatic fire detection systems installed to meet a particular OSHA standard. Only approved devices and equipment may be used. All fire detection systems and components shall be restored to normal operating condition as soon as possible after each test or alarm.

All systems must be maintained in an operable condition except during repairs or maintenance. Fire detectors and fire detection systems (unless factory calibrated) must be tested and adjusted as often as needed to maintain proper reliability and operating condition. Servicing, maintenance, and testing of fire detection systems must be performed by a trained person knowledgeable in the operations and functions of the system.

Fire detection equipment installed outdoors or in the presence of corrosive atmospheres shall be protected from corrosion. Detection equipment must be located and protected from mechanical or physical impact. Fire detection systems installed for the purpose of actuating fire extinguishment or suppression systems shall be designed to operate in time to control or extinguish a fire. Detection systems installed for the purpose of employee alarm and evacuation must be designed and installed to provide a warning for emergency action and safe escape of employees.

The number, location, and spacing of fire detectors must be based on design data obtained from field experience, or tests, engineering surveys, manufacturer's recommendations, or a recognized testing laboratory listing.

15.4.11 Employee Alarm Systems—1910.165

This section applies to all emergency employee alarms installed to meet a particular OSHA standard. The employee alarm system shall provide warning for necessary emergency action as called for in the emergency action plan, or for the reaction time for safe escape of employees. The employee alarm shall be capable of being perceived above ambient noise or light levels by all employees in the affected portions of the workplace. The alarm must be distinctive and recognizable as a signal to evacuate the work area or to perform actions designated under the emergency action plan.

The employer shall explain to each employee the preferred means of reporting emergencies, such as manual pull box alarms, public address systems, radio, or telephones and should be so designed that the alarm alerts those who are hearing impaired (not an OSHA requirement).

All devices, components, and systems installed to comply with this standard must be approved. All employee alarm systems must be restored to normal operating condition as promptly as possible after each test or alarm. All employee alarm systems shall be maintained in operating condition except when undergoing repairs or maintenance.

A test of the reliability and adequacy of non-supervised employee alarm systems must be made every 2 months. A different actuation device shall be used in each test of a multi-actuation device system so that no individual device is used for two consecutive tests.

All supervised employee alarm systems must be tested at least annually for reliability and adequacy. Servicing, maintenance, and testing of systems must be done by persons trained in the designed operation and functions necessary for reliable and safe operation of the system. Manually operated actuation devices for use in conjunction with employee alarms shall be unobstructed, conspicuous, and readily accessible.

15.5   FIRE PROTECTION SUMMARY

The causes of fires in an industrial environment are often

- Electrical causes (22%)—Lax maintenance in wiring, motors, switches, lamps, and heating elements
- Matches and smoking (18%)—Near flammable liquids, stored combustibles, etc.
- Friction (11%)—Hot bearings, misaligned or broken machine parts, choking or jamming materials, and poor adjustment of moving parts
- Hot surfaces (9%)—Exposure of combustibles to furnaces, hot ducts or flues, electric lamps or heating elements, and hot metal
- Overheated materials (7%)—Abnormal process temperatures, materials in dryers, and overheating of flammable liquids
- Open flames (6%)—Gasoline or other torches and gas or oil burners
- Foreign substances (5%)—Foreign material in stock
- Spontaneous heating (4%)—Deposits in ducts and flues, low-grade storage, scrap waste, oily waste, and rubbish
- Cutting and welding (4%)—Highly dangerous in areas where sparks can ignite combustibles

- Combustion sparks (4%)—Burning rubbish; foundry cupolas, furnaces, and fireboxes
- Miscellaneous (10%)—Including incendiary cases, fires spreading from adjoining buildings, molten metal or glass, static electricity near flammable liquids, chemical action, and lighting

Spotting fire hazards in your work area is a matter of being familiar with the causes listed previously. Fire inspections should be conducted on a daily, weekly, monthly, etc., basis.

When you spot a fire hazard, eliminate it immediately, if capable of doing so and have the authority to do so. Fill out a fire hazard report form and bring it to the supervisor's attention. If a fire has started, notify the appropriate personnel (company fire brigade, a supervisor, safety director, etc.) or turn in a general alarm.

If the fire is not out of control, attempt to extinguish it with the appropriate fire extinguishing equipment if the employer has trained employees to fight fires. Use the proper extinguisher for the type of fire as follows:

- Plain water, gas propelled (Class A fires)
- Soda-acid (Class A fires)
- Foam (Class A and B fires)
- Loaded steam (Class A and B fires)
- Vaporizing liquid (Class B and C fires)
- Carbon dioxide (Class B and C fires)
- Dry chemical (Class B and C fires)

Fixed extinguishing systems throughout the workplace are among the most reliable firefighting tools. These systems detect fires, sound an alarm, and send water to the fire and heat. To meet OSHA standards employers who have these systems must

- Substitute (temporarily) a fire watch of trained employees to respond to fire emergencies when a fire suppression system is out of service
- Ensure that the watch is included in the fire prevention plan and the EAP

If the fire is out of control, follow evacuation procedures recommended by the company.

15.6   FIRE PROTECTION TECHNIQUES

Certain types of workplace factors that are most likely to cause fires are flammable and combustible liquids, electricity, housekeeping, and hot work.

15.6.1   FLAMMABLE AND COMBUSTIBLE LIQUIDS

One of the common factors in industrial fires are flammable and combustible liquids. The following guidelines should be followed:

- Keep these liquids away from open flames and motors that might spark
- When you transfer them, bond the containers to each other and ground the one that is dispensed from, to prevent sparks from static electricity
- Clean up spills right away, and put oily rags in a tightly covered metal container
- Change clothes immediately if you get oil or solvents on them
- Watch out for empty containers that held flammable or combustible liquids; vapors might still be present
- Store these liquids in approved containers in well-ventilated areas away from heat and sparks
- Be sure all containers for flammable and combustible liquids are clearly and correctly labeled

15.6.2 ELECTRICITY

Electricity is a cause of many fires and the following recommendations should be adhered to

- Check for frayed insulation and damaged plugs on power cords or extension cords, damp or wet wires, oil and grease on wires
- A cord that is warm to the touch when current is passing through should warn you of a possible overload or hidden damage
- Do not overload motors; watch for broken or oil-soaked insulation, excessive vibration or sparks; keep motors lubricated to prevent overheating
- Defective wiring, switches, and batteries on plant vehicles should be replaced immediately
- Electric lamps need bulb guards to prevent contact with combustibles and to help protect the bulbs from breakage
- Do not try to fix electrical equipment yourself if you are not a qualified electrician

15.6.3 HOUSEKEEPING

Housekeeping is one of the important considerations when trying to prevent fires. The following are some housekeeping practices the will help with prevention:

- Your work area must be kept clean
- Passageways and fire doors should be kept clear and unobstructed
- Material must not obstruct sprinkler heads or be piled around fire extinguisher locations or sprinkler controls

Combustible materials should be present in work areas only in quantities required for the job and should be removed to a designated storage area at the end of each workday.

15.6.4 HOT WORK

Hot work is welding and cutting operations that have potential to cause fires: Some guidance when doing hot work are

- Never do cutting or welding without supervision or a hot work permit.
- Watch out for molten metal; it can ignite combustibles or fall into cracks and start a fire that might not erupt until hours after the work is done.
- Portable cutting and welding equipment is often used where it is unsafe; keep combustibles at least 30 ft from a hot work area.
- Be sure that tanks and other containers that held flammable or combustible liquids are completely neutralized and purged before you do any hot work on them.
- Have a fire watch (another employee) on hand to put out a fire before it can get out of control.

15.6.5 CHECKLIST

The fire protection checklist should help reinforce the aspects of fire prevention (see Figure 15.5).

Fire protection checklist

Yes ☐ No ☐ Are all smoking materials extinguished before entering a No Smoking area?

Yes ☐ No ☐ Are all oil-soaked rags, shavings, and waste placed in self-closing metal containers?

Yes ☐ No ☐ Have all empty boxes, cartons, excelsior, or loose paper been disposed of?

Yes ☐ No ☐ Are all flammable liquids stored in metal containers away from excessive heat?

Yes ☐ No ☐ Is the right cleaning agent used for a job? Remember benzene, gasoline, or any other flammable liquid should not be used as a cleaning agent.

Yes ☐ No ☐ Are clothing hung in a metal locker? Clothes hanging on walls or behind doors are an invitation to fire.

Yes ☐ No ☐ Are all fire doors, stairways, and aisles clear? These are not storage areas and should not be used as such.

Yes ☐ No ☐ Is the work area cleaned before prior to leaving work? Make sure floors, ceiling, walls, ledges, beams, and equipment in the immediate work area are free of dust, lint, and loose paper.

Yes ☐ No ☐ Have power cords been checked for frayed insulation or damaged plugs? If a cord being using gets warm, stop using it-this could be a sign of overload or a clue that something else is wrong.

Yes ☐ No ☐ Are electric lamps protected by guards?

Yes ☐ No ☐ Is any motor being used running smoothly? If it is vibrating excessively or throwing sparks, stop using it. Make sure it is well lubricated and not overloaded.

Yes ☐ No ☐ Have maintenance been informed of any electrical equipment in need of repair?

Yes ☐ No ☐ Defective equipment should not be left lying around -although you know not to touch it, others may not. And do not try to fix electrical equipment unless qualified to do so.

Yes ☐ No ☐ Are fire extinguishers are located in a known area, and employees trained how to use them—just in case?

Yes ☐ No ☐ Do workers know where the nearest exits are located so that you can get out fast if necessary?

Yes ☐ No ☐ Do workers know how to turn in an alarm if you discover a fire?

Yes ☐ No ☐ Do workers know the company's evacuation procedure?

FIGURE 15.5 Fire protection checklist.

16 Hot Processes

Hot ovens present a potential for burns to workers.

Hot processes usually bring to mind high temperatures and fire or enough heat to cause a fire. It is not possible to discuss the hazards of each specific hot process. It is more logical to generically iterate the hazards and their potential to cause trauma or illness related to exposure during hot processes.

Two types of harm arise from exposure to thermal energy. First, the physiological effect that transpires from exposure to hot environments such as heat exhaustion. Second, the injuries that are caused by the release of heat energy (i.e., burns). From utility workers to warehouse employees, Americans work in a wide variety of hot or hot and humid environments. Some examples of processes that generate heat or result in exposure to high temperatures are as follows:

- Outdoor operations in hot weather (including surveying crew, waste collection, utility facilities, servicing work, non-air conditioned facilities, maintenance work outdoors, outside activities, food preparation, and other construction-like activities)
- Electrical utilities (particularly boiler rooms)
- Bakeries
- Confectioneries
- Restaurant kitchens
- Laundries (see Figure 16.1)
- Steam tunnels
- Welding and cutting

FIGURE 16.1   Steam pressing machines in cleaners are burn hazards.

Most of these processes also possess many if not most of the inherent hazards present in most other industrial environments along with thermal energy issues. These are the common slips, falls, electrical hazards, machinery, equipment, tools, toxic or hazardous chemicals, ergonomic, etc., which result in lacerations, broken bones, sprains, and strains. Being uncomfortable is not the major problem with working in high temperatures and humidity. Workers who are suddenly exposed to working in hot environments face additional and generally avoidable hazards to their safety and health. The employer should provide detailed instructions on preventive measures and adequate protection necessary to prevent heat stress.

16.1   PHYSIOLOGY OF HOT PROCESSES

The human body, being warm blooded, maintains a fairly constant internal temperature, even though it is exposed to varying environmental temperatures. To keep internal body temperatures within safe limits, the body must get rid of its excess heat, primarily through varying the rate and amount of blood circulation through the skin and the release of fluid onto the skin by the sweat glands. These automatic responses usually occur when the temperature of the blood exceeds 98.6°F and are kept in balance and controlled by the brain. In this process of lowering internal body temperature, the heart begins to pump more blood, blood vessels expand to accommodate the increased flow, and the microscopic blood vessels (capillaries) that thread through the upper layers of the skin begin to fill with blood. The blood circulates closer to the surface of the skin, and the excess heat is lost to the cooler environment.

If heat loss from increased blood circulation through the skin is not adequate, the brain continues to sense overheating and signals the sweat glands in the skin to shed large quantities of sweat onto the skin surface. Evaporation of sweat cools the skin, eliminating large quantities of heat from the body.

As environmental temperatures approach normal skin temperature, cooling of the body becomes more difficult. If air temperature is as warm as or warmer than the skin, blood brought to the body surface cannot lose its heat. Under these conditions, the heart continues to pump blood to the body surface, the sweat glands pour liquids containing electrolytes onto the surface of the skin, and the evaporation of the sweat becomes the principal effective means of maintaining a constant body temperature. Sweating does not cool the body unless the moisture is removed from the skin by evaporation. Under conditions of high humidity, the evaporation of sweat from the skin is decreased and the body's efforts to maintain an acceptable body temperature may be significantly impaired. These conditions adversely affect an individual's ability to work in a hot environment. With so much blood going to the external surface of the body, relatively less goes to the active muscles, the brain, and other internal organs; strength declines; and fatigue occurs sooner than it would otherwise. Alertness and mental capacity also may be affected. Workers who must perform delicate or detailed work may find their accuracy suffering, and others may find their comprehension and retention of information lowered.

16.1.1 SAFETY PROBLEMS

Certain safety problems are common to hot environments. Heat tends to promote accidents due to the slipperiness of sweaty palms, dizziness, or the fogging of safety glasses. Wherever there is molten metal hot surfaces, steam, etc., the possibility of burns from accidental contact also exists.

Aside from these obvious dangers, the frequency of accidents, in general, appears to be higher in hot environments than in more moderate environmental conditions. One reason is that working in a hot environment lowers the mental alertness and physical performance of an individual. Increased body temperature and physical discomfort promote irritability, anger, and other emotional states, which sometimes cause workers to overlook safety procedures or to divert attention from hazardous tasks.

16.1.2 HEALTH PROBLEMS

Excessive exposure to a hot work environment can bring about a variety of heat-induced disorders (Figure 16.2).

Heat stroke is the most serious of health problems associated with working in hot environments. It occurs when the body's temperature regulatory system fails and sweating becomes inadequate. The body's only effective means of removing excess heat is compromised with little warning to the victim that a crisis stage has been reached. A heat stroke victim's skin is hot, usually dry, red, or spotted. Body temperature is usually 105°F or higher, and the victim is mentally confused,

FIGURE 16.2 Working over a hot grill while cooking can be a cause of heat-related problems for workers.

delirious, perhaps in convulsions, or unconscious. Unless the victim receives quick and appropriate treatment, death can occur. Any person with signs or symptoms of heat stroke requires immediate hospitalization. However, first aid should be immediately administered. This includes removing the victim to a cool area, thoroughly soaking the clothing with water, and vigorously fanning the body to increase cooling. Further treatment at a medical facility should be directed at the continuation of the cooling process and the monitoring of complications that often accompany the heat stroke. Early recognition and treatment of heat stroke are the only means of preventing permanent brain damage or death.

Heat exhaustion includes several clinical disorders with symptoms that may resemble the early symptoms of heat stroke. Heat exhaustion is caused by the loss of large amounts of fluid by sweating, sometimes with excessive loss of salt. A worker suffering from heat exhaustion still sweats but experiences extreme weakness or fatigue, giddiness, nausea, or headache. In more serious cases, the victim may vomit or lose consciousness. The skin is clammy and moist, the complexion is pale or flushed, and the body temperature is normal or only slightly elevated.

In most cases, the treatment involves having the victim rest in a cool place and drink plenty of liquids. Victims with mild cases of heat exhaustion usually recover spontaneously with this treatment. Those with severe cases may require extended care for several days. There are no known permanent effects.

Heat cramps are painful spasms of the muscles, which occur among those who sweat profusely in heat, drink large quantities of water, but do not adequately replace the body's salt loss. The drinking of large quantities of water tends to dilute the body's fluids, while the body continues to lose salt. Shortly thereafter, the low salt level in the muscles causes painful cramps. The affected muscles may be part of the arms, legs, or abdomen, but tired muscles (those used in performing the work) are usually the ones most susceptible to cramps. Cramps may occur during or after work hours and may be relieved by taking salted liquids by mouth.

A worker who is not accustomed to hot environments and who stands erect and immobile in the heat may faint. With enlarged blood vessels in the skin and in the lower part of the body due to the body's attempts to control internal temperature, blood may pool there rather than return to the heart to be pumped to the brain. Upon lying down, the worker should soon recover. By moving around, and thereby preventing blood from pooling, the patient can prevent further fainting.

Heat rash, also known as prickly heat, is likely to occur in hot, humid environments where sweat is not easily removed from the surface of the skin by evaporation and the skin remains wet most of the time. The sweat ducts become plugged, and a skin rash soon appears. When the rash is extensive or when it is complicated by infection, prickly heat can be very uncomfortable and may reduce a worker's performance. The worker can prevent this condition by resting in a cool place part of each day and by regularly bathing and drying the skin.

16.2 PHYSICAL EFFECTS OF HOT PROCESSES

Burns are one of the most common types of injury, both on and off the job. Many are minor, with no lasting effects. However, some burns can be life threatening and some can cause permanent skin damage, which has to be treated with difficult skin grafts and plastic surgery. A very bad burn may even damage the tissues beneath the skin. Fortunately, good safety practices—and good sense—can usually prevent burns. There are different ways of getting burned that might occur on the job. This section discusses the potential results, and what can be done to prevent these injuries. This section also discusses the need to know what to do if, by some chance, someone is burned and how to respond, because taking correct action immediately can literally save someone's skin—or life.

16.2.1 GENERAL HAZARDS

Almost every worker has the potential to be exposed to a variety of different burn hazards on and off the job. The most common—and most obvious—burn hazard is heat from flames, boiling water, or direct contact with a very hot item such as a welding or cutting tool or a hot process.

However, these are not the only things that can burn a worker. One of the most dangerous is a hazard we rarely think of in this context: electricity. Workers can get a skin burn from electrical arcs, flashes, and fires and can also get an electrical burn from touching overheated electric wires or equipment.

Another type of electrical burn passes right through the skin and damages the underlying body tissues. That can occur if you touch the ground at the same time you touch a live wire or a tool or machine that is poorly insulated. When that happens, your body can become a conductor for electricity, which means that the electricity goes right through you, resulting in both shock and tissue burns. This is a hazard you really want to avoid, since it can cause serious internal damage and may even kill those who suffer them.

Materials such as metal and glass can be changed into a molten state through a furnace or similar equipment. These operations pose a number of serious safety hazards to exposed workers. Workers may be exposed to hazardous fiber, dust, vapors as well as to extremely high temperatures and noise. Burns, irritation of the skin, eyes, nose, and throat are a few of the acute health issues. Some eye injuries because of infrared radiation from molten material cause changes to the lens and cataracts. Longer-term health effects may include damage to many of the organs of the body. Special precautions are often needed in these circumstances.

Chemicals—especially corrosives—can pose yet another burn hazard. In fact, the central definition of a corrosive is that it can burn skin and eyes on contact.

Many industrial acids are corrosive; some examples are hydrochloric acid, nitric acid, and sulfuric acid. Bases—also known as alkalines and caustics—are also often corrosive. Sodium hydroxide and ammonium hydroxide are both corrosive bases, as are many other substances whose names end with hydroxide, oxide, or amine. Some oxidizers are also corrosive; two examples are fluorine and chlorine.

In addition to recognizing the various potential burn hazards in the workplace, workers should know the meaning of descriptions of a burn's seriousness. There are three categories of burns:

- First-degree burns are the least serious. They usually make the skin red and can be painful. However, they affect only the outside of the skin and are not likely to have long-term effects. Most sunburns are first-degree burns.
- Second-degree burns go deeper. The skin gets very red and often blisters. These burns are also painful, though they usually do not become infected or scar. Scalding hot water can cause a second-degree burn.
- Third-degree burns are the most serious—and that is very serious indeed. They destroy the skin, usually leaving it white or charred. In addition, these burns go through the skin, damaging the tissues below. They also destroy nerve endings, so the burns are not painful. Nevertheless, they are dangerous. There is a great risk of infection with third-degree burns, in addition to bad scars that often require skin grafts.

16.2.2 IDENTIFYING BURN HAZARDS

Because there are so many potential burn hazards on the job, workers always need to be alert to their presence. Take precautions to prevent fires and contact with heat sources when workers encounter any of the following burn hazards:

- Flammable liquids such as oil, solvents, and many chemicals often have invisible vapors that move quickly through the air. Put them together with an ignition source—even a spark—and you could have a fire.
- Smoking is a hazard because lit cigarettes or matches can be an ignition source for paper, flammable liquids, or almost anything that is capable of burning.
- Molten materials such as metals, glass, etc.

FIGURE 16.3 Hot oil from french fryers' splatters can result in burns at restaurants.

- Welding and cutting operations create flames and sparks, so they' are a potential cause of burns and fires.
- Hot machines and processes are another potential burn hazard.
- Space heaters can, if not used properly, cause fires—and, therefore, burns.
- Very hot water is yet another burn hazard.
- Hot liquids, solids, and gases (see Figure 16.3).
- Workers should be aware of electrical hazards that could burn you directly or cause electrical fires. These hazards include
  - Wiring with frayed or worn insulation
  - Overloaded circuits, fuses, motors, or outlets
  - Loose ground connections
  - Lights or machinery that comes in contact with combustible materials
  - Direct contact with power lines or other live wires

To determine when chemical burns are a potential hazard, basic chemical information tools are used: labels and material safety data sheets. If they state that a substance can catch fire under certain circumstances or that it can burn the skin on contact, this signals that precautions should be taken.

16.3 PREVENTION OF HOT PROCESS ACCIDENTS

One of the best ways to reduce heat stress on workers is to minimize heat in the workplace. However, there are some work environments where heat production is difficult to control, such as when furnaces or sources of steam or water are present in the work area or when the workplace itself is outdoors and exposed to varying warm weather conditions.

Humans are, to a large extent, capable of adjusting to the heat. This adjustment to heat, under normal circumstances, usually takes about 5–7 days, during which time the body will undergo a series of changes that will make continued exposure to heat more endurable.

On the first day of work in a hot environment, the body temperature, pulse rate, and general discomfort will be higher. With each succeeding daily exposure, all of these responses will gradually decrease, while the sweat rate will increase. When the body becomes acclimated to the heat, the worker will find it possible to perform work with less strain and distress.

Gradual exposure to heat gives the body time to become accustomed to higher environmental temperatures. Heat disorders in general are more likely to occur among workers who have not been given time to adjust to working in the heat or among workers who have been away from hot environments and who have gotten accustomed to lower temperatures. Hot weather conditions of the summer are likely to affect workers who are not acclimatized to heat. Likewise, workers who return to work after a leisurely vacation or extended illness may be affected by the heat in the work environment. Whenever such circumstances occur, the worker should be gradually reacclimatized to the hot environment.

16.3.1 Lessening Hot Process Conditions

Many industries have attempted to reduce the hazards of heat stress by introducing engineering controls, training workers in the recognition and prevention of heat stress, and implementing work–rest cycles. Heat stress depends, in part, on the amount of heat the worker's body produces while a job is performed. The amount of heat produced during hard, steady work is much higher than that produced during intermittent or light work. Therefore, one way of reducing the potential for heat stress is to make the job easier or lessen its duration by providing adequate rest time. Mechanization of work procedures can often make it possible to isolate workers from the heat sources (perhaps in an air-conditioned booth) and increase overall productivity by decreasing the time needed for rest. Another approach to reducing the level of heat stress is the use of engineering controls, which include ventilation and heat shielding.

Rather than be exposed to heat for extended periods of time during the course of a job, workers should, wherever possible, be permitted to distribute the workload evenly over the day and incorporate work–rest cycles. Work–rest cycles give the body an opportunity to get rid of excess heat, slow down the production of internal body heat, and provide greater blood flow to the skin.

Workers employed outdoors are especially subject to weather changes. A hot spell or a rise in humidity can create overly stressful conditions. The following practices will be helpful:

- Postpone nonessential tasks
- Permit only workers acclimatized to heat to perform the more strenuous tasks
- Provide additional workers to perform the tasks, keeping in mind that all workers should have the physical capacity to perform the task and they should be accustomed to the heat

16.3.2   CONTROLLING THERMAL CONDITIONS

A variety of engineering controls can be introduced to minimize exposure to heat. For instance, improving the insulation on a furnace wall can reduce its surface temperature and the temperature of the area around it. In a laundry room, exhaust hoods installed over sources releasing moisture will lower the humidity in the work area. In general, the simplest and least expensive methods of reducing heat and humidity are

- Opening windows in hot work areas
- Using fans
- Using other methods of creating airflow such as exhaust ventilation or air blowers

Providing cool rest areas in hot work environments considerably reduces the stress of working in those environments. There is no conclusive information available on the ideal temperature for a rest area. However, a rest area with a temperature near 76°F appears to be adequate and may even feel chilly to a hot, sweating worker, until acclimated to the cooler environment. The rest area should be as close to the workplace as possible. Individual work periods should not be lengthened in favor of prolonged rest periods. Shorter but frequent work–rest cycles are of the greatest benefit to the worker.

In the course of a day's work in the heat, a worker may produce as much as 2–3 gal of sweat. Because so many heat disorders involve excessive dehydration of the body, it is essential that water intake during the workday be about equal to the amount of sweat produced. Most workers exposed to hot conditions tend to drink less fluid than needed because of an insufficient thirst drive. A worker, therefore, should not depend on thirst to signal when and how much to drink. Instead, the worker should drink 5–7 oz of fluids every 15–20 min to replenish the necessary fluids in the body. There is no optimum temperature of drinking water, but most people tend not to drink warm or very cold fluids as readily as they will cool ones. Whatever the temperature of the water, it must be palatable and readily available to the worker. Individual drinking cups should be provided—never use a common drinking cup.

Heat acclimatized workers lose much less salt in their sweat than do workers who are not adjusted to the heat. The average American diet contains sufficient salt for acclimatized workers even when sweat production is high. If, for some reason, salt replacement is required, the best way to compensate for the loss is to add a little extra salt to the food. Salt tablets should not be used.

16.3.3   PROTECTIVE CLOTHING

Clothing inhibits the transfer of heat between the body and the surrounding environment. Therefore, in hot jobs where the air temperature is lower than skin temperature, wearing clothing reduces the body's ability to lose heat into the air.

When air temperature is higher than skin temperature, clothing helps to prevent the transfer of heat from the air to the body. However, this advantage may be nullified if the clothes interfere with the evaporation of sweat.

In dry climates, adequate evaporation of sweat is seldom a problem. In a dry work environment with very high air temperatures, protective clothing could be an advantage to the worker. The proper type of clothing depends on the specific circumstance. Certain work in hot environments may require insulated gloves, insulated suits, reflective clothing, respirators, or infrared reflecting face shields. For extremely hot conditions, thermally conditioned clothing is available. One such garment carries a self-contained air conditioner in a backpack, while another is connected to a compressed air source, which feeds cool air into the jacket or coveralls through a vortex tube. Another type of garment is a plastic jacket, which has pockets that can be filled with dry ice or containers of ice.

The key to preventing excessive heat stress is educating the employer and worker on the hazards of working in heat and the benefits of implementing proper controls and work practices. The employer should establish a program designed to acclimatize workers who must be exposed to hot environments and provide necessary work–rest cycles and water to minimize heat stress.

16.3.4 SPECIAL CONSIDERATIONS

During unusually hot weather conditions lasting longer than 2 days, the number of heat illnesses usually increases. This is due to several factors, such as progressive body fluid deficit, loss of appetite (and possible salt deficit), buildup of heat in living and work areas, and breakdown of air-conditioning equipment. Therefore, it is advisable to make a special effort to adhere rigorously to the above preventive measures during these extended hot spells and to avoid any unnecessary or unusual stressful activity. Sufficient sleep and good nutrition are important for maintaining a high level of heat tolerance. Workers who may be at a greater risk of heat illnesses are the obese, the chronically ill, and older individuals.

When feasible, the most stressful tasks should be performed during the cooler parts of the day (early morning or at night). Double shifts and overtime should be avoided whenever possible. Rest periods should be extended to alleviate the increase in the body's heat load.

The consumption of alcoholic beverages during prolonged periods of heat can cause additional dehydration. Persons taking certain medications (e.g., medications for blood pressure control, diuretics, or water pills) should consult their physicians to determine if any side effects could occur during excessive heat exposure. Daily fluid intake must be sufficient to prevent significant weight loss during the workday and over the workweek.

16.3.5 PROTECTION AGAINST HAZARDS

Now that burn hazards have been identified as potential burn hazards, the next step is to protect workers from those hazards. First is a discussion of how to prevent contact with flames and heat and the fires that often create them. To begin with a protective measure that is almost too obvious: Do not touch anything that is or might be hot. Keep a safe distance from any tool or process that is supposed to be hot. And do not use touch to test temperature. If something is really hot, burned skin is the indicator of the test results. Be especially cautious with processes that use or create hot water, as a splash could badly burn the skin.

Smoking is another dangerous hazard and one of the easiest to prevent. Obey all "No Smoking" signs and limit smoking to permitted areas. Lit cigarettes and matches are ignition sources, and certainly must not be discarded anywhere near flammable and combustible materials. When smoking is allowed in permitted areas or at home, always extinguish cigarettes and matches in ashtrays and make sure that they are really out.

Space heaters are another potential fire source. It is best not to use them at all. However, if they must be used, be sure that they are in areas with good ventilation. Other safety precautions include using the proper fuel for that heater and keeping the heater away from combustible materials. Flammable liquids can, as mentioned earlier, have invisible vapors that may catch fire quickly. Workers could be burned before they know it; therefore, play it safe.

- Keep flammable liquids away from all ignition sources—heat, fire, cigarettes, sparking tools, etc.
- Use flammable liquids only where there is good ventilation.
- Keep containers closed when not in use—and make sure the liquids are in airtight metal containers approved for that use.
- Clean up all spills and leaks immediately.
- Be sure an "empty" container has been tested to be sure that it does not have even a drop of flammable liquid left in it. In addition, do not cut a container that once contained a flammable liquid.

Corrosives and other chemicals can burn the skin and a barrier is required between those with exposure potential and the chemical. Always wear protective clothing when you handle corrosives; and be sure that the clothing and protection fully cover all body parts, including the eyes, with clothing that is approved to protect against the particular substance. Check the MSDS if unsure of what personal protective equipment (PPE) to use. Before putting the PPE on, inspect it to make sure it has no rips, tears, or deterioration that could lessen the protection expected. Moreover, when removing PPE contaminated with corrosives, follow company procedures to the letter to make sure that not even a drop contacts the skin. Be sure, too, to wash carefully after working with corrosives.

Electricity is the most common cause of workplace fires. To prevent those fires, and protect others from electrical burns, keep these basic rules in mind:

- Check wiring on all tools and equipment, including lights and extension cords, before use. If the insulation that covers the wire is worn or frayed, report it immediately.
- Do not overload circuits, fuses, motors, or outlets.
- Match plugs to outlets; do not ever force a three-pronged grounded plug into a two-pronged outlet.
- Make sure that electrical connections are tight.

- Keep machines and tools properly lubricated and do not let grease, dust, or dirt build up on them.
- Dispose of flammable and combustible trash in assigned containers—and keep all such trash away from electrical lights and machinery.
- Do not overheat transmission shafts or bearings, especially in areas where there is dust, lint, or anything that could burn.
- Do not touch anything electrical with wet hands or while standing on a wet surface.
- Wear rubber insulating gloves—and possibly rubber clothing and boots—if you handle electrical equipment.
- Keep electrical tools away from water.

16.3.6 SAFETY PRACTICES AND PROCEDURES

Serious injuries or death can result from contact with hot equipment or uncontrolled releases of hot gases, liquids, and solids. Radiated and convected heat may also cause heat strain, heat stress, and death, and its effects can be exacerbated when wearing PPE. This will be mitigated if the following work practices are followed:

- Understand the process that is used. A job safety analysis should identify potential hazards such as heat, noise, and fumes. Safe work procedures should include: worker permits and any PPE required to eliminate or control the risk of contact or exposure.
- Material safety data sheets should be provided with information on the properties of any chemical substance used or generated during the process.
- Be aware of and report any unusual occurrence of suspect conditions in the facility, such as odors, flames, or leaks.
- Heat acclimatization is very important everyday for a person to adequately adapt to working in the heat generated by the hot process. A formal acclimatization program should be in place for new employees and those returning from leave or absence.
- Wear clothing that breathes and allows moisture transfer when sweating. Long sleeves and trousers reduce the risk of contact types of burns. All cotton clothing breathe better than nylon or other synthetic fabrics, which may also melt on contact with a hot surface.
- Wear appropriate PPE when required by the job task, which may include aprons, gloves, face and respiratory protection. For certain activities, air or ice cooled clothing may be required.

Hopefully, these protective measures will prevent burns (see Figure 16.4). However, if they cannot, it is essential to know how to treat different kinds of burns so that the correct action can be taken quickly. Proper first aid can be a life and death proposition in some burn cases; in others, it can help ensure that a less serious burn heals.

First, here is a sequence to memorize for each worker so that they need not hesitate for a second. If a worker or their clothing catches fire, stop, drop, and roll. In other words, stop what they are doing, drop to the floor or ground, cover their face

FIGURE 16.4   Close quarters and inattention by workers around hot areas can result in burns.

with their hands to protect it and their lungs, and roll over and over to smother the flames. If someone else's clothing is on fire, wrap that person in a rug or blanket to smother the flames. Once the fire is out, cut away any loose clothing, but do not remove clothing if it is stuck to a burn. Soak a first- or second-degree burn in cold water for at least 5 min. Do not, however, apply oil, butter, or lotion to a burn. In addition, do not pack the burn in ice or rub burned skin. After soaking the burned skin, cover it with a clean (preferably sterile), moist cloth. If arms or legs are burned, elevate them. Do not break any skin blisters; if they break on their own, leave them alone.

Burns often require medical attention. Be sure to see a doctor as quickly as possible for all burns that

- Appear to be third-degree (white or charred skin)
- Blister
- Affect the hands, feet, or face
- Cover more than 10% of the body
- Cause pain for 48 h

Before treating an electrical burn, other first-aid measures may need to be taken. Do not touch a person who is in contact with a live electric source, or this could cause that individual to end up in the same position. First, pull the main switch or fuse to turn off the electricity. To remove someone from a live wire, stand on something dry and use a dry stick or board—nothing metal, wet, or damp—to push the person away from the wire.

Someone who has had contact with live electricity may well be in shock and require CPR. Once that has been tended to, treat the burn with the procedures detailed here.

Treat chemical burns the same way you would any skin or eye chemical exposure. Be sure you get medical attention as soon as you have completed these steps:

If a corrosive or other chemical gets in the eyes, go immediately to emergency eyewash. Flush the eye with water for at least 15 min, lifting your eyelids occasionally. If a corrosive or other chemical splashes on the skin, quickly and carefully remove contaminated clothing. Then rinse the skin with water for at least 15 min and follow any other instructions on the MSDS. Finally, if the skin is burned, apply a clean compress.

16.4   HOT PROCESSES IN THE SERVICE INDUSTRY

In the service industry hot processes often occur at power plants, during food preparation, during outdoor activities, and in facilities without heat regulation. As you might suspect hot processes and hot temperature extremes have the possibility of occurring in almost any industry sector of the service industry.

16.5   SUMMARY OF OSHA REGULATIONS

OSHA does not specifically direct us to prevent and respond effectively to burns nor are there regulations that govern the physiological exposure to thermal conditions. However, the agency does build fire prevention into many of its regulations. For example, Subpart L of 29 CFR Part 1910 includes detailed requirements for fire detection and alarm systems as well as for firefighting equipment and firefighting teams.

Other OSHA fire prevention requirements are found in its regulations for flammable, combustible, and other materials that could burn and for hot operations like welding and cutting. Many of these regulations require employers to have fire prevention plans and emergency action plans. To prevent fire hazards from becoming real fires, OSHA explains in 29 CFR 1910.38 that these fire prevention plans must spell out:

- Workplace fire hazards, along with prevention techniques that apply specifically to them
- Potential ignition sources and controls for them
- Fire protection systems and equipment that control fires
- Individuals or job titles responsible for maintaining these prevention and control systems and hazards
- Maintenance procedures and plans to prevent accidental ignition of combustible materials

The plans must also include written procedures to "control accumulations of flammable and combustible waste materials and residues so that they do not contribute to a fire emergency." Finally, this OSHA regulation requires employers to let employees know about "the fire hazards of the materials and processes to which they are exposed," along with what employees must do to prevent fires and how to respond if there is a fire.

Electrical safety is covered in OSHA regulations (29 CFR 1910.331–335), and chemical safety is, of course, covered extensively in rules related to specific substances and the hazard communication standard (29 CFR 1910.1200). As you know, the hazard communication standard gives you the right to know about the hazards of on-the-job chemicals, including burns, as well as how to protect yourself against those hazards and respond to emergencies that are related to them.

There are no regulations that specifically address hot process and the physiological and physical prevention of these types of accidents, but only guidelines. These guidelines come from the National Institute for Occupational Safety and Health (NIOSH) and the American Conference of Governmental Industrial Hygienist (ACGIH). Some vague references to thermal situations exist in some specific OSHA regulations such as 1910.261(pulp, paper, and paperboard mills), 1910.262 (textiles), 1910.263 (Bakery equipment), 1910.264 (laundry machinery and operations), etc.

16.6 CHECKLIST FOR HOT PROCESSES

A checklist for safety and health when involved in hot processes can be found in Figure 16.5.

Hot processes checklist

Although specific OSHA regulations do not exist for hot processes, this checklist will help you determine if you hot process or heat related hazards by simply answering each item yes or no.

☐ Yes ☐ No Is there potential thermal loading from the sun?
☐ Yes ☐ No Is there potential thermal loading to an inside hot environment?
☐ Yes ☐ No Are there sources of heat one that would burn the skin?
☐ Yes ☐ No Are chemicals in use that could cause a heated reaction?
☐ Yes ☐ No Are there chemicals in that could burn the skin such as corrosive chemicals?
☐ Yes ☐ No Is there exposed electrical circuits or conducts that could burn the skin?
☐ Yes ☐ No Are procedures in place to prevent heat-related illnesses?
☐ Yes ☐ No Are first aid supplies available to treat heat-related illness and burn injuries?
☐ Yes ☐ No Is there trained first aiders trained to treat heat-related or hot process related incidents?
☐ Yes ☐ No Is protection equipment and clothing available when they will prevent heat-related injuries or illnesses?
☐ Yes ☐ No Are work/rest cycles in place and other steps taking to prevent heat illnesses?
☐ Yes ☐ No Are controls or guards in place to prevent contact with hot materials?
☐ Yes ☐ No Are controls to decrease temperature in place and functioning (e.g., air conditioning)?
☐ Yes ☐ No Is ventilation available and working as designed?
☐ Yes ☐ No Is fire prevention practiced around hot sources?
☐ Yes ☐ No Are there means to extinguish fired the might start?
☐ Yes ☐ No Are high heat source contained to prevent accidents?
☐ Yes ☐ No Are safe work procedures practiced around hot processes?

FIGURE 16.5 Hot processes checklist.

16.7 SUMMARY

There are many potential burn hazards in most workplaces—and many steps can be taken to ensure that workers do not get burned. In other words, know where and why possible burn hazards are present and build proper precautions into the preparation for and performance of any job tasks. Remember to keep checking for hazards and double-checking to be sure that all the necessary steps have been taken to protect all workers from burns. Another aspect of being prepared is knowing what to do if the worst occurs—in this case, if any worker is burned. Burns can be very serious, and even the smallest burn cannot be treated lightly. But knowing what to do—and doing it immediately—can make sure that failure to perform in these situations does not add to the problem and, in many cases, prevents a burn from causing serious long-term damage.

17 Ionizing Radiation

The most recognized warning symbol for ionizing radiation.

17.1 IONIZING RADIATION

Ionizing radiation (IR) has always been a mystery to most people. Actually, much more is known about ionizing radiation than the hazardous chemicals that constantly bombard the workplace. After all, there are only four major types of radiation (alpha particles, beta particles, gamma rays, and neutrons) rather than thousands of chemicals. There are instruments that can detect each type of radiation and provide an accurate dose received value. This is not so for chemicals, where the best that we could hope for in a real-time situation is a detection of the presence of a chemical and not what the chemical is. With radiation detection instruments, the boundaries of contamination can be detected and set while such boundaries for chemicals are near to impossible except for a solid.

In the service fields of power generation, medicine, research, and technology, radiation has it uses. By understanding it and following good safety and health practices, radiation can be respected while working with it and not working in fear of it. In many ways, ionizing radiation is very similar to working with hazardous chemicals or hazardous waste:

- All can cause acute and chronic effects
- All require special training before doing any work at a job site

- Medical examinations or medical monitoring is required to provided by the employer
- Personal/area monitoring of air, soil, contaminated tools, and equipment is required
- Use of personal protective equipment (PPE) is usually required
- Skills in donning and doffing of PPE/respirators, which are or could be contaminated, are often required
- Training and knowledge of decontamination policies and procedures are needed
- Both radiation and hazardous chemicals can pose a danger and the use of PPE can cause physical and mental stress as well as a real concern regarding heat stress
- All types of this work must consider the workers, and public and environmental protection
- All types of work performed have the potential for workers to encounter mixed materials that contain hazardous chemicals and radioactive material together

There are advantages to working with radiation compared to hazardous chemicals. The advantages are as follows:

- There are only four major types of radiation while there are tens of thousands of hazardous chemicals.
- There are instruments that can detect each of the four types of radiation and provide the amount of present, but no such easily useable detection instruments exist for the thousands of chemicals.
- Precise radioactive contamination can be detected and contamination zones can be identified as well as the boundaries of that contamination. This is at time not possible or difficult related to hazardous chemicals.
- The mechanisms for protection are known. They are time, distance, and shielding. Respirators/PPE provide a degree of shielding.
- There is a better understanding of the dose of radiation that can cause health effects than for hazardous chemicals.
- Most work where exposure is possible will have medical monitoring, environmental/personal worker monitoring, security systems, training requirements, radiation work permits, specific work practices, and well-posted hazard areas. Workers many not even know that they have a potential for exposure to hazardous chemicals, and hazardous chemicals may be considered just an acceptable part of working.

It is possible to maintain a lifetime dose for individuals exposed to radiation. Most workers wear a personal dosimeter, which provides levels of exposure. The same is impossible for chemicals where no standard unit of measurement such as the rem (roentgen equivalent man) exists for chemicals. The health effects of specific doses are well known such as 20–50 rem—minor changes in blood occur, 60–120 rem—vomiting occurs but no long-term illness, or 5,000–10,000 rem—certain death within 48 h.

Ionizing radiation is radiation that has sufficient energy to remove electrons from atoms. One source of ionizing radiation (IR) is the nuclei of unstable atoms. For these radioactive atoms (also referred to as radionuclides or radioisotopes) to become more stable, the nuclei eject or emit subatomic particles and high-energy photons (gamma rays). This process is called radioactive decay. Unstable isotopes of radium, radon, uranium, and thorium, for example, exist naturally. Others are continually made naturally or by human activities such as the splitting of atoms in a nuclear reactor. Either way, they release ionizing radiation. The major types of IR emitted as a result of spontaneous decay are alpha and beta particles and gamma rays. X-rays, another major type of IR, arise from processes outside of the nucleus. Neutrons are also a form of IR that is present around nuclear reactors.

17.2   TYPES OF RADIATION

The four basic types of ionizing radiation of concern in most radiological work situations are alpha particles, beta particles, gamma rays, and neutron particles. These may exist in various amounts, depending on the exact location and nature of the work.

17.2.1   Alpha Particles

Alpha particles are energetic, positively charged particles (helium nuclei) that rapidly lose energy when passing through matter. They are commonly emitted in the radioactive decay of the heaviest radioactive elements such as uranium and radium as well as by some man-made elements. Alpha particles lose energy rapidly in matter and do not penetrate very far; however, they can cause damage over their short path through tissue, causing it to ionize or become electrically charged. These particles are usually completely absorbed by the outer dead layer of the human skin and, so, alpha emitting radioisotopes are not a hazard outside the body. However, they can be very harmful if they are ingested or inhaled. Alpha particles can be stopped completely by a sheet of paper. Information about the physical characteristic, range, shielding, and biological hazards of alpha particles is as follows:

- Physical characteristics—Alpha particles are emitted during the decay of certain types of radioactive materials. Compared to other types, the alpha particle has a relatively large mass. It consists of two protons and two neutrons that result in a positive charge of +2. It is a highly charged particle that is emitted from the nucleus of an unstable atom. The positive charge causes the alpha particle (+) to strip electrons (−) from nearby atoms as it passes through the material, thus ionizing these atoms.
- Range—Alpha particles deposit a large amount of energy in a short distance of travel. This large energy deposit limits the penetrating ability of alpha particles to a very short distance. This range in air is about 1–2 in.

- Shielding—Most alpha particles are stopped by a few centimeters of air, a sheet of paper, or the dead layer (outer layer) of skin on our bodies.
- Biological hazard—Alpha particles are not considered an external radiation hazard. This is because they are easily stopped by the dead layer of skin. If alpha emitting radioactive material is inhaled or ingested, it becomes a source of internal exposure. Internally, the source of the alpha radiation is in close contact with body tissue and can deposit large amounts of energy in a small volume of body tissue. This internal exposure has resulted in lung cancer in uranium miners and concerns regarding radon levels in homes.

17.2.2 Beta Particles

Beta particles are fast moving, positively or negatively charged electrons emitted from the nucleus during radioactive decay. Humans are exposed to beta particles from man-made and natural sources such as tritium, carbon-14, and strontium-90. Beta particles are more penetrating than alpha particles, but are less damaging over equally traveled distances. Some beta particles are capable of penetrating the skin and causing radiation damage; however, as with alpha emitters, beta emitters are generally more hazardous when they are inhaled or ingested. Beta particles travel appreciable distances in air, but can be reduced or stopped by a layer of clothing or by a few millimeters of a substance such as aluminum. Information about the physical characteristics, range, shielding, and biological hazards of beta particles is as follows:

- Physical characteristics—Beta particle is an energetic electron emitted during radioactive decay. Compared to an alpha particle, a beta particle is nearly 8000 times less massive and has half the electrical charge. Beta radiation causes ionization by the same forces at work with alpha radiation, mainly electrical interactions with atoms that are encountered as it travels. However, because it is not as highly charged, the beta particle is not as effective at causing ionization. Therefore, it travels further before giving up all its energy and finally coming to rest.
- Range—Beta particle has a limited penetrating ability. Its typical range in air is up to about 10 ft. In human tissue, the same beta particle would travel only a few millimeters.
- Shielding—Beta particles are easily shielded by relatively thin layers of plastic, glass, aluminum, or wood. Dense materials such as lead should be avoided when shielding beta radiation due to the increase in production of x-rays in the shield.
- Biological hazard—Externally, beta particles are potentially hazardous to the skin and eyes. They cannot penetrate to deep tissue such as the bone marrow or other internal organs. This type of external exposure is called a shallow dose. When taken into the body, materials that emit beta radiation can be a hazard in a similar way to that described from alpha emitters, although comparatively less damage is done in the tissue exposed to the beta emitter.

17.2.3 GAMMA RAYS/X-RAYS

Like visible light and x-rays, gamma rays are weightless packets of energy called photons. Gamma rays often accompany the emission of alpha or beta particles from a nucleus. They have neither a charge nor a mass and are very penetrating. One source of gamma rays in the environment is naturally occurring potassium-40. Man-made sources include plutonium-239 and cesium-137. Gamma rays can easily pass completely through the human body or be absorbed by tissue, thus constituting a radiation hazard for the entire body. Several feet of concrete or a few inches of lead may be required to stop the more energetic gamma rays.

X-rays are high-energy photons produced by the interaction of charged particles with matter. X-rays and gamma rays have essentially the same properties, but differ in origin, i.e., x-rays are emitted from processes outside the nucleus, while gamma rays originate inside the nucleus. They are generally lower in energy and therefore less penetrating than gamma rays. Literally thousands of x-ray machines are used daily in medicine and industry for examinations, inspections, and process controls. X-rays are also used for cancer therapy to destroy malignant cells. Because of their many uses, x-rays are the single largest source of man-made radiation exposure. A few millimeters of lead can stop medical x-rays (see Figure 17.1).

Information about the physical characteristics, range, shielding, and biological hazards of gamma/x-rays is as follows:

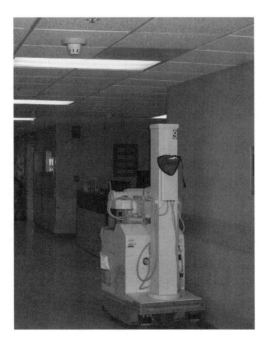

FIGURE 17.1 Example of a portable x-ray machine.

- Physical characteristics—Gamma/x-ray radiation is an electromagnetic wave or photon and has no electrical charge. Gamma rays and x-rays can be thought of as physically identical. The only difference is in the place of origin. These photons have no mass or charge but can ionize matter as a result of direct interactions with orbital electrons. Like all electromagnetic radiations, gamma rays travel at the speed of light.
- Range—Because gamma/x-ray radiation has no charge and no mass, it has a very high penetrating power (i.e., the radiation has a low probability of interacting in matter). Gamma rays have no specific range but are characterized by their probability of interacting in a given material. There is no distinct maximum range in matter, but the average range in a given material can be used to compare materials for their shielding ability.
- Shielding—Gamma/x-ray radiation is best shielded by very dense materials such as lead, concrete, or steel. Shielding is often expressed by thicknesses that provide a certain shielding factor, such as a half-value layer (HVL). An HVL is the thickness of a given material required to reduce the dose rate to one-half the unshielded dose rate.
- Biological hazard—Due to the high penetrating power, gamma/x-ray radiation can result in radiation exposure to the whole body rather than a small area of tissue near the source. Therefore, a photon radiation has the same ability to cause a dose to tissue whether the source is inside or outside the body. This is in contrast to alpha radiation, for example, which must be received internally to be a hazard. Gamma radiation is considered an external hazard.

17.2.4 Neutron Particles

Neutrons are primarily found near nuclear reactors. Thus, exposure is most often expected in those who work around or near a nuclear reactor. The exposure of individuals to neutrons will usually be found in workers who are involved in nuclear research and nuclear electrical power generation. Information about the physical characteristics, range, shielding, and biological hazards of gamma/x-rays is as follows:

- Physical characteristics—Neutron radiation consists of neutrons that are ejected from the nuclei of atoms. A neutron has no electrical charge. Due to their charge, neutrons do not interact directly with electrons in matter. A direct interaction occurs as the result of a collision between a neutron and the nucleus of an atom. A charged particle or other radiation that can cause ionization may be emitted during these interactions. This is called indirect ionization.
- Range—Because neutrons do not experience electrostatic forces, they have a relatively high penetrating ability and are difficult to stop. Like gamma radiation, the range is not absolutely defined. The distance they travel depends on the probability for interaction in a particular material. You can think of neutrons as being scattered as they travel through material, with some energy lost with each scattering event.

- Shielding—Moderate to low-energy neutron radiation is best shielded by materials with a high hydrogen content, such as water (H_2O) or polyethylene plastic (CH_2–CH_2–X). High-energy neutrons are best shielded by more dense materials such as steel or lead. Sometimes a multilayered shield will be used to first slow down very fast neutrons, and then absorb the slow neutrons.
- Biological hazard—Like gamma radiation, neutrons are an external whole-body hazard due to their high penetrating ability.

17.3 SOURCES OF RADIATION

17.3.1 NATURAL RADIATION

Humans are primarily exposed to natural radiation from the sun, cosmic rays, and naturally occurring radioactive elements found in the earth's crust. Radon, which emanates from the ground, is another important source of natural radiation. Cosmic rays from space include energetic protons, electrons, gamma rays, and x-rays. The primary radioactive elements found in the earth's crust are uranium, thorium, potassium, and their radioactive derivatives. These elements emit alpha and beta particles, or gamma rays.

17.3.2 MAN-MADE RADIATION

Radiation is used on an increasing basis in medicine, dentistry, and industry. The main users of man-made radiation include medical facilities such as hospitals and pharmaceutical facilities; research and teaching institutions; nuclear reactors and their supporting facilities such as uranium mills and fuel preparation plants; and federal facilities involved in nuclear weapons production as part of their normal operation.

Many of these facilities generate some radioactive waste, and some release a controlled amount of radiation into the environment. Radioactive materials are also used in common consumer products such as digital and luminous dials on instruments, ceramic glazes, and smoke detectors.

17.3.3 SOURCES OF EXPOSURE

The ionizing radiations of primary concern are alpha and beta particles, gamma rays, and x-rays. Alpha and beta particles and gamma rays can come from natural sources or can be technologically produced. Most of the x-ray exposure people receive is technologically produced. Natural radiation comes from cosmic rays, naturally occurring radioactive elements found in the earth's crust (uranium, thorium, etc.), and radioactive decay products such as radon and its subsequent decay products. The latter group represents the majority of the radiation exposure of the public.

In addition to these natural sources, radiation can come from such wide-ranging sources as hospitals, research institutions, nuclear reactors and their support facilities, certain manufacturing processes, and federal facilities involved in nuclear

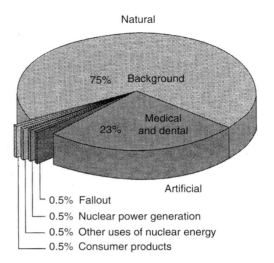

Natural

75% Background

Medical
23% and dental

Artificial

0.5% Fallout
0.5% Nuclear power generation
0.5% Other uses of nuclear energy
0.5% Consumer products

FIGURE 17.2 Average ionizing radiation exposure to individuals in the U.S. population. (Courtesy of the Department of Energy.)

weapons production. Figure 17.2 shows the percentage contribution that various radiation sources make toward the yearly average effective dose received by the U.S. population.

Any release of radioactive material is a potential source of radiation exposure to the population. In addition to exposure from external sources, radiation exposure can occur internally by ingesting, inhaling, injecting, or absorbing radioactive materials. Both external and internal sources may irradiate the whole body or a portion of the body. The amount of radiation exposure is usually expressed in a unit called millirem (mrem). In the United States, the average person is exposed to an effective dose equivalent of approximately 360 mrem (whole-body exposure) per year from all sources.

17.4 PREVENTING EXPOSURES FROM IONIZING RADIATION

As low as reasonably achievable (ALARA) philosophy is intent on maintaining exposures to ionizing radiation at as low a level as possible. A radiation safety program should fully support the concept that all radiation doses should be ALARA. This implies that no dose should be acceptable if it can be avoided or is without benefit. An ALARA program depends on the cooperation of all users of radionuclides and their supervisors involved in areas with IR exposure. The program must include the use of proper equipment and procedures to lower radiation exposure. The radiation department should investigate any whole-body dose in excess of 125 or 1875 mrem to the extremities to any individual in any one quarter. If any worker receives a whole-body dose in excess of 375 or 5625 mrem to the extremities per quarter, direct actions must be taken to minimize any future exposures. These actions may require a change in procedures or an increased application of the principles of personnel protection. Maintain ALARA exposures by practicing the basic principles of radiation protection. Radiation exposure dose limits for adult radiation workers levels are espoused to be

- 5000 mrem (5 rem) per year total effective dose equivalent to the whole body, no one organ of which may exceed 50,000 mrem (50 rem) per year
- 15,000 mrem (15 rem) per year to the lens of the eye
- 50,000 mrem (50 rem) per year to the skin or to any extremities (hands and forearms, or feet and ankles)

Radiation dose limits to minor (under age 18) radiation workers:

- 500 mrem (0.5 rem) per year total effective dose equivalent to the whole body, no one organ of which may exceed 5000 mrem (50 mSv) per year
- 1500 mrem (1.5 rem) per year to the lens of the eye
- 5000 mrem (50 rem) per year to the skin or to any extremities (hands and forearms, or feet and ankles)

Radiation dose limits to members of the general public:

- 100 mrem (0.1 rem) per year total effective dose equivalent to the whole body
- 2 mrem (0.002 rem) total effective dose equivalent to the whole body per hour
- 500 mrem (0.5 rem) total effective dose equivalent to the fetus of a declared pregnant woman per gestation (9 months)

17.4.1 EXPOSURE AND DOSE

When people are exposed to radiation, the energy of the radiation is deposited in the body. This does not make the person radioactive or cause them to become contaminates. An analogy would be to shine a bright light upon the human body. The body absorbs the light (energy), and in some cases the absorption of the light energy may cause noticeable heating in the body tissue, but the body does not become a light source or emit light now that it has absorbed the light.

In a similar way, when exposed to radiation, the body absorbs the radiation energy. As this absorption takes place, the tissue of the body may be damaged by the penetration and conversion of the radiation energy. Again, the body does not become radioactive or emit radiation due to this energy absorption.

Since absorption of radiation can damage tissue, a way to measure that damage and ensure that it is kept to a minimum is necessary. The amount of radiation energy absorbed in a body is known as dose. The special unit for measuring dose in a person (called dose equivalent) is the rem.

Rem is a unit used for equating radiation absorption with biological damage. Since the rem is a fairly large unit, radiation exposure is usually recorded in thousandths of a rem or millirem. Thousand millirem equals one rem. Millirem is usually abbreviated as mrem. For example, if you receive a chest x-ray, the amount of exposure or dose

TABLE 17.1

Summary of Radiation Types and Characteristics

Type	Alpha	Beta	Gamma	Neutron
Penetrating power	Very small	Small	Very great	Very great
Hazard	Internal	Internal/external	External	External
Shielding material	Paper	Plastic, aluminum	Lead, steel, concrete	Water, concrete, steel (high energy)
Quality factor	20	1	1	2–10

Source: Courtesy of the Department of Energy.

would be approximately 10 mrem (0.010 rem). This same amount of dose or biological harm, could be received from making two or three coast to coast airline flights—each round trip involves about 5 mrem (from elevated cosmic radiation levels in the upper atmosphere). The important point is, the source of the exposure is relatively unimportant; once the dose has been measured in a standard unit, it can be compared to other doses, added to other doses, or used in risk comparisons regarding nonradiation risks. Some of these comparisons are made later.

Rad (radiation absorbed dose)—Rad is a unit for measuring absorbed dose in any material. Absorbed dose results in the amount of energy deposited by the radiation. It is defined for any material and applies to all types of radiation. It does not take into account the potential effect that different types of radiation have on the body. Therefore, it can be used as a measure of energy absorbed by the body, but not as a measure of the relative biological effect (harm or risk) to the body.

A measure of the biological effect on the body is the rem. As stated previously, the rem is the unit for measuring the special quantity called dose equivalent. The rem takes into account the energy absorbed (dose) and the relative biological effect on the body due to the different types of radiation (expressed as the quality factor of the radiation). It is therefore a measure of the relative harm or risk caused by a given dose of radiation when compared to any other doses of radiation of any type. Occupational radiation exposure is recorded in rem. The value of rem for an exposure is calculated by multiplying the rad value (dose) by the quality factor equals the amount of biological. The rem can be thought of as the unit of biological hazard. The quality factor is an indicator of the amount of biological damage a type of radiation can cause; for example, alpha particles can cause 20 times the damage that a beta particle can cause (see Table 17.1).

17.4.2 EXTERNAL RADIATION PROTECTION

Maximize the distance from the source. The dose rate for most gamma and x-radiation varies with the inverse square of the distance from a point source. Therefore, the farther workers position themselves from the source of radiation, the smaller the dose they receive. For example, doubling the distance from a radiation source will result in one-fourth the exposure in the same amount of time. One practical implementation of this principle is using remote handling devices such as forceps, tongs, and tube racks, etc., to minimize direct contact with sources and containers. Even a small increase in distance can result in a dramatic decrease in dose rate.

Minimize time of exposure—The less time a worker remains in a radiation field, the smaller the dose they receive. Perform the task or the procedure as quickly and as efficiently as possible without increasing the probability of a spill or other accident. Shield the radiation source. Place shielding between an individual and a source of penetrating radiation to decrease the dose. For low-energy beta emitters (hydrogen-3, carbon-14, phosphorus-33, and sulfur-35), shielding is not necessary. For high-energy beta emitters (phosphorus-32), 3/8 in. acrylic is the shielding material of choice. For gamma or x-ray emitters (chromium-51 and iodine-125) lead is used when exposure rates are significant. Shielding is the most impractical form of personal protection with exposure to radiation. Distance and time are the easiest to implement.

Certainly, radiation can be dangerous but exposures can usually be controlled by one or a combination of three factors: distance, time, and shielding. Distance is the best since the amount of radiation from a source drops off quickly as a factor of the inverse square of the distance; for instance, at 8 ft away the exposure is 164th of radiation occurring at the source. As for time, many radiation workers are allowed to stay in a radiation area only for a certain length of time and then they must leave that area. Shielding often conjures up lead plating or lead suits (similar as to when x-rays are taken by physician or dentist). Wearing a lead suit may seem appropriate but the weight alone would be prohibitive. Lead shielding can be used to protect workers from gamma rays (similar to x-rays). Once they are emitted, they would pass through anything in their path and continue on their way, unless a lead shield is thick enough to protect the worker.

For beta particles, aluminum foil will stop its penetration. Thus, a protective suit will prevent beta particles from reaching the skin. Beta particles can burn the skin and cause surface contamination of the skin. Alpha particles can enter the lungs and cause the tissue to become electrically charged (ionized). Protection for alpha particles can be obtained with the use of air-purifying respirators with proper cartridges that will filter out radioactive particles. Neutrons are found around the core of a nuclear reactor and are absorbed by water and the material in the control rods of the reactor. If a worker is not close to the core of the reactor, then no exposure can occur.

17.4.3 INTERNAL EXPOSURE PROTECTION

Inhalation—A chemical fume hood, which has been certified for radioactive materials work, is highly recommended when using potentially volatile compounds. Certain equipment are capable of generating radioactive aerosols. Use centrifuges, vortex mixers, shakers, and chromatography plate scraping procedures, etc., in such a way that production of and exposure to radioactive aerosols is minimized.

Puncture—Dispose of syringes and pipettes promptly and in appropriate containers. Guard against glass breakage and puncture injury during use and disposal. Do not attempt to recap needles before disposal.

Ingestion—Never introduce any food or drink into a posted restricted area, even for temporary storage. Do not eat or drink in any area where radionuclides are used, never pipette by mouth, and never store food and drinks in a cold room or refrigerator that is designated for radioactive material storage.

Absorption—Use measures that prevent the contamination of skin and eyes. If there is any possibility that the clothes have been contaminated, remove this clothing before leaving the lab. Eye protection (e.g., goggles, face shield) is encouraged to prevent contamination of the eyes. This is particularly important for individuals wearing contact lenses since some lenses will absorb and concentrate radiochemicals. Wear protective gloves at all times when working with radioactive materials. Change gloves frequently during the work, disposing of the used gloves as radioactive waste. Wash hands after using radioactive materials and monitor the hands for contamination, especially before eating or smoking and before leaving the facility.

17.4.4 PREVENTIVE STEPS

Before being ordered or received, approval must be obtained for all radioactive materials, radiation sources, radiation-producing equipment, self-shielded irradiators, gas chromatographs with radioactive sources, medical diagnostic and therapy devices using radioactive sources, or any other equipment or materials containing or capable of producing ionizing radiation. All orders for radioactive materials must be approved by authorized user (AU) supervisor or their alternates before they are forwarded to the procurement system.

All shipments of radioactive materials received must be secured or be under constant surveillance at all times. Shipments of radioactive materials that have not been delivered must be secured at the receiving site by trained personnel.

Any radioactive material in use in a facility must be attended at all times or secured by locking it up when not attended. Radioactive materials may not be left unsecured even momentarily.

Radioactive materials in storage, i.e., not used, must be secured when the room in which it is stored is unoccupied. The required security may be accomplished by locking the room while unoccupied, or alternatively, by locking the radioactive materials within refrigerators, freezers, cabinets, or lock boxes. Wherever possible, lock boxes are recommended for storage of radioactive materials. Only authorized persons may have access to radioactive materials.

All radioactive wastes are considered as radioactive materials. Radioactive wastes, including dry waste, liquid waste, medical pathological waste, and mixed waste, must be secured at all times. Radioactive waste may be placed in lockable containers.

An authorized user (AU) or a designee must keep records of all radioactive materials ordered, received, and disposed. Packing slips must be strictly maintained to keep an accurate inventory.

17.5 HUMAN EXPOSURE OUTCOMES

17.5.1 HEALTH EFFECTS OF RADIATION EXPOSURE

Depending on the level of exposure, radiation can pose a health risk. It can adversely affect individuals directly exposed as well as their descendants. Radiation can affect cells of the body, increasing the risk of cancer or harmful genetic mutations, which can be passed on to future generations; or, if the dosage is large enough to cause massive tissue damage, it may lead to death within a few weeks of exposure.

Ionizing radiation can cause changes in the chemical balance of cells. Some of those changes can result in cancer. In addition, by damaging the genetic material (DNA) contained in all cells of the body, ionizing radiation can cause harmful genetic mutations, which can be passed on to future generations. Exposure to large amounts of radiation, a rare occurrence, can cause sickness in a few hours or days and death within 60 days of exposure. In extreme cases, it can cause death within a few hours of exposure.

17.5.2 Results of Exposure

Ionizing radiation affects people by depositing energy in body tissue, which can cause cell damage or cell death. In some cases there may be no effect. In other cases, the cell may survive but become abnormal, either temporarily or permanently, or an abnormal cell may become malignant. Large doses of radiation can cause extensive cellular damage and result in death. With smaller doses, the person or particular irradiated organ(s) may survive, but the cells are damaged, increasing the chance of cancer. The extent of the damage depends on the total amount of energy absorbed, the period and dose rate of exposure, and the particular organ(s) exposed (see Tables 17.2 and 17.3).

Evidence of injury from low or moderate doses of radiation may not show up for months or even years. For leukemia, the minimum period between the radiation exposure and the appearance of disease (latency period) is 2 years. For solid tumors, the latency period is more than 5 years. The types of effects and their probability of occurrence can depend on whether the exposure occurs over a large part of a person's lifespan (chronic) or during a very short portion of the lifespan (acute). It should be noted that all of the health effects of exposure to radiation can also occur in unexposed people due to other causes. In addition, there is no detectable difference in appearance between radiation-induced cancers and genetic effects and those due to other causes.

17.5.3 Chronic Exposure

Chronic exposure is continuous or intermittent exposure to low levels of radiation over a long period. Chronic exposure is considered to produce only effects that can be observed after some time following initial exposure. These include genetic effects and other effects such as cancer, precancerous lesions, benign tumors, cataracts, skin changes, and congenital defects.

17.5.4 Acute Exposure

Acute exposure is exposure to a large, single dose of radiation, or a series of doses, for a short period. Large acute doses can result from accidental or emergency exposures or from special medical procedures (radiation therapy). In most cases, a large acute exposure to radiation can cause both immediate and delayed effects. For humans and other mammals, acute exposure, if large enough, can cause rapid

TABLE 17.2

Summary of Clinical Symptoms of Radiation Sickness

Time After Exposure	Survival Improbable (700 rem or More)	Survival Possible (300–550 rem)	Survival Probable (100–250 rem)
First hour	Nausea, vomiting, and diarrhea in first few hours	Nausea, vomiting, and diarrhea in first few hours	Possible nausea, vomiting, and diarrhea on first day
First week	No definite symptoms in some cases (latent period)		
		No definite symptoms (latent period)	
Second week	Diarrhea Hemorrhage Purpura (skin hemorrhages) Inflammation of the mouth and throat Fever		No definite symptoms (latent period)
Third week	Rapid emaciation Death (mortality— probably 100% without medical treatment)	Epilation (hair loss) Loss of appetite and general malaise (discomfort) Fever Hemorrhage Purpura (skin hemorrhages) petechaie, (skin spots) nosebleeds, pale skin Inflammation of the mouth and throat Diarrhea Emaciation (extreme malnutrition)	Epilation (hair loss) Loss of appetite and malaise Hemorrhage Purpura Petechaie Pale skin Diarrhea Moderate emaciation
Fourth week		Death in most serious cases (mortality—50% for 450 rem without medical treatment)	Recovery likely in about 3 months unless complication by previous poor health or superimposed injuries or infections

Source: Courtesy of the Department of Energy.

development of radiation sickness, evidenced by gastrointestinal disorders, bacterial infections, hemorrhaging, anemia, loss of body fluids, and electrolyte imbalance. Delayed biological effects can include cataracts, temporary sterility, cancer, and genetic effects. Extremely high levels of acute radiation exposure can result in death within a few hours, days, or weeks.

TABLE 17.3

Effects of Acute Whole Body Radiation Doses

Acute Dose (rem)	Probable Effect
0–20	No obvious effect. No injury or disability. Could be detected by chromosome analysis.
20–50	Minor blood changes.
60–120	Vomiting and nausea for about 1 day in 5%–10% of those exposed. Fatigue, but no serious disability.
130–170	Vomiting and nausea for about 1 day, followed by other symptoms of radiation sickness (increased temperature, blood changes, fatigue) in about 25% of those exposed. No deaths anticipated.
180–220	Vomiting and nausea for about one day followed by other symptoms of radiation sickness in about 50% of those exposed. No or few deaths anticipated.
270–330	Vomiting and nausea in nearly all those exposed on first day, followed by other symptoms of radiation sickness. About 20% of the group will die within 2–6 weeks after exposure; survivors convalesce for about 3 months.
400–500	Vomiting and nausea in all those exposed on the first day, followed by other symptoms of radiation sickness. Bone marrow destruction (reversible). Without medical treatment 50% of the group will die within a month. Survivors convalesce for about 3 months.
550–750	Vomiting and nausea to all those exposed within four hours, followed by other symptoms of radiation sickness. Irreversible destruction of bone marrow. Deaths in 100% of the group is expected.
1,000–5,000	Vomiting, diarrhea, and nausea in all those exposed within 1–2 h. Damage primarily to the digestive system. First and second degree burns of the skin. Death will occur to entire group in 3–10 days. There is no medical treatment to prevent death after this high of a dose.
	Note: These are the doses used to treat cancer but those treatments are kept in small areas of the body.
5,000–10,000	Unconscious minutes after exposure. Death within 48 h due to nervous system damage.

Source: Courtesy of the Department of Energy.

With proper medical attention the chances of survival are excellent even for very high absorbed doses. Radiotherapy treatment for cancer patients often exposes the tumor area to protracted doses of several thousand rem over a period of 6–8 weeks. Recall that the federal dose limit is 3 rem per quarter.

17.5.5 RISKS OF HEALTH EFFECTS

All people are chronically exposed to background levels of radiation present in the environment. Many people also receive additional chronic exposures or relatively small acute exposures. For populations receiving such exposures, the primary concern is that radiation could increase the risk of cancers or harmful genetic effects.

The probability of a radiation-caused cancer or genetic effect is related to the total amount of radiation accumulated by an individual. Based on current scientific evidence, any exposure to radiation can be harmful (or can increase the risk of cancer);

however, at very low exposures, the estimated increases in risk are very small. For this reason, cancer rates in populations receiving very low doses of radiation may not show increases over the rates for unexposed populations.

For information on effects at high levels of exposure, scientists largely depend on epidemiological data on survivors of the Japanese atomic bomb explosions and on people receiving large doses of radiation medically. These data demonstrate a higher incidence of cancer among exposed individuals and a greater probability of cancer as the level of exposure increases. In the absence of more direct information, that data are also used to estimate what the effects could be at lower exposures. Where questions arise, scientists try to extrapolate based on information obtained from laboratory experiments, but these extrapolations are acknowledged to be only estimates. For radon, scientists largely depend on data collected on underground miners. Professionals in the radiation protection field prudently assume that the chance of a fatal cancer from radiation exposure increases in proportion to the magnitude of the exposure and that the risk is as high for chronic exposure as it is for acute exposure. In other words, it is assumed that no radiation exposure is completely risk free.

17.5.6 PERSONAL MONITORING

Dosimeters—Personal monitoring devices (dosimeters) are required for workers who may receive 10% of the maximum dose of external radiation permissible under Nuclear Regulatory Commission (NRC) regulations. To apply for a monitoring device, those exposed must complete the form for request for radiation monitoring badge, and it must be returned to the Radiation Safety Office (RSO). RSO will request the dosimetry records of new radiation workers from other previous workplaces where they used radioactive materials. Old and new dosimetry records will be added to obtain cumulative records. Annual maximum permissible dose equivalent in mrem is

- Whole body 5,000
- Lens of the eye 15,000
- Skin 50,000
- Extremities (hand, forearms, feet, ankles) 50,000

Thermoluminescent dosimeters (TLD's) are used for monitoring gamma and high-energy, beta-emitting radioisotopes, such as phosphorus-32, chromium-51, and iodine-125. The RSO may use radiation dosimeters to monitor levels of radiation in other areas.

An accurate record of an employee's radiation exposure history must be maintained by RSO. Employees must provide information regarding any prior occupational radiation exposure. If a worker is occupationally exposed to radiation elsewhere in addition to being exposed at the present place of employment, the authorized user should report this to the RSO so that an accurate record of the worker's total radiation exposure can be maintained. Two NRC forms are to be maintained and they are Occupational Dose Record for a Monitoring Period (NRC Form 5) and Cumulative Occupational Dose History (NRC Form 4).

Employees must wear dosimeters recommended by the RSO while working in any restricted area. While not being worn, dosimeters should be stored away from all radiation sources in a desk drawer or in some other location where they will not be exposed to excessive heat, sunlight, or moisture (for e.g., never left in a car). They are not to be worn off the facilities' premises. Individuals who do not work directly with radioisotopes or in an area where radioisotopes are used may not be issued dosimeters.

Note: Individuals who wear radiation badges should review their radiation dosimetry records to ascertain their radiation exposures.

Any dosimeter contaminated or exposed to heat, moisture, or medical x-rays should be returned to the RSO for replacement. After any accident or if an overexposure is suspected, the dosimeters should be returned immediately to the RSO to be read. Dosimeters should be worn on a shirt, coat pocket, lapel, or in some other position between the waist and the shoulders, which will be representative of any radiation exposure. If, during a radiological process, a hand might receive a dose, a ring dosimeter should be worn on a finger of the hand under the glove. When both whole-body and hand doses can occur, two dosimeters will be issued so that one is for the whole body and one for a hand.

Authorized user supervisors are responsible for distributing and collecting dosimeters for all personnel under their supervision or authorization. Ring and whole-body dosimeters should be exchanged quarterly through the RSO. The RSO will keep a record of any dose received and will send each worker a copy of his or her exposure record upon request. Maximum permissible doses—federal limits for radiation doses were discussed previously; however, all doses must be maintained ALARA.

17.6 TRAINING

All authorized users and radiation workers under their supervision who work with radioisotopes must receive instruction on radiation safety, biological effects of radiation, regulatory requirements, and laboratory techniques. The RSO should be designed to achieve strict compliance with applicable federal regulations. Title 10, Code of Federal Regulations, Part 19, Section 12 Instructions to Workers (10 CFR 19.12) is the regulation requiring training of all individuals working in or frequenting any portion of a restricted area associated with radioactive materials or radiation.

Radiation workers are those personnel listed on the authorized user form of the supervisor to conduct work with radioactive materials. Radiation workers must

- Attend the Occupational Health and Safety (OHS) radiation safety in the laboratory course prior to beginning work with radioactive materials
- Read the CDC Radiation Safety Manual and be responsible for its contents as applicable to their work duties
- Call the emergency medical service or fire department immediately for any fire, explosion, or major accident and tell the dispatcher that the accident involves radioactive materials

- Next, notify OHS and their immediate supervisor or the authorized user responsible for their work area
- Adhere to all CDC radiation safety regulations, policies, and procedures governing the use of radioactive materials as outlined in this manual and required by the NRC
- Follow the recommendations of the authorized user for those procedures that are specific to their work area for the storage, usage, recording, and disposal of radioactive materials
- Receive necessary on-the-job training from an authorized user as it relates to the duties of the job
- Maintain exposures ALARA through work procedures, shielding, and the use of gloves and other protective clothing

A permitted worker is a worker who does not work with radioactive materials but works in a radiation laboratory. To be a permitted worker, the employee must successfully complete the first day on the job. The duties of permitted workers are as follows:

- Enroll, attend, and complete the first day of the radiation safety course
- Wear issued radiation monitoring badges at all times in the radiation work area
- Use the principles of time, distance, and shielding to protect themselves from radiation exposure
- Confer with radiation workers to find out where radioisotopes are used and stored so that these areas can be avoided
- Report any observed radiation safety infractions, shortcomings, or failures in a timely manner

Records of all training, including documentation of attendance, are required to be maintained by OHS for review by NRC inspectors. Visiting scientists or contract personnel are not to undertake work involving radioactive material until they have taken the appropriate safety training courses and are under the supervision of an authorized user.

17.7 RADIATION AREAS

There are a variety of classifications for areas where radiation sources or potentials for radiation exposure exist. This system of classification indicates the potential risk due to radiation. The following are definitions of radiation areas and used as warnings signs to alert workers as to the hazard present:

- Restricted area—An area to which access is limited by the licensee for the purpose of protecting individuals against undue risks from exposure to radiation and radioactive materials.
- Unrestricted area—An area to which access is neither limited nor controlled by the licensee.
- Radiation area—An area accessible to individuals, in which radiation levels could result in an individual receiving a dose equivalent in excess of 0.005

FIGURE 17.3 High-radiation area warning sign. (Courtesy of the Department of Energy.)

rem in 1 h at 30 cm from the radiation source or from any surface that the radiation penetrates.

- High-radiation area—An area accessible to individuals, in which radiation levels could result in an individual receiving a dose equivalent in excess of 0.1 rem in 1 h at 30 cm from the radiation source or from any surface that the radiation penetrates (see Figure 17.3).
- Airborne radioactivity area—A room, enclosure, or area in which airborne radioactive materials, are composed wholly or partly of licensed material.
- Labeling requirements—Each area or laboratory used to store or contain licensed radioactive material shall be conspicuously posted with a door sign bearing the radiation caution symbol and the words "caution (or danger), radioactive materials." Containers or areas where radioactive material are stored, used, and disposed of shall bear a durable, clearly visible label that identifies the contents. The label must have the radiation symbol and the words "caution (or danger), radioactive materials." Beakers, test tubes, and other glassware that contain radioactive material transiently during an experiment need not be individually labeled. However, containers that will be left unattended must be labeled.

A current notice to employees (Form NRC 3) must be posted so that it can be easily seen by persons entering or leaving a restricted area. Authorized users are responsible for posting all signs required and provided by the OHS. Authorized users must also remove signs that are no longer needed or that have become incorrect or inappropriate for their laboratories.

Active radiation areas where radioactive materials are used are to be surveyed by authorized users or their designees at least once a month by using wipes or a suitable survey meter. A survey using an instrument such as a GM (Geiger–Muller) counter is acceptable as long as it is sensitive enough to detect the nuclides used. In addition to routine surveys, facilities or other potentially contaminated areas must be surveyed in the following cases:

- After any spill, leak, fire, or other disturbances in the facility
- When work with radioactive materials is terminated
- Before and after facility construction modifications
- Before maintenance or removal of any equipment that may have come in contact with radioactive material or that contains radioactive material

Facilities with sealed sources will be surveyed at least biannually. The following sealed sources will be surveyed by the qualified staff, with RSO assistance, for leakage and external contamination at least once every 6 months. The sources will also be surveyed before and after they are moved within a facility or to another facility, after being dropped or otherwise damaged, and before and after maintenance:

- Multi hazard waste—This waste contains any combination of radioactive, biohazardous, and chemically hazardous materials known as mixed waste. Avoid creating such materials, if possible. Disposal of multi-hazard waste is extremely costly and difficult.
- Solid waste—This includes test tubes, beakers, absorbent paper, gloves, pipettes, and other dry items contaminated with radioactive material but not containing liquid radioactive waste. This material must be placed in plastic bags and sealed with tape. Hypodermic needles, capillary pipettes, and other sharp objects must be placed in puncture-proof containers before they are put into the large waste cans.
- Containers bearing a radioactive label, but no longer containing radioactive material, must be disposed of as ordinary waste only after the radioactive label is defaced or removed and after decontaminating.
- Aqueous liquid waste—No liquid radioactive waste shall be disposed of by the sewage system unless (1) the liquid is readily soluble or dispersible in water, and (2) the material is diluted to the concentrations flushed simultaneously with measured amounts of water sufficient to achieve those concentrations.

17.8   SPECIFIC EMERGENCY PROCEDURES

The procedures for spills of radioactive materials are as follows:

- Put on protective clothing, such as shoe covers, and gloves, before starting containment and cleanup of the spills.

- Cover the spill with absorbent material as quickly and as completely as possible to prevent spreading. To localize the contamination, wipe inward toward the center of the spill. Do not wipe back and forth or in a random fashion.
- Have someone who is not contaminated call the OHS immediately.
- If a biological agent is involved, soak the area with a disinfectant for at least 30 min to inactivate the agent, and wash your hands and arms thoroughly with soap or an appropriate disinfectant. Scrub your hands for several minutes and rinse them thoroughly.
- If personnel must leave the contaminated area, remove gloves, shoes, and laboratory coat; segregate them as radioactive waste before leaving the areas of spill.
- After removing protective clothing, wash all contaminated areas of skin thoroughly, without vigorous scrubbing, with cool water and mild soap for 5–10 min. Do this as soon as possible after the accident.
- Remember also to remove all clothing that may have been contaminated.
- Take care not to recontaminate yourself after you have thoroughly washed.
- Do not leave the area until someone from the OHS has determined that you have been successfully decontaminated.

17.9   SUMMARY OF APPLICABLE REGULATIONS

Standards for protection from radiation are published in NRC's Rules and Regulations, Title 10, Chapter 1, Code of Federal Regulations, Part 19, titled Notices, Instructions, and Reports to Workers; Inspections, and Part 20, titled Standards for Protection Against Radiation. They are referred to in this manual as 10 CFR 19 and 10 CFR 20, respectively. These regulations may also be viewed via the Internet at http://www.nrc.gov/NRC/CFR/index.html. Additional requirements are included in the NRC licenses issued to facilities governing the possession and use of radioisotopes. Employees are encouraged to refer to these standards and the current licenses assuring compliance with all ALARA regulations as defined by 10 CFR 20.

17.9.1   SUMMARY

Ionizing radiation is any radiation capable of displacing electrons from atoms or molecules, thereby producing ions. Some examples are alpha, beta, gamma, x-rays, and neutrons. Ionizing radiation is energy that is carried by several types of particles and rays given off by radioactive material, x-ray machines, and fuel elements in nuclear reactors. Ionizing radiation includes alpha particles, beta particles, x-rays, and gamma rays. Alpha and beta particles are essentially small fast moving pieces of atoms. X-rays and gamma rays are types of electromagnetic radiation. These radiation particles and rays carry enough energy to knock out electrons from molecules, such as water, protein, and DNA, with which they interact. This process is called ionization, which is why it is named ionizing radiation. We cannot sense

FIGURE 17.4   New symbol for inizing radiation (proposed). (Courtesy of International Atomic Energy Agency.)

ionizing radiation, so we must use special instruments to learn whether we are being exposed to it and to measure the level of radiation exposure. High doses of ionizing radiation may produce severe skin or tissue damage. The International Atomic Energy Agency (IAEA) has proposed a new symbol to warn about ionizing radiation dangers (see Figure 17.4).

18 Machine/Equipment Safeguarding

A mixer that has no safeguarding.

18.1 MACHINE/EQUIPMENT SAFEGUARDING

Accidents resulting from moving machinery and persons working on or around machinery must be prevented. The installation and maintenance of machinery and machine guards are governing factors in controlling and preventing accidents and injuries. In devising protection against moving machinery and parts, the goal should be to make it as effective as possible. All possible contingencies should be considered, including acts of thoughtlessness and foolhardiness, in guarding machinery to prevent injuries.

18.2 PRINCIPLES FOR MACHINE GUARDING

All companies should develop safeguarding requirements for fixed powered machines and equipment and power transmission devices and equipment. This chapter is not applicable to hand tools. These requirements apply both to point-of-operation safeguarding of machines and to any other danger point where an employee may come into contact with the moving parts of the machine, or material handling equipment.

The intent of this chapter is to provide five principles for machine and equipment safeguarding:

- Employers should conduct a risks assessment for all machines and equipment.
- Employers are to comply with all applicable machine-guarding standards.
- Safeguards shall be provided that prevent the entry of any part of an employee's body into the point of operation of machines or equipment when the machines or equipment are operating.
- Certain types of machines such as presses (mechanical, pneumatic, hydraulic), press brakes, powered crimpers, riveting machines, shears, and staking machines shall be provided with at least two-point operation safeguards, installed in such a manner that the failure of either individual safeguard shall not prevent the proper operation of the other.
- Administrative or disciplinary actions shall be enforced if any individual intentionally removes or bypasses, or permits the removal or bypassing of safeguards, unless it is approved in advance and alternate safeguarding methods have been put in place.

It is recognized that the installation of safeguards alone will not prevent injuries. The training of operators and maintenance personnel and maintaining the equipment in good working order are also essential to a successful machine safety program. All of these elements are an integral part of any type of safety program.

All of the machine-guarding principles should apply where fixed powered machines, equipment, or power transmission devices are used without regard for their location. Machine guarding applies both to production operations and to nonproduction activities such as research, laboratory, experiments, inspection, and maintenance.

It should be the responsibility of any employer to ensure that all new equipment/machines as well as existing equipment and machines are guarded to protect workers and meet applicable Occupational Safety and Health Administration (OSHA) standards.

18.3 CONDUCTING AN ASSESSMENT

An assessment must be made of each powered machine and piece of equipment at all company facilities. This assessment will determine the inherent risks to operator(s), support, or maintenance personnel resulting from the operation, maintenance, and repair of the machine or equipment. The assessment will also evaluate the adequacy of the machine/equipment control system design, and installed safeguards, to meet the requirement to protect the workforce. The assessment process must evaluate the need for redundant safeguards on machines and equipment.

In the planning and design of new parts, equipment, manufacturing, or service operations, an assessment must be conducted to identify, reduce, or eliminate potential hazards created in or by any operations and services of the new part or piece of equipment while complying with all applicable regulations.

18.4 GUIDANCE ON ASSESSING MACHINE/EQUIPMENT SAFEGUARDS

The following sections provide general guidance on identifying safeguarding risks and assessing the effectiveness of machine/equipment safeguards by evaluating the risk and the adequacy of existing guards.

18.4.1 ASSESSING SAFEGUARDING RISKS

Assessments should include a comprehensive review of all machines and equipment to determine point-of-operation hazards and hazards from moving material, machinery, or equipment, including power transmission equipment and devices. The need for redundant safeguards should also be evaluated.

The assessment should include each location on or about the machine or equipment (front, back, top, and sides) where a person may come into contact with moving parts or material as well as each operation done on the machine or equipment. The assessment should also address how the equipment is used, (i.e., for its intended purpose, etc.), the need for redundant safeguards, and the movement of material as it is processed. If possible, the machine or equipment should be designed and configured so that the potential for contact with moving parts is eliminated, without the need for additional safeguards. Other information that should be obtained during the assessment process includes, but is not limited to, the following:

- Are machines fitted with auxiliary equipment such as robotic arms, activators, manipulators, conveyors, dial feeders, augers, hoppers, grinders, indexing tables, etc., designed to feed or unload the machine that may have moving components?
- Is there a hazard to individuals who may have to work on the machine or equipment with safeguards removed? (see Figure 18.1)
- Are there loose electrical fittings, poor grounding, improper fusing, or other electrical hazards on or about the machine or equipment?
- Are workers potentially exposed to harmful substances or excessive noise during the operation of the machine or equipment?
- Are workers exposed to hazards from elevated temperatures in the machines, tools, dies, or equipment?

It would be advisable to develop a form that could be tailored to a specific workplace to summarize the findings on each machine or piece of equipment that is assessed.

18.4.2 EVALUATING EFFECTIVENESS OF SAFEGUARDS

In trying to determine the effectiveness, the appropriateness, or the need for a new or revised guard, the following questions should be answered and the answers evaluated for the action to be taken in ensuring that all machines and equipment are adequately guarded:

FIGURE 18.1   Some machine operations are very open and difficult to guard unless a barrier surrounds the machines. No barrier exists for this printing press.

- Are power transmission devices such as belts and shafts located so that employees cannot come into contact with moving parts? (see Figure 18.2) (i.e., overhead and out of reach)
  - If yes, additional safeguarding may not be required
  - If no, safeguards will be required

FIGURE 18.2   Failure to guard power transmission apparatus can result in incidents like this. (Courtesy of Mine Safety and Health Administration.)

- Can employees place any part of their bodies into the danger points of a machine or equipment during any operation done on the machine or equipment?
  - If yes, safeguards will be required
- Are safeguards firmly secured in place and not easily removed?
- Do the safeguards permit safe, comfortable, and relatively easy operation of the machine?
- Do the safeguards themselves create a hazard to the employees?
  - If yes, additional safeguards will be required
- Do the safeguards allow employees to watch the operation with the safeguard in place? (Required only if employees are required to watch the work performed.)
- If fixed guards are used, can they only be removed by the use of tools?
- If adjustable guards are used, can they be easily adjusted without the use of tools?
- If there is the risk of flying or ejected material or objects, i.e., chips coolant, etc., have safeguards been installed to protect the operator and other employees from being struck by material, parts, cuttings, waste, etc.?

18.4.3 ASSESSING MANAGEMENT OF SAFEGUARDS

There must be policies, procedures, and programs to make sure that the machine safety program is managed efficiently and effectively. The following questions can be used as a guide:

- Is there a formal documented preventive maintenance program to verify the safe operation of the safeguards on a periodic basis?
- Is there a formal program that requires employees to promptly report any suspected problems with machine/equipment safeguards or safe operation to supervision as soon as they are noticed?
- Is there a formal program to require supervision to evaluate the effectiveness of machine/equipment safeguards and safe operation once possible deficiencies have been reported?

18.4.4 FURTHER MACHINE-RELATED ISSUES

The following questions should be answered as part of the ongoing assessment:

- Are machines and equipment secured to the floor or other supports to eliminate the potential for "walking" or tipping over?
- If manual guards (e.g., guards that require the operator to replace them after each use) are used, is a second means of protecting the employee also used?
- Are hand tools or push sticks used as a "safeguard"? Is an acceptable safeguard used with the hand tools or push sticks?

18.4.5 CONTROL AND MACHINES

An integral part of machine guarding is related to the type operational controls and their use. Thus, as part of evaluating a machine or equipment safety, operational controls play a significant part; therefore, the following questions should be addressed:

- Are controls that start or cycle a machine protected against unintended or accidental operation? This applies to hand and foot operated controls as well as any other control with which an employee may inadvertently come into contact.
- Is the reliability of control systems periodically verified and are the control systems properly maintained?
- Is the machine/equipment control system and logic designed, installed, and maintained so that the failure or bypassing of any single component of the control system will not allow the machine to start, recycle, or repeat or prevent the machine from stopping at the completion of its cycle?
- Are the machine/equipment controls clearly visible and well identified?
- Are all control devices located outside of the danger zone of the machine or equipment?
- Are machines equipped with well identified "emergency stop" buttons where necessary, located so that any operator can stop the machine?
- If the emergency stop buttons or other safety devices such as interlocked guards or presence-sensing devices do not immediately stop the machine when activated, i.e., the machine coasts to a stop, are additional safeguards in place to eliminate the potential for employees to access moving parts until all movement has ceased?
- Do the emergency stop devices have priority over other machine or equipment controls?
- Does the emergency stop device stop the process as quickly as possible without creating additional hazards?
- Are control devices and circuits designed so that in the event of a power failure, the machine will not restart automatically upon resumption of power?
- Are machine control systems provided with a positive means of de-energizing and locking out all energy sources (compressed air, electricity, hydraulic, etc.) when the machine is taken out of service for maintenance or repairs?
- If the control system is designed so that safeguards may be bypassed during maintenance or setup work, is a key controlled switch provided?
- If the control system allows the machine to operate in continuous mode, is the mode selector provided with a key controlled switch?

18.4.6 OPERATOR TRAINING

All machine and equipment operators are to be trained. Training should be addressed during any assessment regarding machine or equipment safe guards. Use the following questions to assist in accomplishing answering them:

- Are only specifically trained and authorized employees designated to perform maintenance or setup work, and are additional safeguards such as locking out the equipment, training of equipment operators, etc., in place?
- Is the mode selection done by designated and trained person?
- Are all machine operators trained in how their particular machines and auxiliary equipment are safeguarded?
- Does this training also include the procedures that they should follow for testing, maintaining, and adjusting safeguards?
- Is there an annual retraining program for machine operators, setup and repair personnel?
- Have written setup, operating, and maintenance instructions been developed and posted on each machine?

18.5 TYPICAL MACHINES/EQUIPMENT REQUIRING SAFEGUARDS

The following is a listing of some of the most common machines and equipment that require safeguards:

- Fixed powered machines
  - Boring machines
  - Broaching machines
  - Conveyors
  - Drilling machines
  - Electrical machining units
  - Gear cutting and finishing machines
  - Grinders
  - Honing machines
  - Injection molding machines
  - Joiners
  - Lathes (wood and metal)
  - Milling machines
  - Presses (mechanical, pneumatic, and hydraulic)
  - Press brakes
  - Planers
  - Powered crimpers
  - Polishing/buffing machines
  - Printing presses
  - Riveting machines
  - Robots
  - Rollers
  - Sanders
  - Saws (wood and metal)
  - Shapers

- Shears
- Staking machines
- Welders
- Typical power transmission devices
  - Gears
  - Sprockets and chains
  - Belts/ropes and pulleys (includes drive sheaves)
  - Shafts and couplings
  - Flywheels
  - Screws

18.6 SAFEGUARDS

Based on the assessment, safeguards are to be installed to prevent any part of an employee's body from entering the point of operation of the machine while the machine is in operation.

Based on the assessment, safeguards must be installed to prevent operators, support personnel, or passersby from accessing mechanical pinch points created by machine components, material or material handling equipment, or power transmission elements. High temperature and electrical hazards must be protected likewise.

Manual safeguards (e.g., blocking devices, etc.) that are installed temporarily, and then removed, during the machine operating cycle, should not be used as the sole means of safeguarding the machine.

While push sticks or other hand-held devices used to insert, position, or remove stock during machine operation may be valued safety measures, they are not to be accepted as a machine safeguard.

In addition to the safeguards provided for normal machine or equipment operation, additional safeguards must be provided, if required, to protect personnel involved in setup, maintenance or repair operations. Proper lockout provisions and control system procedures are considered as safeguards under these conditions.

Safeguards interlocked with a machine or piece of equipment should be designed, installed, and maintained so that the failure of the safeguard does not prevent the machine/equipment from being stopped or deactivated safely and will prevent further operation of the machine/equipment until the safeguard failure has been corrected.

The safeguarding assessment is to include the hazards of uncontrolled energy release, either suddenly or gradually. Where necessary, safeguards or other means must be provided to eliminate the possibility of employee contact with material or any hazardous part of the machine until all potential for movement has stopped.

18.7 AREAS TO BE SAFEGUARDED

Certain areas on machines and pieces of equipment present dangers and must be guarded to protect workers. These are danger point or zone, point of operation, and power transmission devices.

FIGURE 18.3 A fixed guard prevents a worker from contacting points of operation.

18.7.1 DANGER POINT OR ZONE

Danger point or zone is the position or point in or about a machine or piece of equipment where part of an employee's body may come into contact with the movement on or about the machine. Movement refers to movement of the machine while it cycles, movement of material, movement of energy through a power transmission device, or movement of any equipment or device attached to a machine. Nip points, pinch points, the point of operation, etc., are components of the danger point or zone (see Figure 18.3).

18.7.2 POINT OF OPERATION

Point of operation is the place/point of actual contact between a machine and the material that is processed. It is where the material is actually inserted, maintained, and withdrawn. Typically, the point of operation is where the saw blade, power press die, press brake die, lathe tool, router bit, or grinding wheel comes into contact with the material that is processed.

18.7.3 POWER TRANSMISSION DEVICE

Power transmission device is the means by which power is transmitted to a machine or piece of equipment. Power transmission devices include belts, pulleys, drive sheaves, chains, flywheels, gears, pumps, and couplings.

18.7.4 Lockout

Lockout is a program or procedure that requires all energy sources to a machine or piece of equipment to be isolated/neutralized and made inoperative by the use of a lock on each energy source. Energy sources include electrical, mechanical including suspended parts or springs, hydraulic, pneumatic, chemical, and thermal as well as any other energy source that provides energy to a machine or operation.

18.8 TYPES OF SAFEGUARDS

Well-designed safeguards cannot be easily removed, allow for normal operation of the machine, and enable the operator to work with a minimum of interference. Safeguards themselves should not create a hazard by virtue of their position or movement. Common types of safeguards include barrier guards, presence-sensing devices i.e., light curtains, mats, and safety controls i.e., two-hand controls, gates, enclosures, distance and location, and use of robots or automatic feed and removal equipment.

18.8.1 Barrier Guards

Barrier guards create a barrier between the employee and the danger point of the machine or equipment. There are four types of barrier guards: fixed, interlocked, adjustable, and self-adjusting. A description of these barrier guards are as follows:

- Fixed guards are used to prevent entry into danger points. A fixed guard becomes a permanent part of a machine or die, or power transmission device or equipment. Fixed guards can be made out of sheet metal, screen, wire cloth, metal or plastic bars, plastic sheets, or any other material that can withstand impact.
- Interlocked guards completely stop the action (movement) of the machine as soon as the guard is opened or removed. An interlocked guard can only be used on a machine or equipment that can be stopped quickly. These guards can be automatic, i.e., powered or manually actuated. The control system for an interlocked guard should be designed such that a failure of the guard or control circuit safely stops the action of the machine. Interlocked guards should be checked by a designated and trained person on a frequent (i.e., at start of shift) basis.
- An adjustable guard is a guard that is readily adjustable to allow different sizes and shapes of material to be cut or shaped. Adjustable guards should be checked by a designated and trained person on a frequent (i.e., each shift) basis for adjustment, operation, and maintenance.
- A self-adjusting guard adjusts automatically to the material used. Adjustable guards should be checked for adjustment, operation, and maintenance on a frequent basis.

18.8.2 Redundant Safeguards

At least two safeguards must be provided for the point of operation of presses (mechanical, pneumatic, hydraulic), press brakes, powered crimpers, riveting machines, shears, and staking machines. However, the assessment process may identify other machines or equipment that also require two-point operation safeguards. The only exception to this requirement is those situations in which it can be demonstrated that a single safeguard provides absolute assurance that no part of the operator's body can enter the point of operation.

In such instances, the single safeguard must be essentially non-removable. All exceptions must be inspected and approved by the local senior operations manager.

In considering the redundancy requirement for safeguards provided for setup, maintenance, and repair functions, the mandatory use of power supply lockout, the proper release of all stored energy, and the use of die blocks will be considered as a redundant system.

18.9 CONTROLS AND CONTROL SYSTEMS

Machine or equipment control systems will be guarded or located so that the machine or equipment may only be started or cycled by the intentional act of an operator. Operating controls, including foot pedals, will be guarded against unintentional activation.

All energy control devices associated with the machine or equipment (e.g., electrical switches and pneumatic or hydraulic supply valves) must be provided with a means to lockout the energy source while maintenance or repair operations are performed. Lockout procedures must include the release of any stored energy (e.g., electrical capacitors, hydraulic/pneumatic pressure tanks, compressed springs, etc.) in the machine or equipment operating system.

Based on the assessment, machines or equipment will be equipped with a clearly marked emergency shutdown switch or control, easily accessible to the machine operator(s) and other employees.

18.10 CONTROL SYSTEM FUNCTION

Machine/equipment control systems will be designed, installed, and maintained so that the failure or bypassing of any single part or component of the control system will not allow the machine/equipment to start or repeat. Failure or bypassing of any single part or component of the control system also shall not prevent the machine/equipment from safely stopping at the completion of its cycle.

Machine and equipment control systems will be designed, installed, and maintained so that in the event of a power failure, the machine/equipment will not create a hazard when power is restored.

If a machine can be optionally operated in a mode where additional hazards may exist, e.g., inching, jogging, purging, etc., the mode selection device shall be key controlled with the keys kept under the control of designated competent personnel.

18.11 TYPES OF CONTROL DEVICES

Control devices protect against entry by stopping a machine or piece of equipment if a hand or other body part enter the point of operation or danger zone (presence-sensing device), preventing employees from placing any part of their body in the point of operation or danger point (restraint), automatically withdrawing employees' hands (pullbacks), requiring employees to keep their hands on controls outside of the danger zone during the hazardous portion of the machine cycle (two-hand controls), or monitoring the braking performance of the machine on a continuous and automatic basis (brake monitor).

18.11.1 PRESENCE-SENSING DEVICES

Electric eyes, light curtains, pressure mats, safety wires, or optical devices are used to detect body parts (hands, arms, fingers, etc.) before they enter into the danger zone. Activation of these devices stops the action of the machine. They must be located such that the operator cannot reach the point of operation or danger point before the hazardous motion of the machine or equipment has stopped. These devices must be installed so that any failure or fault in the device will prevent the machine from operating. Presence-sensing devices should be tested at the beginning of each shift and whenever the machine is stopped for maintenance, lubrication, adjustment, or setup (see Figure 18.4).

18.11.2 TWO-HAND CONTROL DEVICES

Two-hand control devices require the operator to keep his or her hands on the controls through the hazardous portion of the machine cycle. The control device should be designed so that only intentional use of employees' hands may be used to

FIGURE 18.4   This power press for cutting paper is equipped with a presence-sensing device that shuts it down when the beam is broken by a worker's hand or body part.

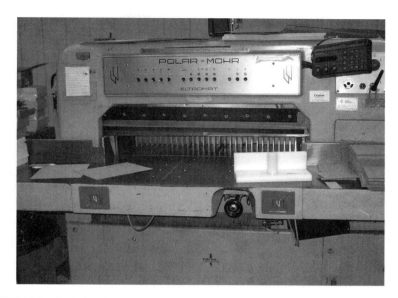

FIGURE 18.5 Both hands must be used to depress the two separate operation buttons at the same time to actuate the power press.

cycle the machine. A two-hand control device requires simultaneous activation of the controls by both hands. The controls shall have an anti-repeat and anti-tie down feature so that one or both control devices cannot be wedged or forced in the cycle or run position. The machine must stop immediately if the operator removes one or both hands. The control devices must be located at a predetermined safe distance from the danger zone of the machine or equipment (see Figure 18.5).

Two-hand controls must be tested to verify that they are in good operating condition at the start of each shift and after the machine has been stopped for maintenance, lubrication, adjustment, or setup.

18.11.3 BRAKE MONITOR

A brake monitor is a sensor, system, or device that monitors the performance of the machine's braking systems during each cycle. If the braking systems do not meet the design criteria, the monitor automatically prevents the machine from being cycled. The operation of the brake monitor should be identified on the control panel of the machine.

18.11.4 RESTRAINT/PULL BACK DEVICE

A restraint device typically attaches to the operator's hands and connects to the machine in such a fashion that it either prevents the operator from reaching into a danger point or withdraws the operator's hands from a danger point as the machine operates. Restraint devices must be checked for proper adjustment and operation at the start of each shift and after the machine is stopped for maintenance, adjustment, lubrication, setup, or operator change. Examples of restraint devices include pull-backs and holdbacks.

18.11.5 LOCATION/DISTANCE

Protecting danger points from contact by location or distance is an acceptable machine safeguarding means under certain well-defined, conditions. Very often power transmission devices are safeguarded by location in that the hazards are located above where an employee may come into contact with them e.g., locating fans above 7 ft, putting guardrails around entire pieces of equipment such as air compressors, or locating moving equipment or machines in locked rooms. If locked rooms are used, personnel who can enter the room need to be trained in the hazards of the machines/equipment and how the hazards can be controlled, i.e., lockout.

18.12 TRAINING

Any individual required to install, set up, operate, or otherwise use, service, etc., machines, equipment, or safeguards, shall be trained by a knowledgeable person in the safe operation of that machine or equipment including the function and use of each safeguard. Management, engineering support, and supervisory personnel also must be trained. Refresher training must be provided on an annual basis.

For each machine or piece of equipment covered by this standard, written setup, operating, and maintenance instructions must be developed. The instructions must be reviewed with all appropriate employees, e.g., operators, maintenance personnel, setup personnel, etc., during the training and located on or about the machine or piece of equipment. Training records should be documented and maintained in the company's personnel policies.

18.13 MAINTENANCE

A program that ensures that required and necessary preventive scheduled maintenance is performed and documented on machines, equipment, and safeguards should be established and in place.

18.14 ENFORCEMENT

Administrative or disciplinary actions must be enforced when any individual intentionally removes or bypasses, or permits the removal or bypassing of safeguards, unless it is approved in advance and alternate safeguarding methods have been established.

18.15 PERIODIC REVIEW

Procedures must be in place to ensure that all new, modified, relocated, etc., machines, equipment, or power transmission equipment are reviewed for conformance safeguard requirements before operation.

18.16 APPLICABLE OSHA REGULATIONS

18.16.1 Machine Guarding (29 CFR 1910.212 and 219)

Moving machine parts have the potential for causing severe workplace injuries, such as crushed fingers or hands, amputations, burns, and blindness, just to name a few. Safeguards are essential for protecting workers from these needless and preventable injuries. Machine guarding and related machinery violations continuously rank among the top 10 of OSHA citations issued. In fact, "Mechanical Power Transmission" (29 CFR 1910.219) and "Machine Guarding: General Requirements" (29 CFR 1910.212) were the No. 6 and No. 7 top OSHA violations for FY 1997, with 3077 and 3050 federal citations issued, respectively. Mechanical power presses have also become an area of increasing concern. In April 1997, OSHA launched a National Emphasis Program on mechanical power presses (CPL 2–1.24). This program targets industries that have high amputation rates and includes both education and enforcement efforts.

Machine guarding is to be provided to protect employees in the machine area from hazards such as those created by point of operation, nip points, rotating parts, flying chips, and sparks. The guard itself should not create an accident hazard. The point-of-operation guarding device is to be designed to prevent an operator from placing any part of their body in the danger zone during the operating cycle. Special supplemental hand tools for placing and removing material are to be designed so that operators can perform without their hands in the danger zone.

Some of the machines that usually require point-of-operation guarding are guillotine cutters, shears, alligator shears, power presses, milling machines, power saws, jointers, portable power tools, and forming rolls and calenders.

Machines designed for a fixed location are to be securely anchored to prevent walking or moving, or designed in such a manner that they will not move during normal operation.

It is also a good practice to have a training program to instruct employees on safe methods of machine operation. Supervision should ensure that employees follow safe machine operating procedures. A regular program of safety inspection of machinery and equipment should exist, which will ensure that all machinery and equipment are kept clean and properly maintained. Sufficient clearance provided around and between machines facilitates safer operations, setup, servicing, material handling, and waste removal.

As a part of safeguarding, the power shut-off switch should be within reach of the operator's position at each machine, and electric power to each machine must be locked out for maintenance, repair, or security. Noncurrent-carrying metal parts of electrically operated machines are to be bonded and grounded. Foot-operated switches need to be guarded or arranged to prevent accidental actuation by personnel or falling objects. Any manually operated valves and switches controlling the operation of equipment and machines are to be clearly identified and readily accessible. Provisions must exist to prevent machines from automatically starting when power is restored after a power failure or shutdown. All emergency stop buttons shall be colored red.

All exposed moving chains and gears must be properly guarded. All pulleys and belts that are within 7 ft of the floor or working level are to be properly guarded. Fan blades are to be protected with a guard that has openings no larger than 1/2 in., when operating within 7 ft of the floor.

Splashguards should be mounted on machines that use a coolant to prevent the coolant from reaching employees. Revolving drums, barrels, and containers are required to be guarded by an enclosure that is interlocked with the drive mechanism, so that revolution cannot occur unless the guard enclosure is in place, and therefore guarded.

Machine guarding is an integral part of machine safety and should be addressed without question based on the number of injuries that transpire.

18.16.2   MECHANICAL POWER PRESSES (29 CFR 1910.217)

Occupational Safety and Health Administration has identified power presses for special emphasis based on the number of amputations that occur on a routine basis. The employer must provide and ensure the use of point-of-operation guards or properly applied and adjusted point-of-operation devices to prevent entry of hands or fingers into the point of operation by reaching through, over, under, and around the guard on every operation performed on a mechanical power press. This requirement shall not apply when the point-of-operation opening is 1/4 in. or less. Hand and foot operations are to be provided with guards to prevent inadvertent initiation of the press.

All dies should be stamped with the tonnage and stroke requirements or be otherwise recorded and readily available to the die setter. The employer is to provide and enforce the use of safety blocks whenever dies are adjusted or repaired in the press. Brushes, swabs, or other tools are to be provided for lubrication so that employees cannot reach into the point of operation.

Presence-sensing devices may not be used to initiate the slide motion except when used in total conformance with paragraph (h) or 29 CFR 1910.217, which requires certification of the control system. Any machine using full-revolution clutches must incorporate a single-stroke mechanism. A main disconnect capable of being locked in the off position is to be provided with every power press control system.

To ensure safe operating conditions and to maintain a record of inspections and maintenance work, the employer must establish a program of regular inspections of the power presses to include the date, serial number of the equipment, as well as the signature of the inspector. OSHA requires that all point-of-operation injuries must be reported to its Office of Standards and Development or the state agency administering an occupational safety and health program within 30 days.

18.16.3   POWER TRANSMISSION EQUIPMENT GUARDING (29 CFR 1910.219)

All belts, pulleys, sprockets, chains, flywheels, shafting and shaft projections, gears, and couplings, or other rotating or reciprocating parts, or any moveable portion within 7 ft of the floor or working platform are to be effectively guarded.

All guards for inclined belts are to conform to the standards for construction of horizontal belts and shall be arranged in such a manner that a minimum clearance of 7 ft is maintained between the belt and floor at any point outside the guard. Where both runs of horizontal belts are 7 ft or less from the floor or working surface, the guard is to extend at least 15 in. above the belt or to a standard height except that where both runs of a horizontal belt are 42 in. or less form the floor, the belt must be fully enclosed by guards made of expanded metal, perforated or solid sheet metal, wire mesh on a frame of angle iron, or iron pipe securely fastened to the floor to the frame of the machine.

Flywheels located so that any part is 7 ft or less above the floor or work platform are to be guarded with an enclosure of sheet, perforated or expanded metal, or woven wire. Flywheels protruding through a working floor shall be entirely enclosed by a guardrail and toeboard.

Gears, sprocket wheels, and chains are to be enclosed unless they are higher than 7 ft above the floor or the mesh points are guarded. Couplings with bolts, nuts, or setscrews extending beyond the flange of the coupling are to be guarded by a safety sleeve.

18.17 SUMMARY

Machine guarding is visible evidence of an employer's interest in promoting safety, and it is to the employer's benefit, for unguarded machinery is a principle source of accidents that cause injuries for which compensation must be paid. A guarded machine is a safer machine; and when operators have no fear of a machine, they can better contribute their effort to production.

FIGURE 18.6 If the gate is lifted during operation, an interlocking system will de-energize it.

An effective machine guard should have certain characteristics in design and construction. Such a guard should

- Be considered a permanent part of the machine or equipment
- Afford maximum protection
- Prevent access to the danger zone during operation (see Figure 18.6)
- Be convenient: it must not interfere with efficient operation
- Be designed for the specific job and specific machine, with provisions made for oiling, inspection, adjusting, and repairing machine parts
- Be durable and constructed strongly enough to resist normal wear
- Not present a hazard in itself

Safe machine operating conditions depend on the detection of existing and potential hazards and on taking immediate actions to remedy them. Management should develop and keep its own machine-guarding checklists. Employees should follow safety precautions and be trained on the machines that they are required to operate.

19 Nonionizing Radiation

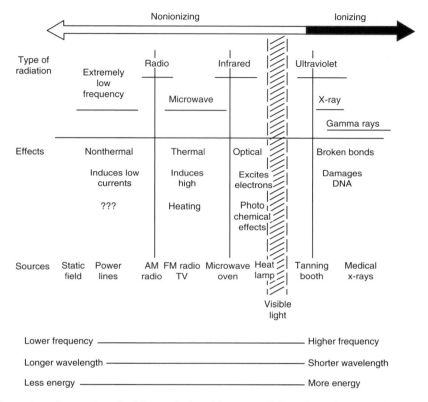

The various forms of nonionizing radiation. (Courtesy of the US Environmental Protection Agency.)

19.1 NONIONIZING RADIATION

Nonionizing radiation (NIR) is a form of electromagnetic radiation and it has varying effects on the body, depending largely on the particular wavelength of the radiation involved. In the following paragraphs, in approximate order of decreasing wavelength and increasing frequency, are some hazards associated with different regions of the nonionizing electromagnetic radiation spectrum. NIR is covered in detail by 29 CFR 1910.97, but not in detail.

Nonionizing radiation, in contrast to ionizing radiation, is electromagnetic radiation that does not have sufficient energy to remove electrons from an atom or molecules to form an ion (or charged particle) during a collision. Instead, it imparts energy to other particles, which typically results in heating. NIR includes frequencies of the electromagnetic spectrum ranging from 1 Hz up to 3×10^{10} Hz (300 GHz) and its wavelengths range from 10^9 m down to 10^{-7} m. As the frequency of a wave decreases and the wavelength increases, the energy decreases. Included within this category of nonionizing radiation are—in order of decreasing energy—the lower-frequency portion of ultraviolet (UV) radiation, visible light, infrared radiation (IR), microwave radiation, radio frequency radiation, and extremely low-frequency (ELF) radiation.

Some types of NIR have both beneficial and harmful effects on human health. For example, exposure to UV radiation facilitates the synthesis of vitamin D in the human body, and vitamin D plays an important role in intestinal calcium absorption. Lack of vitamin D can result in lead overdosing, kidney damage, and elevated serum cholesterol levels. UV is also used as an antimicrobial agent—it can penetrate food packaging and sterilize the contents—and it is used in tanning beds and salons. On the other hand, acute UV exposure can cause eye and skin damage in humans and long-term exposure has been found to cause elastosis (loss of skin elasticity) and skin cancer in humans. The sun is the major source of UV radiation.

19.1.1 Understanding and Evaluating Nonionizing Radiation Hazards

The properties and hazards of NIR can best be understood by considering the following.

Electromagnetic (EM) spectrum has three broad categories:

- Optical radiation (100 nm to 1 mm)
- Microwave radiation (300 GHz to 300 MHz)
- Radiofrequency and lower-frequency radiation (300 MHz to static fields)

Basic characteristics of optical radiation (ultraviolet/visible light/infrared):

- Possesses small wavelengths, large frequencies, and substantial energy (extreme UV approaches the photon energy of ionizing radiation)
- Optical theory can be applied to an analysis of the radiation field
- Both thermal and photochemical (biological) effects are possible from exposures (depending on wavelength)
- Exposures normally occur in the far field where the E (electric) and H (magnetic) fields are strongly coupled
- The inverse square law applies to any analysis of the radiation field.
- Only power density (S) measurements are normally considered in the hazard analysis
- The radiation interacts readily with surfaces and can easily deposit energy in human tissues

Basic characteristics of microwave radiation (300 GHz to 300 MHz):

- Possesses intermediate wavelengths (1 mm to 1 m), frequencies, and moderate photon energy
- Microwave theory can be applied to an analysis of the radiation field
- Both thermal and induced current (biological) effects are possible from exposures
- Exposures may occur in both the near and far fields
- In hazard analysis, both E (electrical field) and H (magnetic field) measurements must be considered in addition to the power density (S) measurements
- This type of radiation resonates (forms standing waves) in tissue dimensions with multiples of 1/2 wavelength (depending on the tissue orientation to the wave plane)

Basic characteristics of radiofrequency and lower-frequency (ELF, static) fields:

- Possess large wavelengths (>1 m), small frequencies, and very low energy
- Circuit theory can be applied to an analysis of the radiation field
- In general, there is poor energy deposition in human tissue but thermal and induced current (biological) effects are possible
- Exposures usually occur in the near field where the E and H fields are not coupled
- The inverse square law may not apply
- The E and H measurements must be considered separately for a hazard analysis (of RF)
- At ELF and static fields, the magnetic field dominates the hazard analysis
- This type of radiation can easily penetrate, but rarely deposit energy in tissue

19.1.2 BIOLOGICAL EFFECTS

Nonionizing radiation although perceived not to be as dangerous as ionizing radiation does have its share of adverse health effects accompanying it. The types of biological or health effects that might occur are eye injury, thermal injury, skin hyperpigmentation and erythema or carcinogenesis (see Table 19.1).

19.1.3 OTHER AREAS OF CONCERN WITH NIR

Biological effects depend on the frequency and intensity of the electromagnetic radiation. Known biological hazards are

- Static magnetic fields (0 Hz) with normal strengths can produce a variety of symptoms including nausea, metallic taste in the mouth, and vertigo.
- Electromagnetic fields can induce current flow in the body. The threshold for perception and discomfort from such current flow is frequency dependent. At frequencies from 0 Hz to 100 MHz, a serious electrical shock can

TABLE 19.1

Summary of Basic Biological Effects of Nonionizing Radiation

Photobiological Spectral Domain	Eye Effects	Skin Effects
Ultraviolet C (0.200–0.280 μm)	Photokeratitis	Erythema (sunburn)
		Skin cancer
Ultraviolet B (0.280–315 μm)	Photokeratitis	Accelerated skin aging
		Increased pigmentation
Ultraviolet A (0.315–0.400 μm)	Photochemical	Pigment darkening
	UV cataract	Skin burn
Visible (0.400–0.780 μm)	Photochemical and	Photosensitive reactions
	Thermal retinal injury	Skin burn
Infrared A (0.780–1.400 μm)	Cataract, retinal burns	Skin burn
Infrared B (1.400–3.00 μm)	Corneal burn	Skin burn
	Aqueous flare	
	IR cataract	
Infrared C (3.00–1000 μm)	Corneal burn only	Skin burn

occur if the induced current flow in the body is great enough and there is a current path from the body to the ground.

- Radiofrequency and microwave energy can cause heating equivalent to that in a microwave oven. Heating becomes significant at frequencies in the MHz and GHz range, especially between about 30 and 300 MHz. Microwave energy is known to cause cataracts and skin burns in humans.
- Static- and lower-frequency fields are known to induce malfunctions in medical electronic implants, such as pacemakers, which can malfunction at field strengths well below applicable occupational exposure limits. As a result, in many cases the only precautions required for elevated static- and lower-frequency fields are
  - To warn pacemaker users to stay out of the area
  - To keep tools and magnetizable objects out of places where elevated static magnetic fields are present
  - Tools and compressed-gas cylinders can become uncontrollable and fly like missiles toward magnets in areas where strong static fields and strong field gradients (changes in field strength over distance) exist
  - These same hazards also apply to individuals with metallic prosthetic implants (e.g., aneurysm clips, pins, or hip replacements)

Mechanical hazards depend on the field strength and the field gradient, and on how rapidly the magnetic field strength changes with distance. This means a field-strength measurement alone is insufficient to adequately identify a hazardous situation. A supplemental survey method is described in the following text. Static fields can also erase data stored on magnetic media or on the strips of credit or debit cards and badges.

Concerns have been expressed about a possible link between cancer and all forms of nonionizing radiation, especially cell phones and power lines (60 Hz). The issue has not been resolved and probably will remain an open issue for some time to come. The preponderance of available information suggests that power-line fields are either unrelated to cancer or, at worst, a very mild hazard. Similarly, there are no replicated studies linking cell phones to cancer. The exposure criteria in this document are intended to prevent known or readily predictable effects and are not based on the risk of cancer.

19.1.4 PERSONAL PROTECTION FOR NIR

Proper eye protection must be worn when working with sources such as ultraviolet, laser, and high intensity visible light, which are capable of producing eye damage. Protective clothing must be worn with sources such as ultraviolet, which are capable of producing skin burns.

19.2 VISIBLE LIGHT RADIATION

Visible radiation, which is about midway in the electromagnetic spectrum, is important because it can affect both the quality and accuracy of work. Good lighting conditions generally result in increased product quality with less spoilage and increased production. Lighting should be bright enough for easy and efficient viewing and directed so that it does not create glare. Illumination levels and brightness ratios recommended for manufacturing and service industries are found in Table 19.2.

All visible light (400–780 nm) entering the human eye is focused on the sensitive cells of the retina where human vision occurs. The retina is the part of the eye normally considered at risk from visible light hazards.

Any very bright visible light source causes a human aversion response (we either blink or turn our head away). Although we may see a retinal afterimage, which can last for several minutes, the aversion response time (about 0.25 s) will normally protect our vision. This aversion response should be trusted and obeyed. Never stare at any bright light source for an extended period. Overriding the aversion response

TABLE 19.2
Recommended Ratios of Illumination for Various Tasks

Conditions	Office	Industrial
Between tasks and adjacent darker surroundings	3:1	3:1
Between tasks and adjacent lighter surroundings	1:3	1:3
Between tasks and more remote darker surface	5:1	20:1
Between tasks and more remote lighter surfaces	20:1	1:20
Between luminaires (or windows, skylights) and surfaces adjacent to them	20:1	NC[a]
Anywhere within normal field of view	40:1	NC[a]

[a] NC means not controllable in practice.

by forcing yourself to look at a bright light source may result in permanent injury to the retina. This type of injury can occur during a single prolonged exposure. Welders and other persons working with plasma sources are especially at risk for this type of injury. Note: The aversion response cannot be relied on to protect the eye from Class 3b or 4 laser exposure.

Visible light sources that are not bright enough to cause retinal burns are not necessarily safe to view for an extended period. In fact, any sufficiently bright visible light source viewed for an extended period will eventually cause degradation of both night and color vision. Appropriate protective filters are needed for any light source that causes viewing discomfort when viewed for an extended period of time.

For these reasons, prolonged viewing of bright light sources (plasma arcs, flash lamps, etc.) should be limited by the use of appropriate filters. Traditionally, welding goggles or shields of the appropriate "shade number" have been used to provide adequate protection for limited viewing of such sources.

Blue light wavelengths (400–500 nm) present a unique hazard to the retina by causing photochemical effects similar to those found in UV radiation exposure. Visible light sources strongly weighted towards blue should be evaluated to determine whether special protective eyewear is needed.

19.3 LASERS

Lasers are a unique application of the use of a light form of nonionizing radiation for a multitude of applications. The term laser is an acronym for light amplification by stimulated emission of radiation. Light can be produced by atomic processes that generate laser light. A laser consists of an optical cavity, a pumping system, and an appropriate lasing medium. It is a form of nonionizing radiation (see Figure 19.1).

The optical cavity contains the media to be excited with mirrors to redirect the produced photons back along the same general path. The pumping system uses photons from another source such as a xenon gas flash tube (optical pumping) to transfer energy to the media, electrical discharge within the pure gas or gas mixture media (collision pumping), or relies on the binding energy released in chemical reactions to raise the media to the metastable or lasing state. The laser medium can be a solid (state), gas, dye (in liquid), or semiconductor. Lasers are commonly designated by the type of lasing material employed.

The wavelength output from a laser depends on the medium that is excited. The uses of a laser are quite varied as can be seen by the major use categories in Table 19.3.

Some of the potential hazards that can be expected are the potential hazards associated with compressed gases, cryogenic materials, toxic and carcinogenic materials, noise, explosion, radiation, electricity, and flammability.

Lasers and laser systems are assigned one of four broad Classes (I–IV) depending on their potential for causing biological damage. Class I cannot emit laser radiation at known hazardous levels (typically continuous wave: cw 0.4 W at visible wavelengths). Users of Class I laser products are generally exempt from radiation hazard controls during operation and maintenance.

Since lasers are not classified on beam access during service, most Class I industrial lasers consist of a higher-class (high power) laser enclosed in a properly

FIGURE 19.1   A laser scanning device for inventorying.

TABLE 19.3
Major Uses of Lasers

Alignment	Heat treating	Spectroscopy
Annealing	Holography	Special photography
Balancing	Information handling	Surgery
Biomedical	Laboratory instruments	Surveying
Cellular research	Marking	Typesetting
Communications	Metrology	Videodisk
Copying	Military	Weaponry
Cutting	Ophthalmology	Welding
Dental	Plate making	
Dermatology	Printing	
Diagnostics	Reading	
Displays	Scanning	
Drilling	Scanning microscopy	
Entertainment	Soldering	

interlocked and labeled protective enclosure. In some cases, the enclosure may be a room (walk-in protective housing), which requires a means to prevent operation when operators are inside it.

Class IA is a special designation that is based on a 1000 s exposure and applies only to lasers that are "not intended for viewing" such as a supermarket laser scanner. The upper power limit of Class IA is 4.0 mW. The emission from a Class IA laser is defined to mean that the emission is not to exceed the Class I limit for an emission duration of 1000 s (see Figure 19.2).

Class II A lasers are low-power visible lasers that emit above Class I levels but at a radiant power not above 1 mW. The concept is that the human aversion reaction to bright light will protect a person. Only limited controls are specified.

Class IIIA lasers are intermediate-power lasers (cw: 1–5 mW), and are only hazardous for intrabeam viewing. Some limited controls are usually recommended.

Class IIIB lasers are moderate-power lasers (cw: 5–500 mW or the diffuse reflection limit, whichever is lower). In general, Class IIIB lasers will not be a fire hazard, nor are they generally capable of producing a hazardous diffuse reflection. Specific controls are recommended.

Class IV are high-power lasers (cw: 500 mW or the diffuse reflection limit), which are hazardous to view under any condition (directly or diffusely scattered), are a potential fire hazard, and a skin hazard. Significant controls are required in Class IV laser facilities.

Accident data on laser usage have shown that Class I, Class II, Class IA, and Class IIIA lasers are normally not considered hazardous from a radiation standpoint unless illogically used. Direct exposure on the eye by a beam of laser light should always be avoided with any laser, no matter how low the power.

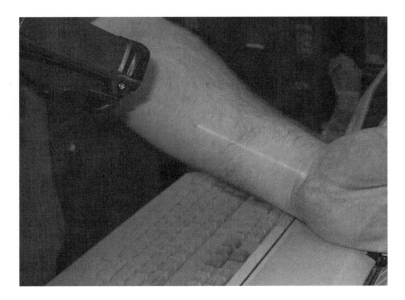

FIGURE 19.2   Low-level lasers are not harmful to the skin as seen in this picture.

19.3.1   LASER HEALTH ISSUES

Laser sources exist with outputs from the infrared to the UV. The eye is generally the most sensitive organ and the one that must be protected. However, ultraviolet and high-power lasers may also be hazardous to the skin. In addition, high-power lasers may become fire hazards. The eye is transparent to radiation from 400 to 1400 nm and is termed the ocular hazard region since the radiation incident on the cornea is focused on the retina. The classes of lasers and their potential to create health-related problems are as follows:

- Class I denotes exempt lasers or laser systems that cannot under normal operating conditions produce a hazard. This includes lasers that are completely enclosed and interlocked or so low in power than no hazard exists. No warning label is required.
- Class II denotes low-power visible lasers or laser systems, which because of the normal human aversion responses (<0.25 s) do not normally present a hazard but may present some potential for hazard if viewed directly for extended periods of time (like many conventional light sources). If manufactured after 1976 the laser will normally have a sign attached stating "Caution—Laser Radiation—Do not stare into beam." This label should be clearly visible (see Figure 19.3).
- Class IIIA denotes lasers or laser systems that normally would not produce a hazard if viewed for only a momentary period with the unaided eye. They may present a hazard if viewed using collecting optics. A sign with the words "Caution—Laser Radiation—Do not stare into beam or view directly with optical instruments" should be clearly visible. Eye protection should be worn.

FIGURE 19.3   Symbol used to indicate that a laser exists or presents a hazard. (Courtesy of the Department of Energy.)

- Class IIIB denotes lasers or laser systems that can produce a hazard if viewed directly. This includes viewing of reflections from various smooth reflective surfaces. Diffuse reflections are not hazardous. A sign that reads "Danger—Laser Radiation—Avoid Direct Exposure to Beam" should be clearly visible. Eye protection must be worn. Access to the area must be controlled while the laser is used. Hazards can be reduced by removing highly reflective surfaces, enclosing the beam path, operating in a closed, shielded room, and permitting only trained personnel to operate the laser. A warning light or buzzer should indicate when the laser is in operation. Written operating procedures must be visibly posted and approved by the Health and Safety Branch.

- Class IV denotes lasers or laser systems that can produce a hazard not only from direct sources or reflections but also from diffuse reflections. In addition, such lasers may produce fire hazards and skin hazards. A sign that reads "Danger—Laser Radiation—Avoid eye or skin exposure to direct or scattered radiation" should be clearly visible. Lasers in this class are capable of producing serious eye injury. These lasers should be enclosed if possible and operated as Class I lasers. If this in not possible, eye protection is required and access must be controlled when the laser is in operation. The skin should also be protected. The laser should be operated remotely if possible. Highly reflective surfaces must be removed or painted. Only qualified and trained personnel should operate the laser.

19.3.2 LASER PROTECTIVE EQUIPMENT

Protective equipment (eye protection, temporary barriers, clothing or gloves, respirators, etc.) would be recommended; for example, only if the hazard analysis indicates a need or if the SOP requires periods of beam access such as during setup or infrequent maintenance activities. Temporary protective measures for service can be handled in a manner similar to the service of any embedded Class IV laser.

When the entire beam path from a Class IIIB or Class IV laser is not sufficiently enclosed or baffled to ensure that radiation exposures will not exceed the maximum permissible exposure (MPE), a laser-controlled area is required. During periods of service, a controlled area may be established on a temporary basis. The controls required for both Class IIIB and Class IV installations are as follows:

- Posting with appropriate laser warning signs
- Operation by qualified and authorized personnel

Control measures recommended for Class IIIB but required for Class IV lasers are as follows:

- Workers are supervised directly by an individual knowledgeable in laser safety.
- Entry of any noninvolved personnel requires approval.

- A beam stop of an appropriate material must be used to terminate all potentially hazardous beams.
- Diffusely reflecting materials must be used near the beam, where appropriate.
- Appropriate laser protective eyewear must be provided to all personnel within the laser-controlled area.
- The beam path of the laser must be located and secured above or below eye level for any standing or seated position in the facility.
- All windows, doorways, open portals, etc., of an enclosed facility should be covered or restricted to reduce any escaping laser beams below appropriate ocular MPE level.
- Storage or disabling of lasers should be required when not in use.

In addition, there are to be specific controls required at the entryway to a Class IV laser-controlled area. These can be summarized as follows:

- All personnel entering a Class IV area shall be adequately trained and provided proper laser protective eyewear (see Table 19.4)
- All personnel shall follow all applicable administrative and procedural controls
- All Class IV area and entryway controls shall allow rapid entrance and exit under all conditions
- The controlled area shall have a clearly marked "Panic Button" (nonlockable disconnect switch), which allows rapid deactivation of the laser

Class IV areas also require some form of area and entryway controls such as

- Nondefeatable entryway controls
- Defeatable entryway controls
- Procedural entryway controls
- Entryway warning systems

Administrative and procedural controls are standard operating procedures, alignment procedures, limitations on spectators, and protective equipment. Protective equip-

TABLE 19.4
Chart for the Proper Selection of Protective Laser Goggles

Intensity, CW Maximum Power Density (W/cm^3)	Optical Density (OD)	Attenuation Factor
10^{-2}	5	10^5
10^{-1}	6	10^6
1.0	7	10^7
10^{+1}	8	10^8

Source: Courtesy of the Department of Energy.
Selecting laser safety glass.

ment for laser safety generally means eye protection in the form of goggles or spectacles, clothing, and barriers and other devices designed for laser protection.

Engineering controls are normally designed and built into the laser equipment to provide for safety. In most instances, these will be included on the equipment (provided by the laser manufacturer). Some of these items are protective housing, master switch control, optical viewing system safety, beam stop or attenuated laser activation warning system, service access panels, protective housing interlock requirements, and a remote interlock connector.

Since lasers are highly collimated (has a small divergence angle), they can have a large energy density in a narrow beam. Direct viewing of the laser source or its reflections should be avoided.

The work area should contain no reflective surfaces (such as mirrors or highly polished furniture) for even a reflected laser beam can be hazardous. Suitable shielding to contain the laser beam should be provided. OSHA covers protection against laser hazards in its construction regulations.

The eye is the organ that is most vulnerable to injury induced by laser energy. The reason for this is the ability of the cornea and lens to focus the parallel laser beam on a small spot on the retina. The fact that infrared radiation of certain lasers may not be visible to the naked eye contributes to the potential hazard. Lasers generating in the ultraviolet range of the electromagnetic spectrum can produce corneal burns rather than retinal damage, because of the way the eye handles ultraviolet light. Other factors that have a bearing on the degree of eye injury induced by laser light are

- Pupil size—the smaller the pupil diameter, the less the amount of laser energy permitted to the retina
- Ability of the cornea and lens to focus the incident light on the retina
- Distance from the source of energy to the retina
- Energy and wavelength of the laser
- Pigmentation of the eye of the subject
- Place on the retina where the light is focused
- Divergence of the laser light
- Presence of scattering media in the light path

As a further assurance that safety is practiced in the use of lasers, use the following audit instrument to check compliance with laser safety procedures (see Figure 19.4).

19.3.3 LASER RADIATION

Laser sources exist with outputs from the infrared to the UV. The eye is generally the most sensitive organ and the one that must be protected. However, ultraviolet and high-power lasers may also be hazardous to the skin. In addition, high-powered lasers may become fire hazards. The eye is transparent to radiation from 400 to 1400 nm and is termed the ocular hazard region since the radiation incident on the cornea is focused on the retina.

Laser room monthly safety audit

- Are warning labels visible on the housings of all class III and IV lasers?
- Are warning labels visible on the housings of all class III and IV lasers?
- Are class III and IV lasers enclosed?
- Do class III and IV laser enclosures have appropriate signs posted?
- Do class III and IV lasers have audible or visible alarms over the posted signs?
- Is eye protection worn during laser beam operation?
- Are employees working with lasers cleared through the medical department?
- Is there documentation that employees working with lasers are properly trained?
- Are there SOP's available to the employees working with lasers?
- Is there a specific SOP for alignment of the laser beam?
- Is access to the area limited to trained employees?

FIGURE 19.4   Laser room monthly safety audit.

19.4   NONCOHERENT LIGHT SOURCE SAFETY

Many devices (or sources) produce either broadband or discrete wavelength radiation between 100 nm and 1 mm. Under certain conditions, these sources may present a health hazard. Factors affecting the potential hazard include the specific wavelength (s) produced, the irradiance value, the source dimensions, and whether the radiation can access the eye or skin. Sources include (but are not limited to) the following:

- Lamps (filament, discharge, fluorescent, arc, solid state, etc.)
- Plasma sources (welding devices, surface deposition, etc.)
- Heat sources (furnaces, molten glass, open flames, etc.)

Whenever practical, sources of noncoherent light not intended for illumination purposes should be shielded to prevent exposure to the eye or skin. For sources intended to produce exposure (lamps), any unneeded wavelengths (example: ultraviolet emitted from mercury vapor lamps) should be removed with appropriate filtration.

19.5   ULTRAVIOLET RADIATION

Ultraviolet (UV) radiation is defined as having a wavelength between 10 and 400 nm. Specific wavelength "bands" are defined as follows:

- Physical definition
  - Extreme UV (10–100 nm)
  - Vacuum UV (10–200 nm)
  - Far UV (200–300 nm)
  - Near UV (300–400 nm)

- Photobiologic definition
  - UV-C (100–280 nm)
  - UV-B (280–315 nm)
  - UV-A (315–400 nm)

19.5.1 ULTRAVIOLET SKIN HAZARDS

Ultraviolet radiation is a known carcinogen for human skin. In addition to cancer induction, erythema (sunburn) and skin aging are also known hazards of ultraviolet skin exposure. Because the biological effects are dependent on the time of exposure, the specific UV wavelength, and the susceptibility of the individual exposed, it is considered prudent to prevent any unnecessary skin exposure to UV sources. Elimination of unnecessary skin exposure is reinforced by the fact that most individuals receive substantial UV exposure from the sun during normal outdoor activities over a human lifetime.

UV radiation causes biological effects primarily through photochemical interactions. The UV wavelengths that produce the greatest biological effects fall in the UV-B, but other wavelengths can also be hazardous. Skin protection is not difficult in theory, as most clothing tends to absorb some of the UV wavelengths. However, in practice, it is often difficult to properly motivate individuals to use appropriate skin protection unless they know they are receiving an erythema (sunburn) dose (see Figure 19.5).

Protection of the skin from UV radiation hazards is best achieved though the use of clothing, gloves, and face shields. The use of UV skin blocks (creams or lotions) is considered inadequate for protection against the high irradiance of man-made UV radiation sources.

FIGURE 19.5 Most of the medical community consider tanning beds that produce ultraviolet radiation as dangerous to skin health.

19.5.2 ULTRAVIOLET EYE HAZARDS

Various components of the human eye are susceptible to damage from extended exposure to direct/reflected UV from photochemical effects. The UV wavelength is the determining factor as to which part(s) of the eye may absorb the radiation and suffer biological effects.

The cornea is like the skin in that it can be sunburned by exposure to too much UV radiation. This is called keratoconjunctivitis (snow blindness or welder's flash), a condition where the corneal (epithelial) cells are damaged or destroyed. This condition usually does not present until 6–12 h following the UV exposure. Although very painful (often described as having sand in the eyes), this condition is usually temporary (a few days) because the corneal cells grow back. In very severe cases, the cornea may become clouded and corneal transplants may be needed to restore vision. Exposure to UV-C and UV-B present the greatest risk to the cornea.

The lens of the eye are unique in that they are formed early in human development and are not regenerated should they become damaged. For normal vision, it is essential that the lens remain clear and transparent. Unfortunately, UV-A exposure is suspected to be a cause of cataracts (clouding of the lens).

To protect the human eye from exposure to UV wavelengths, all that is usually needed is a pair of polycarbonate safety glasses or a polycarbonate face shield. This protective eyewear should be worn whenever there is a potential for ongoing UV radiation exposure.

19.6 INFRARED RADIATION

Infrared (or heat) radiation is defined as that with a wavelength between 780 nm and 1 mm. Specific biological effectiveness bands have been defined as follows:

- IR-A (near IR) (780–1400 nm)
- IR-B (mid IR) (1400–3000 nm)
- IR-C (far IR) (3000 nm to 1 mm)

19.6.1 INFRARED RADIATION HAZARDS

Infrared radiation in the IR-A that enters the human eye reaches (and can be focused upon) the sensitive cells of the retina. For high irradiance sources in the IR-A, the retina is the part of the eye that is at risk. For sources in the IR-B and IR-C, both the skin and the cornea may be at risk from flashburns. In addition, the heat deposited in the cornea may be conducted to the lens of the eye. This heating of the lens is believed to be the cause of the so-called "glass blowers" cataracts because the heat transfer may cause clouding of the lens.

- Retinal IR hazards (780–1400 nm)—possible retinal lesions from acute high irradiance exposures to small dimension sources
- Lens IR hazards (1400–1900 nm)—possible cataract induction from chronic lower-irradiance exposures

- Corneal IR hazards (1900 nm to 1 mm)—possible flashburns from acute high irradiance exposures
- Skin IR hazards (1400 nm to 1 mm)—possible flashburns from acute high irradiance exposures

The potential hazard is a function of the following:

- Exposure time (chronic or acute)
- Irradiance value (a function of both the image size and the source power)
- Environment (conditions of exposure)

Evaluation of IR hazards can be difficult, but reduction of eye exposure is relatively easy through the use of appropriate eye protection. As with visible light sources, the viewing of high irradiance IR sources (plasma arcs, flash lamps, etc.) should be limited by the use of appropriate filters. Traditionally, welding goggles or shields of the appropriate "shade number" have been used to provide adequate protection for limited viewing of such sources. Specialized glassblowers goggles may be needed to protect against chronic exposures.

19.7 MICROWAVE AND RADIOFREQUENCY RADIATION SAFETY

19.7.1 MICROWAVE/RF RADIATION SOURCES

The workplace contains many potential sources of microwave/RF radiation exposure. Some of these sources (primarily antennas) are designed to emit microwave/RF radiation into the environment. Other types of sources (coaxial cables, waveguides, transmission generators, heaters, and ovens) are designed to produce or safely contain the microwave/RF radiation, but may present a hazard should they leak for some reason. A third type of source (primarily power supplies) may create microwave/RF radiation as a by-product of their operation.

19.7.2 FACTORS AFFECTING EXPOSURE TO MICROWAVE/RF RADIATION

The hazards from exposure to microwave/RF radiation are related to the following parameters:

- Frequency of the source
- Power density at the point of exposure
- Accessibility to the radiation field
- Whether the exposure occurs in the near or far field
- Orientation of the human body to the radiation field

This combination of factors is used in both evaluating and mitigating the hazard.

19.7.3 POTENTIAL BIOLOGICAL EFFECTS OF EXPOSURE TO MICROWAVE/RF RADIATION

In general, most biological effects of exposure to microwave/RF radiation are related to the direct heating of tissues (thermal effects) or the flow of current through tissues (induced current effects). Nonthermal effects resulting in carcinogenesis, teratogenesis, etc., have been demonstrated in animals but have not been proven by epidemiological studies on humans. The following biological effects have been demonstrated in humans:

- Cataract formation (from eye exposure)
- RF (induction) burns
- Burns from contact with metal implants, spectacles, etc.

19.7.4 STANDARDS FOR MICROWAVE/RF RADIATION EXPOSURE PROTECTION

A large number of standards have been developed for use in protecting individuals against overexposure to microwave/RF radiation. These standards often address only specific frequency bands or exposure conditions. There is not one definitive standard. Because of the difficulties of performing actual microwave/RF radiation surveys (near-field measurements, cost of equipment, etc.), it is often necessary to use calculations or computer models to replace actual measurements in evaluating the hazard and set limits of exposure.

19.7.5 ANTENNAS AND ANTENNA ARRAYS

Operation of radio, television, microwave, and other related communication systems using electromagnetic radiation and carrier-current systems fall under FCC regulations and the appropriate exposure model to use in a hazard assessment process.

19.7.6 WIRELESS LOCAL AREA NETWORKS

Radio frequency-based wireless local area networks (WLAN) are becoming available and used widely in the workplace. Wireless LAN systems (indoor and outdoor) are very safe when properly installed and used. WLAN systems operate on extremely low power (less than that of a cell phone). It is important that only approved equipment be used to build an operating WLAN. The placement of base station antennas should be high on a wall or on the ceiling. This not only increases the useful range of the system, but also allows for a separation distance of 50 cm, which is sufficient of safe operation. In general, persons should avoid direct contact with antennas attached to computer cards. A separation distance of 10 cm is sufficient for safe operation.

19.7.7 OTHER POTENTIAL MICROWAVE/RF RADIATION SOURCES (LEAKAGE SOURCES)

For waveguides, coaxial cables, generators, sealers, and ovens, probably the most important aspect of controlling microwave/RF radiation hazards is a careful physical inspection of the source. Leaking sources will normally show misalignment of doors or plates, missing bolts, or physical damage to plane surfaces. Sources, which are suspected of leaking, should be repaired and then surveyed with appropriate instrumentation to verify they are no longer leaking.

19.7.8 MICROWAVE OVENS

A large number of microwave ovens are used and their presence is found in many workplaces today (see Figure 19.6). Specific guidance on microwave oven safety is as follows:

- Do not operate the oven if it is damaged or does not operate properly. It is imperative that the oven door seals properly and that there is no damage to the door seal, hinges, latches, or oven surfaces.
- Clean ovens used for food preparation on a regular basis to prevent biological contamination, fire potential, and door seal damage.
- Do not use ovens used for laboratory applications for food preparation. Conversely, food preparation ovens should never be used for other applications.
- Do not use aluminum foil or any metal containers, metal utensils, metal objects, or objects with metal or foil trim in the oven. Such items can cause arcing, damaging the oven and creating a fire or burn hazard. A classic item, which is often overlooked, is the metal handle on the paper Chinese food box.
- Do not heat objects that are sealed as they may explode, damaging the oven and blowing off the door.

FIGURE 19.6 Example of a common microwave oven.

- Never heat any flammable or combustible liquid in the oven. A fire or explosion may result.
- Be careful when removing containers from the microwave. Containers or their contents may be very hot, resulting in burns or spills of hot materials.
- If a fire should start inside the oven, leave the door closed, disconnect the power cord, and call the fire department at 911.
- Never make adjustments to or tamper with any component of the oven. Do not try to perform repairs on your own. The oven operates on high voltage and amperage that can be lethal if improperly handled.

Generally, commercially available microwave ovens are very safe and reliable, regardless of the manufacturer. All ovens produced for sale in the United States must meet a strict FDA/CDRH product performance requirement that limits their microwave leakage during service to <5 mW/cm at 5 cm from any oven surface.

It is very unusual for a commercial microwave oven to leak, but misuse, damage, and interlock failures have caused ovens to leak. Any microwave oven suspected of leaking should be tagged with a warning sign and removed from service.

19.7.9 POWER SUPPLIES

Many high-voltage power supplies operate in the microwave or radiofrequency regions. If damaged, or not properly shielded, these sources can leak, producing unintended microwave/RF radiation exposure. Most of the time, the leakage from these sources is minimal and does not present a hazard. However, if you have an indication of microwave/RF radiation leakage (RF interference with other equipment or documentation warning of interference), remove from service and contact appropriate maintenance or installation personnel.

19.8 EXTREMELY LOW-FREQUENCY RADIATION SAFETY

19.8.1 ELF RADIATION

Whenever a current passes through a wire, a magnetic field is produced. Because electric power generation in the United States changes polarity at 60 Hz (cycles per second), the magnetic fields generated also alternate at 60 Hz. Since about 1980, these 60 Hz magnetic fields (and their frequency harmonics) have been suspected of causing various types of negative health effects. These magnetic fields are commonly called ELF fields.

19.8.2 ELF POTENTIAL HUMAN HEALTH HAZARDS

No one has a definitive answer regarding whether a potential health hazard exists with ELF exposure.

Although some health effects have been statistically related to ELF exposure, these effects are poorly understood and may exist only as statistical or scientific errors. Some research studies, which originally indicated ELF health effects, could not be duplicated.

Much of the data supporting effects are from epidemiological studies, and the effects found were slightly outside the boundaries of statistical error. What can be determined from this information is that any real effect (and the corresponding hazard) must be relatively small.

19.8.3 PROTECTION STANDARDS FOR ELF EXPOSURE

In the absence of conclusive data, the International Radiation Protection Association/International Nonionizing Radiation Committee (IRPA/INIRC) have produced an interim exposure guideline. The U.S. government has no regulations on exposure to ELFs, and has adapted the IRPA/INIRC guidelines to address employee concerns.

19.8.4 NORMAL ELF FIELD

Since even the wiring for electric lights will generate ELF magnetic fields, these fields are present in virtually every room of every building in every workplace. The ELF field intensity is a function of the amperage passing through the wiring. In general, transformers and large motors will produce the most intense fields. Mechanical spaces and machine shops normally have the most intense fields and these may (rarely) exceed the nonoccupational exposure limit. Laboratories and offices usually do not have intense fields and a reasonable average value for these areas has been measured at about 3–5 mG (milliGauss). Therefore, normally, the ELF fields employees are exposed to are less than 1% of the nonoccupational exposure limit.

19.9 STATIC MAGNETIC FIELD SAFETY

Many sources (devices) produce static magnetic fields. Static magnetic fields result from either fixed magnets or the magnetic flux resulting from the flow of direct current (DC). Sources producing these fields include (but are not limited to) the following:

- Nuclear magnetic resonance (NMR) imaging and spectroscopy devices
- Electron paramagnetic resonance (EPR, ESR, EMR) devices
- Helmholtz coils, solenoids, DC motors, etc.

19.9.1 FACTORS AFFECTING STATIC MAGNETIC FIELD HAZARDS

Under certain conditions, sources of static magnetic fields can present health hazards. Factors affecting the potential hazards include

- Magnetic flux intensity associated with the source
- Design of the magnetic field source
- Accessibility of the magnetic field
- Equipment/hazardous materials associated with the magnetic field source

Sources of large static magnetic fields may require appropriate controls to mitigate potential hazards. For sources intended to produce human exposure to the magnetic field (such as MRI devices), it is critical that safety precautions cover not only the user of the device, but also the research subject.

19.9.2 BIOLOGICAL EFFECTS OF EXPOSURE TO STATIC MAGNETIC FIELDS

There are no known adverse biological effects for flux densities within the ACGIH (American Conference of Governmental Industrial Hygienists) exposure limits. Implanted medical devices present a potential hazard to individuals exposed to fields above the ACGIH limits (see Section 19.9.4).

19.9.3 KINETIC ENERGY HAZARDS

Due to the large fields associated with NMR magnets, ferrous objects can be accelerated toward the magnet with sufficient energy to seriously injure persons and damage the magnet. As a precaution, even small metal objects (screws, tools, razor blades, paper clips, etc.) should be kept at least 1.5 m from the magnet (or anywhere the field exceeds 30 G). Large ferrous objects (equipment racks, tool dollies, compressed-gas cylinders, etc.) should be moved with care whenever the field approaches 300 G. In many recorded instances, large objects have been drawn toward and even into the bore of the magnet.

19.9.4 STANDARDS FOR EXPOSURE TO STATIC MAGNETIC FIELDS

The ACGIH and International Council on Non-Ionizing Radiation Protection (ICNIRP) have set guidelines for continuous exposure to static electromagnetic fields as indicated in the table below. (Note: 1 Gauss (G) = 0.1 millitesla (mT)):

- 5 G—highest allowed field for implanted cardiac pacemakers
- 10 G—watches, credit cards, magnetic tape, computer disks damaged
- 30 G—small ferrous objects present a kinetic energy hazard
- 600 G—allowed TWA for routine exposure (whole body)
- 6000 G—allowed TWA for routine exposure (extremities)
- 20,000 G—(2T) whole body ceiling limit (no exposure allowed above this limit)
- 50,000 G—(5T) extremity ceiling limit (no exposure allowed above this limit)

Note: TWA exposure time is normally only a concern for extremely high field exposures to the whole body.

19.9.5 MAGNETIC FIELD MEASUREMENTS

Nuclear magnetic resonance magnets commonly produce core fields from 0.2 to 20 T. These fields decrease in intensity as the distance from the core increases. A field strength map of the area surrounding the magnet should be developed and posted for use by staff.

19.9.6 POSTING OF MAGNETIC FIELD HAZARDS

All access points to rooms containing magnetic fields in excess of 5 G shall be marked with magnetic field hazard signs. The 5 G threshold line shall be clearly identified with floor tape or equivalent markings. The location of the 5 G line will vary with the operating frequency and resulting magnetic fields. As an example, one vendor indicates the following values for their NMR spectroscopy equipment:

- Operating frequency of 200 MHz—5 G threshold line at 1.3 m
- Operating frequency of 500 MHz—5 G threshold line at 3.5 m
- Operating frequency of 800 MHz—5 G threshold line at 6.0 m

19.9.7 ACCESS RESTRICTIONS

Persons with cardiac pacemakers or other implanted medical devices shall be restricted to areas outside the 5 G threshold line. Security (locked doors) and proper door markings shall be maintained to prevent unauthorized access to the magnet area.

19.9.8 USE OF NIR HAZARD SIGNS AND WARNING LABELS

Depending on the accessibility and level of NIR hazards, it may be necessary to mark rooms or other areas with appropriate warning signs (static magnetic fields, UV light, etc.).Warning labels should be placed on equipment to indicate the presence of specific NIR hazards (UV light, RF fields, etc.).

19.10 SUMMARY

The intent of this chapter is to ensure that nonionizing radiation sources are identified and posted, users are properly trained to work with and around these sources, and measurements are taken to evaluate worker exposures. Controls to mitigate hazards are implemented when surveys indicate that exposures can exceed acceptable limits.

Nonionizing radiation, although perceived not to be as dangerous as ionizing radiation, does have its share of adverse health effects accompanying it. We take advantage of the properties of nonionizing radiation for common tasks:

- Microwave radiation—telecommunications and heating food
- Infrared radiation—infrared lamps to keep food warm in restaurants
- Ultraviolet radiation—disinfecting and tanning beds
- Radio waves—broadcasting
- Lasers—medicine and industry with multiple applications (see Figure 19.7)

Nonionizing radiation ranges from ELF radiation, shown on the far left through the audible, microwave, and visible portions of the spectrum into the ultraviolet range.

Lower-frequency nonionizing radiation, such as microwave, radio frequency, and ELF radiation, have many beneficial uses—such as tracking radars, weather radars, microwave ovens, radio navigation, satellite communication, broadcast radio and television, and a variety of other communications devices including two-way

FIGURE 19.7   Example of a laser level being used.

radios and cellular phones. Acute effects from direct exposure to high levels of this type of radiation can include severe burns, electric shocks, and even death. The human health effects of chronic exposure to these types of radiation are less clear. The higher frequencies in this range, such as microwaves, may cause adverse heating effects. The majority of studies, which have focused on exposure to ELF radiation,

---

NIR emergency response procedure

In the event of an accident involving an NIR source, immediately do the following:

- Using caution first shut down the source of the NIR radiation and then lock out/tag out the power supply.
- Provide for the safety of personnel (first aid, evacuation, etc.) as needed. Note: If an eye injury is suspected, have the injured person keep their head upright and still to restrict any bleeding in the eye. A physician should evaluate eye injuries as soon as possible.
- Obtain medical assistance for anyone who may be injured. Contact the appropriate medical personnel.
- If there is a fire, leave the area, pull the fire alarm, and contact the fire department by calling 911. Do not fight the fire unless it is very small and you have been trained in fire fighting techniques.
- Inform the appropriate organizational personnel regarding the nature of the emergency.
- After an accident, do not resume use of the NIR source until after an investigation of the proper releases or resume operation approvals are obtained.

FIGURE 19.8   NR emergency response procedure.

have examined cancer, adverse reproductive outcomes, neurodegenerative diseases, and cardiac abnormalities. However, difficulties in conducting these studies, including defining and measuring the biologically relevant exposure and variation in subjects' responses, have left substantial uncertainty. A recommended response to an NIR emergency is found in Figure 19.8.

The most effective means of preventing exposure to most types of nonionizing radiation is to maintain a safe distance from the source. Other means of preventing or limiting exposure, such as using shields, is far more difficult and costly, particularly as the size of the source increases. Unlike ionizing radiation, nonionizing radiation penetrates through most materials relatively unchanged. Special metallic grids can be designed to exclude radiation of particular frequencies, but this is not currently practical, physically, for handheld devices such as cellular phones and two-way radios, and is very onerous for large-scale implementation, such as around houses or buildings.

20 Visitor/Client Safety and Health

Providing a safe and health environment for all those who enter a workplace open to people other than workers is an integral part of conducting business.

All businesses have a number of individuals who enter the workplace for various reasons. They may be clients, patients, sales representatives, family members, groups, students, vendors, and other business representatives. Employers assume a certain amount of liability for individuals who enter their work area or facility. Employers are also responsible for the safety and health of these visitors or clients.

Thus, employers need to institute a policy and plan to handle those that enter their place of business. If an employer has adequately protected his/her own workforce, the foundation exists to protect visitors and others who enter the workplace. However, the procedures to protect outsiders who enter an employer's workplace need to be formalized.

The first step is to develop a policy that addresses the following elements:

- Responsibility of visitors
- Rules and procedures visitor must follow
- Statement of employer policy on violence and weapons
- Acknowledgement of receipt of guidelines for visitors
- Right of the employer to remove the visitor from the workplace for failure to abide by the policy and rules

Each business entity should develop a one page written document that addresses unique hazards of the business, special safety and health issues, guidance for emergencies, and directions for the facility layout. Employers may include other details specific to their business.

20.1 WALK THROUGH

In order to ensure that the workplace is as visitor friendly as possible, a walk through of the visitor's journey through the facility from their entrance to their exit should be performed.

The following steps are a suggested procedure and should be tailored or personalized by each individual business. The process should begin in the following manner:

- Is the parking lot for visitors well marked and reasonably accessible?
- Is the parking area secure or guarded?
- Are there clear directions to the facility entrance?
- Is the visitor an expected guest (by appointment)?
- Is there a reception area separate from the employee work areas?
- Is there a guard or receptionist on the front desk?
- Is there a sign-in procedure that denotes date, time-in, and time-out?
- Are all visitors issued a visitor's badge?
- Is the visitor given a visitor safety and health sheet to read and sign as to receiving it?
- If visitors need special training or indoctrination, is this accomplished and documented?
- Are visitors met by an employee who is to escort the visitor while at the facility?
- Has the visitor been issued any needed personal protective equipment?
- Is the visitor wearing the issued personal protective equipment?
- Upon completion of the visit, is the visitor logged out and the badge collected as he/she departs the facility?

20.2 HAZARD IDENTIFICATION

Employers need to safeguard their visitors while minimizing their liability. A well-thought-out process for handling visitors has the effect of protecting employees, assets, company's property, trade secrets, capital investments, proprietary information, and intellectual property. However, since potential hazards that may be unknown to visitors could be present special care must be taken to address these. Some of the potential hazards that face many businesses are

- Weather
- Traffic
- Parking

- Hazardous processes
- Emergencies
- Chemicals
- Radiation
- Excess noise
- Fire
- Violence/weapons

20.2.1 WEATHER

Adverse weather can always be a factor affecting visitors. Special effort needs to be taken to remove any ice or snow from walkways and steps both for visitors and for employees' safety. Special procedures should exist to shelter or evacuate visitors and workers during earthquakes, tornados, hurricanes, or other severe weather conditions.

20.2.2 TRAFFIC

During times of high traffic volume within or around the business facility due to employees and visitors coming and going or due to poor traffic patterns or control lights, a security guard or assigned personnel might be needed to direct traffic flow.

20.2.3 PARKING

At times parking facilities do not have adequate lighting or are isolated from view, which can cause concern for the well-being of both visitors and employees. If this is the case and no roving security guards or panic buttons are available, then both employees and visitors may need an escort to their vehicle to ensure safety (see Figure 20.1).

20.2.4 HAZARDOUS PROCESSES

Some companies may have hazardous processes ongoing during the normal working day. In most cases, it is best to limit access of visitors to manufacturing processes, laboratories, workshops, or other hazardous areas. These areas are best off limit to visitors or restricted to visitors with special reasons to be in these areas and those who have been provided with training or have been briefed on the dangers and provided appropriate personal protective equipment. Keep visitors away from dangerous processes and areas if possible.

20.2.5 EMERGENCIES

The employer must brief visitors on emergency procedures and indicate that they are to follow these procedures as directed by supervisors or designated employees. Even if a drill is conducted during a visit, the visitor is to follow the procedures as though it was an actual emergency. There should be a mechanism in place to account for all visitors during an emergency or a drill (see Figure 20.2).

FIGURE 20.1   Modern parking facilities are more secure and less hazardous in most cases.

The employer should have adequate fully supplied first-aid kits available. Automatic defibrillators are found in more and more locations at present. Employers should have all security staff, supervisors, and other staff trained in first aid, use of automatic defibrillators, and cardiopulmonary resuscitation (CPR).

20.2.6 CHEMICALS

Many companies use hazardous chemicals as part of their processes or procedures. Special care must be taken to not allow visitors in areas where they might be exposed to these chemicals, which might cause sickness or cancer. Visitors should be informed if they will have the potential to be exposed to these chemicals used by the company.

20.2.7 RADIATION

If the business uses radioactive sources or generates radiation in some form, visitors should be aware of the possibility of exposure. Unless the visitors have been specifically trained in radiation safety and health, they should not be allowed into

FIGURE 20.2   Emergency and fire warning systems are needed to protect visitors and clients.

areas where the possibility of exposure exists unless the radiation is used on them as part of a medical diagnostic test or treatment as a patient.

20.2.8   Excessive Noise

If visitors are taken to areas where high levels of noise exist, they should be given ear protection and shown how to use it and must wear it during the visit to the excessive noise area. Failure to wear the proper protective equipment can result in damage to an individuals hearing.

20.2.9   Fire

Since fire is one of the most frequent causes of emergency action and evacuation, visitors should be informed of their role in case of a fire. For liability reasons and safety, visitors should evacuate and leave any firefighting to trained personnel. Visitors should evacuate when any fire drill occurs.

20.2.10   Violence/Weapons

Employers should have a zero tolerance policy regarding acts of violence for visitors and employees, and no weapons should be permitted at any time for both visitors and employees. This policy is a good deterrent for any type of violent acts and is a critical protection for employees from within and from outside sources such as visitors (see Figure 20.3).

FIGURE 20.3   Uniformed security is a excellent deterrent against violence and crime.

20.2.11   OTHER HAZARDS

Any other hazard identified by the employer should be addressed in a similar fashion with the intent of protecting the workforce and any visitors or guests.

20.3   OTHER CONTROLS

There are other controls that are not hazard specific, which will put a company in good standing and offer a positive image to visitors to sense the company's commitment to the well being of visitors and employees.

20.3.1   HOUSEKEEPING

Workplaces free from litter and waste with uncluttered aisles and walkways are a visible commitment of the company to safety and health (see Figure 20.4).

20.3.2   SIGNAGE

Well-placed direction signs make for easier access and egress within the facility and during evacuations. In addition, the use of appropriate warning signs is necessary to prevent visitors from coming into contact or entering a restricted area where they might come into harm's way. Signs, decals, or company logo should be placed on glass walls and doors or plants or furniture should be placed in front of them so that no one will attempt to walk through them.

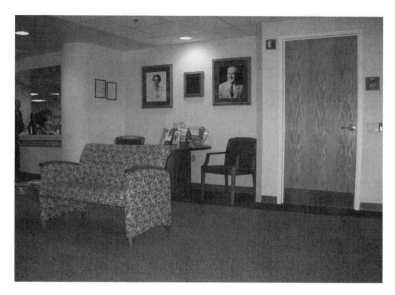

FIGURE 20.4   Good housekeeping is a positive safety image for waiting areas/lobbies.

20.3.3   HANDRAILS

The presence of maintained handrails on steps, ramps, and areas where fall through could occur are vital to safety.

20.3.4   MAINTENANCE

Preventive maintenance and continuous maintenance are an integral part of preventing hazards from occurring and thus an accident-prevention technique that is well worth the effort. This is especially essential for elevators, escalators, and automatically opening doors.

20.4   TRAINING

Security staff and employees should be trained regarding the treatment and limitations placed upon visitors, clients, patients, or other outside business representatives. Security should not be the only one to challenge visitors who appear not to be where they are expected to be. Explain to them that they have ventured into non-visitor areas. If they do not have a visitor's badge and they refuse to leave the area or have no escort, call security immediately for assistance. If they become agitated or belligerent, summon help; employees should not take action on their own.

20.5   OSHA APPLICABLE REGULATIONS

Occupational Safety and Health Administration (OSHA) has no specific regulations that are directly applicable to visitors on an employer's worksite. It is just good safety and health practice to have procedures in place to address visitor safety and health issues.

21 Personal Protective Equipment

Example of a bump cap to provide minimal head protection, often used where no hazard for falling objects exist.

Personal protective equipment (PPE) is the least favorite of workers to wear for protection purposes. This is for a number of reasons including

- No liking for PPE
- Perception that it is not needed
- Lacking proper fitting
- Lacking comfort to wear
- Cosmetics (How it looks when worn)
- Failure to fit women or not made for women

However, PPE is considered the last choice of safeguards or controls. But, when nothing else can be used to effectively protect workers then it is the best solution available.

PPE is used to protect the body including hands, arms, feet, legs, trunk, eyes, head, ears, and lungs. PPE protects from blunt forces, crushes, amputations, cuts, and bruises as well as exposure to hazardous chemicals, airborne hazards, and noise. No better alternative exists when the need for PPE is a part of doing the job. PPE does the job that it was intended for.

If an employer is not sure of the best PPE to provide adequate protection from a hazard, the employer should contact a qualified health and safety professional or an industrial hygienist to advise him or her regarding selecting the right PPE for the hazard that is present.

If an employer is not sure whether PPE is needed in the workplace, then they need to conduct a hazard assessment to identify all the potential hazards. This assessment will help in determining whether PPE is needed to protect the workforce effectively. Previous accident history can be helpful in identifying the potential hazards in the employer's workplace.

21.1 PERSONAL PROTECTIVE EQUIPMENT AND HAZARD PREVENTION

Hazards exist in every workplace in many different forms: sharp edges, falling objects, flying sparks, chemicals, noise, and a myriad of other potentially dangerous situations. Occupational Safety and Health Administration (OSHA) requires that employers protect their employees from workplace hazards that can cause injury.

Controlling a hazard at its source is the best way to protect employees. Depending on the hazard or workplace conditions, OSHA recommends the use of engineering or work practice controls to manage or eliminate hazards to the greatest extent possible. For example, building a barrier between the hazard and the employees is an engineering control; changing the way in which employees perform their work is a work practice control.

When engineering, work practice, and administrative controls are not feasible or do not provide sufficient protection, employers must provide PPE to their employees and ensure its use. PPE is equipment worn to minimize exposure to a variety of hazards. Examples of PPE include such items as gloves, foot and eye protection, protective hearing devices (earplugs, muffs), hard hats, respirators, and full body suits.

This chapter will help both employers and employees do the following:

- Understand the types of PPE
- Know the basics of conducting a hazard assessment of the workplace
- Select appropriate PPE for a variety of circumstances
- Understand what kind of training is needed in the proper use and care of PPE

The information in this chapter is general in nature and does not address all workplace hazards or PPE requirements. The information, methods, and procedures in this chapter are based on the OSHA requirements for PPE as set forth in the Code of Federal Regulations (CFR):

- 29 CFR 1910.132 (General requirements)
- 29 CFR 1910.133 (Eye and face protection)
- 29 CFR 1910.134 (Respiratory protection)
- 29 CFR 1910.135 (Head protection)

- 29 CFR 1910.136 (Foot protection)
- 29 CFR 1910.137 (Electrical protective equipment)
- 29 CFR 1910.138 (Hand protection)

21.1.1 REQUIREMENT FOR PPE

To ensure the greatest possible protection for employees in the workplace, the cooperative efforts of both employers and employees will help in establishing and maintaining a safe and healthful work environment.

In general, employers are responsible for

- Performing a hazard assessment of the workplace to identify and control physical and health hazards
- Identifying and providing appropriate PPE for employees
- Training employees in the use and care of PPE
- Maintaining PPE, including replacing worn or damaged PPE
- Periodically reviewing, updating, and evaluating the effectiveness of the PPE program

In general, employees should

- Properly wear PPE
- Attend training sessions on PPE
- Care for, clean, and maintain PPE
- Inform a supervisor of the need to repair or replace PPE

Specific requirements for PPE are presented in many different OSHA standards, published in 29 CFR. Some standards require that employers provide PPE at no cost to the employee while others simply state that the employer must provide PPE.

21.1.2 HAZARD ASSESSMENT

A first critical step in developing a comprehensive safety and health program is to identify physical and health hazards in the workplace. This process is known as a hazard assessment. Potential hazards may be physical or health-related and a comprehensive hazard assessment should identify hazards in both categories. Examples of physical hazards include moving objects, fluctuating temperatures, high-intensity lighting, rolling, or pinching objects, electrical connections, and sharp edges. Examples of health hazards include overexposure to harmful dusts, chemicals, or radiation.

The hazard assessment should begin with a walk-through survey of the facility to develop a list of potential hazards in the following basic hazard categories:

- Impact
- Penetration
- Compression (roll-over)

- Chemical hazard
- Heat/cold
- Harmful dust
- Light (optical) radiation
- Biological hazard

In addition to noting the basic layout of the facility and reviewing any history of occupational illnesses or injuries, things to look for during the walk-through survey include

- Sources of electricity
- Sources of motion such as machines or processes where movement may exist that could result in an impact between personnel and equipment
- Sources of high temperatures that could result in burns, eye injuries, or fire
- Types of chemicals used in the workplace
- Sources of harmful dusts
- Sources of light radiation, such as welding, brazing, cutting, furnaces, heat treating, high-intensity lights, etc.
- The potential for falling or dropping objects
- Sharp objects that could poke, cut, stab, or puncture
- Biological hazards such as blood or other potentially infected material

When the walk-through is complete, the employer should organize and analyze the data so that they may be efficiently used in determining the proper types of PPE required at the worksite. The employer should be aware of the different types of PPE available and the levels of protection offered. It is definitely a good idea to select PPE that will provide a level of protection greater than the minimum required to protect employees from hazards.

The workplace should be periodically reassessed for any changes in conditions, equipment, or operating procedures that could affect occupational hazards. This periodic reassessment should also include a review of injury and illness records to spot any trends or areas of concern and taking appropriate corrective action. The suitability of existing PPE, including an evaluation of its condition and age, should be included in the reassessment.

Documentation of the hazard assessment is required through a written certification that includes the following information:

- Identification of the workplace evaluated
- Name of the person conducting the assessment
- Date of the assessment
- Identification of the document certifying completion of the hazard assessment

21.2 SELECTING PPE

All PPE clothing and equipment should be of safe design and construction and should be maintained in a clean and reliable fashion. Employers should take the fit and comfort of PPE into consideration when selecting appropriate items for their workplace. PPE that fits well and is comfortable to wear will encourage employees to use it. Most protective devices are available in multiple sizes and care should be taken to select the proper size for each employee. If several different types of PPE are worn together, make sure they are compatible. If PPE does not fit properly, it can make the difference between being safely covered or dangerously exposed. It may not provide the level of protection desired and may discourage employee use.

Occupational Safety and Health Administration (OSHA) requires that many categories of PPE meet or be equivalent to standards developed by the American National Standards Institute (ANSI). ANSI has been preparing safety standards since the 1920s, when the first safety standard was approved to protect the heads and eyes of industrial workers. Employers who need to provide PPE in the categories listed below must make certain that any new equipment procured meets the cited ANSI standard. Existing PPE stocks must meet the ANSI standard in effect at the time of its manufacture or provide protection equivalent to PPE manufactured to the ANSI criteria. Employers should inform employees who provide their own PPE of the employer's selection decisions and ensure that any employee-owned PPE used in the workplace conforms to the employer's criteria, based on the hazard assessment, OSHA requirements, and ANSI standards. OSHA requires PPE to meet the following ANSI standards:

- Eye and face protection: ANSI Z87.1–1989 (U.S. Standard for Occupational and Educational Eye and Face Protection).
- Head protection: ANSI Z89.1–1986.
- Foot protection: ANSI Z41.1–1991.
- For hand protection, there is no ANSI standard for gloves but OSHA recommends that selection be based on the tasks to be performed and the performance and construction characteristics of the glove material. For protection against chemicals, glove selection must be based on the chemicals encountered, the chemical resistance, and the physical properties of the glove material.

21.3   TRAINING EMPLOYEES IN THE PROPER USE OF PPE

Employers are required to train each employee who must use PPE. Employees must be trained to know at least the following:

- When PPE is necessary?
- What PPE is necessary?
- How to properly put on, take off, adjust, and wear the PPE?
- Limitations of the PPE.
- Proper care, maintenance, useful life, and disposal of PPE.

Employers should make sure that each employee demonstrates an understanding of the PPE training as well as the ability to properly wear and use PPE before they are allowed to perform work requiring the use of the PPE. If an employer believes that a previously trained employee does not demonstrate the proper understanding and skill level in the use of PPE, that employee should receive retraining. Other situations that require additional training or retraining of employees include the following circumstances: changes in the workplace or in the type of required PPE that make prior training obsolete.

Employers should make sure that each employee demonstrates an understanding of the PPE training as well as the ability to properly wear and use PPE before they are allowed to perform work requiring the use of the PPE. If an employer believes that a previously trained employee does not demonstrate the proper understanding and skill level in the use of PPE, that employee should receive retraining. Other situations that require additional or retraining of employees include the following circumstances: changes in the workplace or in the type of required PPE that make prior training obsolete.

The employer must document the training of each employee required to wear or use PPE by preparing a certification containing the name of each employee trained, the date of training, and a clear identification of the subject of the certification.

Employees can be exposed to a large number of hazards that pose danger to their eyes and face. OSHA requires employers to ensure that employees have appropriate eye or face protection if they are exposed to eye or face hazards from flying particles, molten metal, liquid chemicals, acids or caustic liquids, chemical gases or vapors, potentially infected material, or potentially harmful light radiation.

21.4 EYE AND FACE PROTECTION

Many occupational eye injuries occur because workers do not wear any eye protection while others result from wearing improper or poorly fitting eye protection. Employers must ensure that their employees wear appropriate eye and face protection and that the selected form of protection is appropriate to the work performed and properly fits each worker exposed to the hazard.

21.4.1 Prescription Lenses

Everyday use of prescription corrective lenses will not provide adequate protection against most occupational eye and face hazards, so employers must make sure that employees with corrective lenses either wear eye protection that incorporates the prescription into the design or wear additional eye protection over their prescription lenses. It is important to ensure that the protective eyewear does not disturb the proper positioning of the prescription lenses so that the employee's vision will not be inhibited or limited. In addition, employees who wear contact lenses must wear eye or face PPE when working in hazardous conditions.

21.4.2 Eye Protection for Exposed Workers

Occupational Safety and Health Administration (OSHA) suggests that eye protection be routinely considered for use by carpenters, electricians, machinists, mechanics, millwrights, plumbers and pipe fitters, sheet metal workers and tinsmiths, assemblers, sanders, grinding machine operators, sawyers, welders, laborers, chemical process operators and handlers, and timber cutting and logging workers. Employers of workers in other job categories should decide whether there is a need for eye and face PPE through a hazard assessment.

Examples of potential eye or face injuries include

- Dust, dirt, metal, or wood chips entering the eye from activities such as chipping, grinding, sawing, hammering, use of power tools or even strong wind forces
- Chemical splashes from corrosive substances, hot liquids, solvents, or other hazardous solutions
- Objects swinging into the eye or face, such as tree limbs, chains, tools, or ropes
- Radiant energy from welding and harmful rays from the use of lasers or other radiant light (as well as heat, glare, sparks, splash, and flying particles)

21.4.3 Types of Eye Protection

Selecting the most suitable eye and face protection for employees should take into consideration the following elements:

- Should be able to protect against specific workplace hazards
- Should fit properly and be reasonably comfortable to wear
- Should provide unrestricted vision and movement
- Should be durable and cleanable
- Should allow unrestricted functioning of any other required PPE

The eye and face protection selected for employee use must clearly identify the manufacturer. Any new eye and face protective devices must comply with ANSI Z87.1-1989 or be at least as effective as this standard requires. Any equipment purchased before this requirement took effect on July 5, 1994, must comply with the earlier ANSI Standard (ANSI Z87.1-1968) or be shown to be equally effective (see Table 21.1 and Figure 21.1).

An employer may choose to provide one pair of protective eyewear for each position rather than individual eyewear for each employee. If this is done, the employer must make sure that employees disinfect shared protective eyewear after each use. Protective eyewear with corrective lenses may only be used by the employee for whom the corrective prescription was issued and may not be shared among employees.

Some of the most common types of eye and face protection include the following:

- Safety spectacles—These protective eyeglasses have safety frames constructed of metal or plastic and impact-resistant lenses. Side shields are available on some models.
- Goggles—These are tight-fitting eye protection that completely cover the eyes, eye sockets, and the facial area immediately surrounding the eyes and

TABLE 21.1

Eye and Face Protection Selection Guide

Applications

Operation	Hazards	Recommended Protectors (see Figure 21.1)
Acetylene-burning Acetylene-cutting Acetylene-welding	Sparks, harmful rays, molten metal flying particles	7, 8, 9
Chemical handling	Splash, acid burns, fumes	2, 10 (For severe exposure add 10 over 2)
Chipping	Flying particles	1, 3, 4, 5, 6, 7A, 8A
Electric (arc) welding	Sparks, intense rays, molten metal	9, 11 (11 in combination with 4,5,6 in tinted lenses advisable)
Furnace operations	Glare, heat, molten metal	7, 8, 9 (For severe exposure add 10)
Grinding-light	Flying particles	1, 3, 4, 5, 6, 10
Grinding-heavy	Flying particles	1, 3, 7A, 8A (For severe exposure add 10)
Laboratory	Chemical splash, glass	2 (10 when in breakage combination with 4, 5, 6)
Machining	Flying particles	1, 3, 4, 5, 6, 10
Molten metals	Heat, glare, sparks, splash	7, 8 (10 in combination with 4, 5, 6 in tinted lenses)
Spot welding	Flying particles, sparks	1, 3, 4, 5, 6, 10

Source: Courtesy of the Department of Energy.

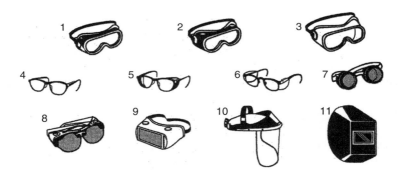

FIGURE 21.1 Recommended eye and face protection. 1. Goggles, flexible fitting, regular ventilation; 2. goggles, flexible fitting, hooded ventilation; 3. goggles, cushioned fitting, rigid body; 4. spectacles, metal frame, without sideshields; 5. spectacles, plastic frame, with sideshields; 6. spectacles, metal–plastic frame, with flat fold sideshields; 7. welding goggles, eyecup type, tinted lenses; 7a. chipping goggles, eyecup type, clear safety lenses (not illustrated); 8. Welding goggles, eyecup type, tinted plate lens; 8a. chipping goggles, coverspec type, clear safety lenses (not illustrated); 9. welding goggles, coverspec type, tinted plate lens; 10. faceshield (available with plastic or mesh window, tinted/transparent); 11. welding helmets. *Note:* These are also available without sideshields for limited use requiring only frontal protection. (Courtesy of the Department of Energy.)

provide protection from impact, dust, and splashes. Some goggles will fit over corrective lenses.

- Welding shields—Constructed of vulcanized fiber or fiberglass and fitted with a filtered lens, welding shields protect eyes from burns caused by infrared or intense radiant light; they also protect both the eyes and face from flying sparks, metal spatter, and slag chips produced during welding, brazing, soldering, and cutting operations. OSHA requires filter lenses to have a shade number appropriate to protect against the specific hazards of the work performed in order to protect against harmful light radiation.
- Laser safety goggles—These specialty goggles protect against intense concentrations of light produced by lasers. The type of laser safety goggles an employer chooses will depend on the equipment and operating conditions in the workplace.
- Face shields—These transparent sheets of plastic extend from the eyebrows to below the chin and across the entire width of the employee's head. Some are polarized for glare protection. Face shields protect against nuisance dusts and potential splashes or sprays of hazardous liquids but will not provide adequate protection against impact hazards. Face shields used in combination with goggles or safety spectacles will provide additional protection against impact hazards.

Each type of protective eyewear is designed to protect against specific hazards. Employers can identify the specific workplace hazards that threaten employees' eyes and faces by completing a hazard assessment as outlined in Section 21.4.

21.5 UNIQUE EYE PROTECTION

21.5.1 WELDING OPERATIONS

The intense light associated with welding operations can cause serious and sometimes permanent eye damage if operators do not wear proper eye protection. The intensity of light or radiant energy produced by welding, cutting, or brazing operations varies according to a number of factors including the task producing the light, the electrode size, and the arc current. In most cases, the worker should not be able to see through the tint shield.

21.5.2 LASER OPERATIONS

Laser light radiation can be extremely dangerous to the unprotected eye and direct or reflected beams can cause permanent eye damage. Laser retinal burns can be painless, so it is essential that all personnel in or around laser operations wear appropriate eye protection.

Laser safety goggles should protect for the specific wavelength of the laser and must be of sufficient optical density for the energy involved. Safety goggles intended for use with laser beams must be labeled with the laser wavelengths for which they are intended to be used, the optical density of those wavelengths, and the visible light transmission.

21.6 HEAD PROTECTION

Protecting employees from potential head injuries is a key element of any safety program. A head injury can impair an employee for life or it can be fatal. Wearing a safety helmet or hard hat is one of the easiest ways to protect an employee's head from injury. Hard hats can protect employees from impact and penetration hazards as well as from electrical shock and burn hazards.

Employers must ensure that their employees wear head protection if any of the following apply:

- Objects might fall from above and strike them on the head
- They might bump their heads against fixed objects, such as exposed pipes or beams
- There is a possibility of accidental head contact with electrical hazards

Some examples of occupations in which employees should be required to wear head protection include construction workers, carpenters, electricians, linemen, plumbers and pipe fitters, timber and log cutters, welders, among many others. Whenever there is a danger of objects falling from above, such as working below others who are using tools or working under a conveyor belt, head protection must be worn. Hard hats must be worn with the bill forward to protect employees properly (see Figure 21.2).

In general, protective helmets or hard hats should do the following:

- Resist penetration by objects.
- Absorb the shock of a blow.
- Be water resistant and slow burning.
- Have clear instructions explaining proper adjustment and replacement of the suspension and headband.
- Have a hard outer shell and a shock-absorbing lining that incorporates a headband and straps that suspend the shell from 1 to 1¼ in. (2.54–3.18 cm)

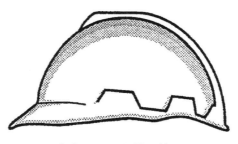

Safety-approved hard hat

FIGURE 21.2   Design of most common hard hats. (Courtesy of the Department of Energy.)

away from the head. This type of design provides shock absorption during an impact and ventilation during normal wear.

- Protective headgear must meet ANSI Standard Z89.1-1986 (Protective Headgear for Industrial Workers) or provide an equivalent level of protection. Helmets purchased before July 5, 1994, must comply with the earlier ANSI Standard (Z89.1-1969) or provide equivalent protection.

21.6.1 TYPES OF HARD HATS

There are many types of hard hats available in the marketplace today. In addition to selecting protective headgear that meets ANSI standard requirements, employers should ensure that employees wear hard hats that provide appropriate protection against potential workplace hazards. It is important for employers to understand all potential hazards when making this selection, including electrical hazards. This can be done through a comprehensive hazard analysis and an awareness of the different types of protective headgear available.

Hard hats require a hard outer shell and a shock-absorbing lining. The lining should incorporate a headband and straps that suspend the shell from 1 to 1¼ in. (2.54–3.18 cm) away from the user's head. This design provides shock absorption during impact and ventilation during wear. As with devices designed to protect eyes, the design, construction, testing, and use of protective helmets must meet standards established by ANSI. Protective helmets purchased after July 5, 1994, must comply with ANSI Z89.1-1986 and they fall into Type I under the new ANSI Z89.1-1997. Type II helmets must meet new requirements for (1) impact resistance from blows to the front, back, sides, and top of the head; (2) off-center penetration resistance; and (3) chinstrap retention. The old classes listed below have been changed to Class G (general), Class E (electrical), and Class C (conductive—no electrical protection). Hard hats are divided into three industrial classes:

- Class A (now Class G)—These helmets are for general service. They provide good impact protection but limited voltage protection (up to 2200 V). They are used mainly in mining, building construction, shipbuilding, lumbering, and manufacturing.
- Class B (now Class E)—They protect against falling objects and high-voltage shocks (up to 20,000 V) and burns. Choose Class B helmets if your employees are engaged in electrical work.
- Class C (now Class C)—Designed for comfort, these lightweight helmets offer limited protection. They protect workers from bumping against fixed objects but do not protect against falling objects or electric shock. These are often called bump hats.

21.6.2 SIZE AND CARE CONSIDERATIONS

Head protection that is either too large or too small is inappropriate for use, even if it meets all other requirements. Protective headgear must fit the body and head size of each individual appropriately. Most protective headgear come in a variety of sizes with adjustable headbands to ensure a proper fit (many adjust in 1/8 in. increments). A proper fit should allow sufficient clearance between the shell and the suspension system for ventilation and distribution of an impact. The hat should not bind, slip, fall off, or irritate the skin.

Some protective headgear allows for the use of various accessories to help employees deal with changing environmental conditions, such as slots for earmuffs, safety glasses, face shields, and mounted lights. Optional brims may provide additional protection from the sun, and some hats have channels that guide rainwater away from the face. Protective headgear accessories must not compromise the safety elements of the equipment.

Periodic cleaning and inspection will extend the useful life of protective headgear. A daily inspection of the hard hat shell, suspension system, and other accessories for holes, cracks, tears, or other damages that might compromise the protective value of the hat is essential. Paints, paint thinners, and some cleaning agents can weaken the shells of hard hats and may eliminate electrical resistance. Consult the helmet manufacturer for information on the effects of paint and cleaning materials on their hard hats. Never drill holes, paint, or apply labels to protective headgear as this may reduce the integrity of the protection. Do not store protective headgear in direct sunlight, such as on the rear window shelf of a car, since sunlight and extreme heat can damage them.

Hard hats with any of the following defects should be removed from service and replaced:

- Perforation, cracking, or deformity of the brim or shell.
- Indication of exposure of the brim or shell to heat, chemicals, or ultraviolet light and other radiation (in addition to a loss of surface gloss, such signs include chalking or flaking).
- Always replace a hard hat if it sustains an impact, even if damage is not noticeable. Suspension systems are offered as replacement parts and should be replaced when damaged or when excessive wear is noticed. It is not necessary to replace the entire hard hat when deterioration or tears of the suspension systems are noticed.

21.7 FOOT AND LEG PROTECTION

Employees who face possible foot or leg injuries from falling or rolling objects or from crushing or penetrating materials should wear protective footwear. In addition, employees whose work involves exposure to hot substances or corrosive or poisonous materials must have protective gear to cover exposed body parts, including legs and feet. If an employee's feet may be exposed to electrical hazards, nonconductive footwear should be worn. On the other hand, workplace exposure to static electricity may necessitate the use of conductive footwear (see Figure 21.3).

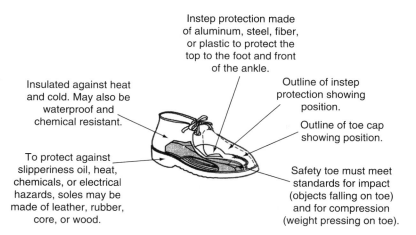

Instep protection made
of aluminum, steel, fiber,
or plastic to protect the
top to the foot and front
of the ankle.

Insulated against heat
and cold. May also be
waterproof and
chemical resistant.

Outline of instep
protection showing
position.

Outline of toe cap
showing position.

To protect against
slipperiness oil, heat,
chemicals, or electrical
hazards, soles may be
made of leather, rubber,
core, or wood.

Safety toe must meet
standards for impact
(objects falling on toe)
and for compression
(weight pressing on toe).

FIGURE 21.3 Protective footwear must meet minimum protective standards. (Courtesy of the Department of Energy.)

Examples of situations in which an employee should wear foot or leg protection include:

- When heavy objects such as barrels or tools might roll onto or fall on the employee's feet
- Working with sharp objects such as nails or spikes that could pierce the soles or uppers of ordinary shoes
- Exposure to molten metal that might splash on feet or legs
- Working on or around hot, wet, or slippery surfaces
- Working when electrical hazards are present

Safety footwear must meet ANSI minimum compression and impact performance standards in ANSI Z41–1991 (American National Standard for Personal Protection-Protective Footwear) or provide equivalent protection. Footwear purchased before July 5, 1994, must meet or provide equivalent protection to the earlier ANSI Standard (ANSI Z41.1–1967). All ANSI ANSI-approved footwear have a protective toe and offer impact and compression protection. However, the type and amount of protection are not always the same. Different footwear protects in different ways. Check the product's labeling or consult the manufacturer to make sure the footwear protects the users from the hazards they face.

Foot and leg protection choices include the following:

- Leggings protect the lower legs and feet from heat hazards such as molten metal or welding sparks. Safety snaps allow leggings to be removed quickly.
- Metatarsal guards protect the instep area from impact and compression. Made of aluminum, steel, fiber, or plastic, these guards may be strapped to the outside of shoes.

- Toeguards fit over the toes of regular shoes to protect the toes from impact and compression hazards. They may be made of steel, aluminum, or plastic.
- Combination foot and shin guards protect the lower legs and feet, and may be used in combination with toeguards when greater protection is needed.
- Safety shoes have impact-resistant toes and heat-resistant soles that protect the feet against hot work surfaces common in roofing, paving, and hot metal industries. The metal insoles of some safety shoes protect against puncture wounds. Safety shoes may also be designed to be electrically conductive to prevent the buildup of static electricity in areas with the potential for explosive atmospheres or nonconductive to protect workers from workplace electrical hazards.

21.8 SPECIAL PURPOSE SHOES

Electrically conductive shoes provide protection against the buildup of static electricity. Employees working in explosive and hazardous locations such as explosives manufacturing facilities or grain elevators must wear conductive shoes to reduce the risk of static electricity buildup on the body, which could produce a spark and cause an explosion or fire. Foot powder should not be used in conjunction with protective conductive footwear because it provides insulation, reducing the conductive ability of the shoes. Silk, wool, and nylon socks can produce static electricity and should not be worn with conductive footwear. Conductive shoes must be removed when the task requiring their use is completed. Note: Employees exposed to electrical hazards must never wear conductive shoes.

Electrical hazard, safety-toe shoes are nonconductive and will prevent the wearers' feet from completing an electrical circuit to the ground. These shoes can protect against open circuits of up to 600 V in dry conditions, and they should be used in conjunction with other insulating equipment and additional precautions to reduce the risk of a worker becoming a path for hazardous electrical energy. The insulating protection of electrical hazard, safety-toe shoes may be compromised if the shoes become wet, the soles are worn through, metal particles become embedded in the sole or heel, or workers touch conductive, grounded items. Note: Nonconductive footwear must not be used in explosive or hazardous locations.

21.8.1 CARE OF PROTECTIVE FOOTWEAR

As with all protective equipment, safety footwear should be inspected before each use. Shoes and leggings should be checked for wear and tear at reasonable intervals. This includes looking for cracks or holes, separation of materials, broken buckles, or laces. The soles of shoes should be checked for pieces of metal or other embedded items that could present electrical or tripping hazards. Employees should follow the manufacturers' recommendations for cleaning and maintenance of protective footwear.

21.9   HAND AND ARM PROTECTION

If a workplace hazard assessment reveals that employees face potential injury to hands and arms, which cannot be eliminated through engineering and work practice controls, employers must ensure that employees wear appropriate protection. Potential hazards include skin absorption of harmful substances, chemical or thermal burns, electrical dangers, bruises, abrasions, cuts, punctures, fractures, and amputations. Protective equipment includes gloves, finger guards, and arm coverings or elbow-length gloves.

Employers should explore all possible engineering and work practice controls to eliminate hazards and use PPE to provide additional protection against hazards that cannot be eliminated through other means. For example, machine guards may eliminate a hazard. Installing a barrier to prevent workers from placing their hands at the point of contact between a table saw blade and the item that is cut is another method.

21.9.1   TYPES OF PROTECTIVE GLOVES

There are many types of gloves available today to protect against a wide variety of hazards. The nature of the hazard and the operation involved will affect the selection of gloves. The variety of potential occupational hand injuries makes selecting the right pair of gloves challenging. It is essential that employees use gloves specifically designed for the hazards and tasks found in their workplace because gloves designed for one function may not protect against a different function even though they may appear to be an appropriate protective device.

The following are examples of some factors that may influence the selection of protective gloves for a workplace:

- Type of chemicals handled
- Nature of contact (total immersion, splash, etc.)
- Duration of contact
- Area requiring protection (hand only, forearm, arm)
- Grip requirements (dry, wet, oily)
- Thermal protection
- Size and comfort
- Abrasion/resistance requirements

Gloves made from a wide variety of materials are designed for many types of workplace hazards. In general, gloves fall into four groups:

1. Gloves made of leather, canvas, or metal mesh
2. Fabric and coated fabric gloves
3. Chemical- and liquid-resistant gloves
4. Insulating rubber gloves (See 29 CFR 1910.137 and the following section on electrical protective equipment for detailed requirements on the selection, use, and care of insulating rubber gloves)

21.9.2   LEATHER, CANVAS, OR METAL MESH GLOVES

Sturdy gloves made from metal mesh, leather, or canvas provide protection against cuts and burns. Leather or canvas gloves also protect against sustained heat. Some examples are

- Leather gloves protect against sparks, moderate heat, blows, chips, and rough objects.
- Aluminized gloves provide reflective and insulating protection against heat and require an insert made of synthetic materials to protect against heat and cold.
- Aramid fiber gloves protect against heat and cold, are cut- and abrasive-resistant and wear well.
- Synthetic gloves of various materials offer protection against heat and cold, are cut- and abrasive-resistant and may withstand some diluted acids. These materials do not withstand alkalis and solvents.

21.9.3   FABRIC AND COATED FABRIC GLOVES

Fabric and coated fabric gloves are made of cotton or other fabric to provide varying degrees of protection.

- Fabric gloves protect against dirt, slivers, chafing, and abrasions. They do not provide sufficient protection for use with rough, sharp, or heavy materials. Adding a plastic coating will strengthen some fabric gloves.
- Coated fabric gloves are normally made from cotton flannel with napping on one side. By coating the unnapped side with plastic, fabric gloves are transformed into general-purpose hand protection offering slip-resistant qualities. These gloves are used for tasks ranging from handling bricks and wire to chemical laboratory containers.

When selecting gloves to protect against chemical exposure hazards, always check with the manufacturer or review the manufacturer's product literature to determine the gloves' effectiveness against specific workplace chemicals and conditions. See Appendix B for an example of parameters for glove selection and a selection chart.

21.9.4   CHEMICAL- AND LIQUID-RESISTANT GLOVES

Chemical-resistant gloves are made with different kinds of rubber: natural, butyl, neoprene, nitrile, and fluorocarbon (viton); or various kinds of plastic: polyvinyl chloride (PVC), polyvinyl alcohol, and polyethylene. These materials can be blended or laminated for better performance. As a general rule, the thicker the glove material, the greater the chemical resistance but thick gloves may impair grip and dexterity, having a negative impact on safety. Some examples of chemical-resistant gloves:

- Butyl gloves are made of a synthetic rubber and protect against a wide variety of chemicals, such as peroxide, rocket fuels, highly corrosive acids (nitric acid, sulfuric acid, hydrofluoric acid, and red-fuming nitric acid), strong bases, alcohols, aldehydes, ketones, esters, and nitrocompounds. Butyl gloves also resist oxidation, ozone corrosion, and abrasion, and remain flexible at low temperatures. Butyl rubber does not perform well with aliphatic and aromatic hydrocarbons and halogenated solvents.
- Natural (latex) rubber gloves are comfortable to wear, which makes them a popular general-purpose glove. They feature outstanding tensile strength, elasticity, and temperature resistance. In addition to resisting abrasions caused by grinding and polishing, these gloves protect workers' hands from most water solutions of acids, alkalis, salts, and ketones. Latex gloves have caused allergic reactions in some individuals and may not be appropriate for all employees.
- Hypoallergenic gloves, glove liners, and powderless gloves are possible alternatives for workers who are allergic to latex gloves.
- Neoprene gloves are made of synthetic rubber and offer good pliability, finger dexterity, high density, and tear resistance. They protect against hydraulic fluids, gasoline, alcohols, organic acids, and alkalis. They generally have chemical and wear resistance properties superior to those made of natural rubber.
- Nitrile gloves are made of a copolymer and provide protection from chlorinated solvents such as trichloroethylene and perchloroethylene. Although intended for jobs requiring dexterity and sensitivity, nitrile gloves stand up to heavy use even after prolonged exposure to substances that cause other gloves to deteriorate. They offer protection when working with oils, greases, acids, caustics, and alcohols but are generally not recommended for use with strong oxidizing agents, aromatic solvents, ketones, and acetates.

Protective gloves should be inspected before each use to ensure that they are not torn, punctured, or made ineffective in any way. A visual inspection will help detect cuts or tears but a more thorough inspection by filling the gloves with water and tightly rolling the cuff toward the fingers will help reveal any pinhole leaks. Gloves that are discolored or stiff may also indicate deficiencies caused by excessive use or degradation from chemical exposure.

Any gloves with impaired protective ability should be discarded and replaced. Reuse of chemical-resistant gloves should be evaluated carefully, taking into consideration the absorptive qualities of the gloves. A decision to reuse chemically exposed gloves should take into consideration the toxicity of the chemicals involved and factors such as duration of exposure, storage, and temperature.

21.10 BODY PROTECTION

Employees who face possible bodily injury of any kind, which cannot be eliminated through engineering, work practice, or administrative controls, must wear appropriate body protection while performing their jobs. In addition to cuts and radiation, the following are examples of workplace hazards that could cause bodily injury:

- Temperature extremes
- Hot splashes from molten metals and other hot liquids
- Potential impacts from tools, machinery, and materials
- Hazardous chemicals

There are many varieties of protective clothing available for specific hazards. Employers are required to ensure that their employees wear PPE only for the parts of the body exposed to possible injury. Examples of body protection include laboratory coats, coveralls, vests, jackets, aprons, surgical gowns, and full body suits.

If a hazard assessment indicates a need for full body protection against toxic substances or harmful physical agents, the clothing should be carefully inspected before each use, it must fit each worker properly, and it must function properly and for the purpose for which it is intended.

Protective clothing comes in a variety of materials, each effective against particular hazards, such as

- Paper-like fiber used for disposable suits provides protection against dust and splashes.
- Treated wool and cotton adapt well to changing temperatures, are comfortable and fire-resistant and protect against dust, abrasions, and rough and irritating surfaces.
- Duck is a closely woven cotton fabric that protects against cuts and bruises when handling heavy, sharp, or rough materials.
- Leather is often used to protect against dry heat and flames.
- Rubber, rubberized fabrics, neoprene, and plastics protect against certain chemicals and physical hazards. When chemical or physical hazards are present, check with the clothing manufacturer to ensure that the material selected provides protection against the specific hazard.

21.11 HEARING PROTECTION

Determining the need to provide hearing protection for employees can be challenging. Employee exposure to excessive noise depends on a number of factors, including

- Loudness of the noise as measured in decibels (dB)
- Duration of each employee's exposure to the noise
- Whether employees move between work areas with different noise levels
- Whether noise is generated from one or multiple sources

Generally, the louder the noise, the shorter the exposure time before hearing protection is required. For instance, employees may be exposed to a noise level of 90 dB for 8 h per day (unless they experience a standard threshold shift) before hearing protection is required. On the other hand, if the noise level reaches 115 dB, hearing protection is required if the anticipated exposure exceeds 15 min.

TABLE 21.2
Permissible Noise Exposure Limits for Workers

Duration Per Day (h)	Sound Level (dBA) Slow Response
8	90
6	92
4	95
3	97
2	100
1–1/2	102
1	105
1/2	110
1/4 or less	115

Source: Courtesy of the Department of Energy.

For a more detailed discussion of the requirements for a comprehensive hearing conservation program, see OSHA Publication 3074 (2002), *Hearing Conservation* or refer to the OSHA standard at 29 CFR 1910.95, Occupational Noise Exposure, section (c).

Table 21.2 shows the permissible noise exposures that require hearing protection for employees exposed to occupational noise at specific decibel levels for specific time periods. Noises are considered continuous if the interval between occurrences of the maximum noise level is 1 s or less. Noises not meeting this definition are considered impact or impulse noises (loud momentary explosions of sound) and exposures to this type of noise must not exceed 140 dB. Examples of situations or tools that may result in impact or impulse noises are powder-actuated nail guns, a punch press, or drop hammers.

If engineering and work practice controls do not lower employee exposure to workplace noise to acceptable levels, employees must wear appropriate hearing protection. It is important to understand that hearing protectors reduce only the amount of noise that gets through to the ears. The amount of this reduction is referred to as attenuation, which differs according to the type of hearing protection used and how well it fits. Hearing protectors worn by employees must reduce an employee's noise exposure to within the acceptable limits noted in Table 21.2. Refer to Appendix B of 29 CFR 1910.95, Occupational Noise Exposure, for detailed information on methods to estimate the attenuation effectiveness of hearing protectors based on the device's noise reduction rating (NRR). Manufacturers of hearing protection devices must display the device's NRR on the product packaging. If employees are exposed to occupational noise at or above 85 dB averaged over an 8 h period, the employer is required to institute a hearing conservation program that includes regular testing of employees' hearing by qualified professionals. Refer to 29 CFR 1910.95(c) for a description of the requirements for a hearing conservation program.

FIGURE 21.4   A canula is a cross between a disposable earplug and a earmuff.

Some types of hearing protection include

- Single-use earplugs are made of waxed cotton, foam, silicone rubber, or fiberglass wool. They are self-forming and, when properly inserted, they work as well as most molded earplugs.
- Preformed or molded earplugs must be individually fitted by a professional and can be disposable or reusable. Reusable plugs should be cleaned after each use (see Figure 21.4).
- Earmuffs require a perfect seal around the ear. Glasses, facial hair, long hair, or facial movements such as chewing may reduce the protective value of earmuffs.

21.12 RESPIRATORY PROTECTION

Respiratory protection has been covered in detail in Chapter 18 of *Industrial Safety and Health for Infrastructure Services.*

22 Workplace Security and Violence

Bulletproof glass acts as a secure barrier to protect workers from outside harm.

In recent years, it has become apparent that workplace security is a major emphasis of occupational safety and health. Workers should feel that they can come to work and work their jobs without the threat that they may come to harm in some way from violence during their work shift.

Thus, with the escalation of workplace violence in the past two decades violence in the workplace has reared it ugly head as a workplace issue, with homicide being the third leading cause of occupational death among all workers in the United States from 1980 to 1988, and the leading cause of fatal occupational injuries among women from 1980 to 1985. Higher rates of occupational homicides were found in the retail and service industries, especially among sales workers. This increased risk may be explained by contact with the public and the handling of money.

Research into the causes of the increasing incidence of death and serious injury to health care workers has led to the theory that exposure to the public may be an important risk. The risk is increased particularly in emotionally charged situations with mentally disturbed persons or when workers appear to be unprotected.

Employers have the responsibility of providing a workplace free from hazards that could cause death or serious physical harm and this includes workplace violence. Thus, the employers of today must take into consideration the security of his/her workplace to assure that their employees can perform their work without the interference of outside sources of danger.

22.1 RISK FACTORS

Some of the common risk factors for workers who could be affected by workplace violence are

- Contact with the public
- Exchange of money
- Delivery of passengers, goods, or services
- Working in a mobile workplace such as a taxi or police cruiser
- Working with unstable or volatile persons in health care, social service, or criminal justice settings
- Working alone or in small numbers
- Working late at night or during early morning hours
- Working in high-crime areas
- Guarding valuable property or possessions
- Working in a community-based setting

Risk factors may be viewed from the standpoint of (1) the environment, (2) administrative controls, and (3) behavior strategies.

22.2 PREVENTION STRATEGIES

Usually three main areas must be considered when considering attempts to provide security and safety for your workforce due to violent occurrences within and without your workplace. Prevention strategies are a good starting point.

22.2.1 ENVIRONMENTAL DESIGNS

Commonly implemented cash-handling policies in retail settings include procedures such as using locked drop safes, carrying small amounts of cash, and posting signs and printing notices that limited cash is available. It may also be useful to explore the feasibility of cashless transactions in taxicabs and retail settings through use of debit or credit cards, especially late at night. These approaches can be used in any setting where cash is currently exchanged between workers and customers.

Physical separation of workers from customers, clients, and the general public through the use of bullet-resistant barriers or enclosures has been proposed for retail settings, such as gas stations and convenience stores, hospital emergency departments, and social service agency claims areas. The height and depth of the counters (with or without bullet-resistant barriers) are also important considerations in protecting workers, since they introduce physical distance between workers and potential attackers. Consideration must nonetheless be given to the continued ease of conducting business: a safety device that increases frustration for workers, customers, clients, or patients may be self-defeating.

Visibility and lighting are also important environmental design considerations. Making high-risk areas visible to more people and installing good external lighting should decrease the risk of workplace assaults (see Figure 22.1).

FIGURE 22.1   Glass and openly visible stairways provide security to those using a parking garage.

Access to and egress from the workplace are also important areas to assess. The number of entrances and exits, the ease with which nonemployees can gain access to work areas because doors are unlocked, and the number of areas where potential attackers can hide are issues that should be addressed. This issue has implications for the design of buildings and parking areas, landscaping, and the placement of garbage areas, outdoor refrigeration areas, and other storage facilities that workers must use during a work shift.

Numerous security devices may reduce the risk for assaults against workers and facilitate the identification and apprehension of perpetrators. These include closed-circuit cameras, alarms, two-way mirrors, card-key access systems, panic-bar doors locked from the outside only, and trouble lights or geographic locating devices in taxicabs and other mobile workplaces.

Personal protective equipment such as body armor has been used effectively by public safety personnel to mitigate the effects of workplace violence. For example, the lives of more than 1800 police officers have been saved by Kevlar vests.

22.2.2   Administrative Controls

Staffing plans and work practices (such as escorting customers and visitors and prohibiting unsupervised movement within and between work areas) are issues that need to be addressed to provide security. Increasing the number of staff on duty may also be appropriate in any number of services and retail settings. The use of security guards or receptionists to screen persons entering the workplace and controlling access to actual work areas has also been suggested by security experts.

Work practices and staffing patterns during the opening and closing of establishments and during money drops and pickups should be carefully reviewed for the increased risk of assault they pose to workers. These practices include having workers take out garbage, dispose of grease, store food or other items in external storage areas, and transport or store money.

Policies and procedures for assessing and reporting threats allow employers to track and assess threats and violent incidents in the workplace. Such policies clearly indicate a zero tolerance of workplace violence and provide mechanisms by which incidents can be reported and handled. In addition, such information allows employers to assess whether prevention strategies are appropriate and effective. These policies should also include guidance on recognizing the potential for violence, methods for defusing or de-escalating potentially violent situations, and instruction about the use of security devices and protective equipment. Procedures for obtaining medical care and psychological support following violent incidents should also be addressed. Training and education efforts are clearly needed to accompany such policies.

22.2.3 Behavioral Strategies

Training employees in nonviolent response and conflict resolution has been suggested to reduce the risk that volatile situations will escalate to physical violence. Also critical is training that addresses hazards associated with specific tasks or worksites and relevant prevention strategies. Training should not be regarded as the sole prevention strategy but as a component in a comprehensive approach to reducing workplace violence. To increase vigilance and compliance with stated violence prevention policies, training should emphasize the appropriate use and maintenance of protective equipment, adherence to administrative controls, and increased knowledge and awareness of the risk of workplace violence.

22.2.4 Perpetrator and Victim Profile

Further, clearly only a small percentage of violence is perpetrated by the mentally ill. Gang members, distraught relatives, drug users, social deviants, or threatened individuals are often aggressive or violent. A history of violent behavior is one of the best indicators of future violence by an individual. This information, however, may not be available, especially for new workers, patients, or clients. Even if this information were available, workers not directly involved with these individuals would not have access to it.

Workers who make home visits or do community work cannot control the conditions in the community and have little control over the individuals they may encounter in their work. The victims of assault are often untrained and unprepared to evaluate escalating behavior or to know and practice methods of defusing hostility or protecting themselves from violence. Training, when provided, is often not required as part of the job and may be offered infrequently. However, using training as the sole safety program element creates an impossible burden on the employee for safety and security for him or herself, coworkers, or other clients. Personal protective measures may be needed and communication devices are often lacking.

22.3 COST OF VIOLENCE

Little has been done to study the cost to employers and employees of work-related injuries and illnesses, including assaults. A few studies have shown an increase in assaults over the past two decades. In one reported situation of 121 workers sustaining 134 injuries, 43% involved lost time from work with 13% of those injured missing more than 21 days from work. In this same investigation, an estimate of the costs of assault was that the 134 injuries from patient violence cost $766,000 and resulted in 4291 days lost and 1445 days of restricted duty.

Additional costs may result from security or response team time, employee assistance program or other counseling services, facility repairs, training and support services for the unit involved, modified duty, and reduction of effectiveness of work productivity in all staff due to a heightened awareness of the potential for violence.

The cost of not developing and providing security at your workplace could be disastrous to your business. This is why it is imperative that a part of your occupational safety and health effort should be directed toward security and the prevention of workplace violence.

22.4 PREVENTION EFFORTS

Although it is difficult to pinpoint specific causes and solutions for the increase in violence in the workplace, recognition of the problem is a beginning. Some solutions to the overall reduction of violence in this country may be found in actions such as eliminating violence in television programs, implementing effective programs of gun control, and reducing drug and alcohol abuse. All companies should investigate programs recently instituted by several convenience store chains or robbery deterrence strategies such as increased lighting, closed-circuit TV monitors, and visible money handling locations. If sales are involved, consider limiting access and egress and providing security staff. You might want to construct a response plan. Although it may not help to prevent incidents, a response plan should be incorporated into an overall plan of prevention. Training employees in the management of assaultive behavior or professional assault response has been shown to reduce the incidence of assaults. Administrative controls and mechanical devices are recommended and gradually implemented.

Some safety measures may seem expensive or difficult to implement but are needed to adequately protect the health and well-being of workers. It is also important to recognize that the belief that certain risks are part of the job contributes to the continuation of violence and possibly the shortage of trained workers.

The guidelines provided in this chapter, while not exhaustive, include philosophical approaches as well as practical methods to prevent and control assaults.

The potential for violence may always exist for workers. The cooperation and commitment of employers are necessary, however, to translate these guidelines into an effective program for the occupational health and safety of their workforce.

22.5   PROGRAM DEVELOPMENT AND ESSENTIAL ELEMENTS

In order to supplement information in the earlier chapters in this book related to safety and health program development, and considering the need to include the four critical elements in such a program, this chapter will discuss how to put the safety and health program development to use by using security as the subject. These elements are as follows.

22.5.1   MANAGEMENT COMMITMENT AND EMPLOYEE INVOLVEMENT

Commitment and involvement are essential elements in any safety and health program. Management provides the organizational resources and motivating forces necessary to deal effectively with safety and security hazards. Employee involvement, both individually and collectively, is achieved by encouraging participation in the worksite assessment, developing clear effective procedures, and identifying existing and potential hazards. Employee knowledge and skills should be incorporated into any plan to abate and prevent safety and security hazards.

22.5.1.1   Commitment by Top Management

The implementation of an effective safety and security program includes a commitment by the employer to provide the visible involvement of everyone, so that all employees, from managers to line workers, fully understand that management has a serious commitment to the program. An effective program should have a team approach with top management as the team leader, and should include the following:

- The demonstration of management's concern for employee emotional and physical safety and health by placing a high priority on eliminating safety and security hazards.
- A policy that places employee safety and health on the same level of importance as customer/patient/client safety. The responsible implementation of this policy requires management to integrate issues of employee safety and security to ensure that this protection is part of the daily functioning of the workplace.
- Employer commitment to security through the philosophical refusal to tolerate violence in the workplace and the assurance that every effort will be made to prevent its occurrence.
- Employer commitment to assign and communicate the responsibility for various aspects of safety and security to supervisors, forepersons, lead workers, and other employees involved so that they know what is expected of them. Also, to ensure that recordkeeping is accomplished and used to aid in meeting program goals.

- Employer commitment to provide adequate authority and resources to all responsible parties so that assigned responsibilities can be met.
- Employer commitment to ensure that each manager, supervisor, professional, and employee is responsible for the security and safety program in the workplace and is accountable for carrying out those responsibilities.
- Employer develops and maintains a program of medical and emotional health care for employees who are assaulted or suffer abusive behavior.
- Development of a safety committee, which evaluates all reports and records of assaults and incidents of aggression. When this committee makes recommendations for correction, the employer reports back to the committee in a timely manner on actions taken on the recommendation.

22.5.1.2 Employee Involvement

An effective program includes a commitment by the employer to provide for and encourage employee involvement in the safety and security program and in the decisions that affect worker safety and health, as well as client well-being. Involvement may include the following:

- An employee suggestion/complaint procedure, which allows workers to bring their concerns to management and receive feedback without fear of reprisal or criticism of ability.
- Employees follow a procedure that requires prompt and accurate reporting of incidents with or without injury. If injury has occurred, prompt first aid or medical aid must be sought and treatment provided or offered.
- Employees participate in a safety and health committee that receives information and reports on security problems, makes facility inspections, analyzes reports and data, and makes recommendations for corrections.
- Employees participate in incident review to identify problems, which may help employees to identify potentially violent behavior patterns and discuss safe methods of managing difficult situations.
- Employees participate in security response teams, which are trained and possess required professional assault response skills.
- Employees participate in training and refresher courses in professional assault response training to learn techniques such as recognizing escalating agitation, deflecting or controlling undesirable behavior and, if necessary, of controlling assaultive behavior, and protecting customers and staff members.
- Training in dealing with hostile individuals or the police department program on "personal safety" should be provided and required to be attended by all involved employees.

Effective implementation requires a written program for job safety, health, and security that is endorsed and advocated by the highest level of management. This program should outline the employer's goals and objectives. The written program should be suitable for the size, type, and complexity of the workplace and its operations and should permit these guidelines to be applied to the specific hazardous situations of each operation.

The written program should be communicated to all personnel regardless of the number of workers or work shift. The program should establish clear goals and objectives that are understood by all members of the company or organization. The communication needs to be extended to all levels of the workforce.

22.5.2 HAZARD IDENTIFICATION AND ANALYSIS

Worksite hazard identification/analysis identifies existing hazards and conditions, operations and situations that create or contribute to hazards, and areas where hazards may develop. This includes close scrutiny and tracking of injury/illness and incident records to identify patterns that may indicate causes of aggressive behavior and assaults. The objectives of worksite hazard identification/analyses are to recognize, identify, and to plan to correct security hazards. Analysis utilizes existing records, and work site evaluations should include record review and identification of security hazards.

22.5.2.1 Record Review

Analyze medical, safety, and insurance records, including the OSHA 300 log and information compiled for incidents or near incidents of assaultive behavior from workers or visitors. This process should ensure confidentiality of records of employees and others. This information should be used to identify incidence and severity and to establish a base line for identifying change.

Identify and analyze any apparent trends in injuries relating to particular departments, units, job titles, unit activities or workstations, activities or time of day. Identification of sentinel events such as threatening of workers or identification and classification of where aggressive behavior could be anticipated and by whom might be included.

22.5.2.2 Identification of Security Hazards

Worksite hazard identification/analysis should use a systematic method to identify those areas needing in-depth scrutiny of security hazards. This analysis should do the following:

- Identify those work positions in which workers are at risk of assaultive behavior.
- Use a checklist for identifying high-risk factors that includes components, such as type of people contacts, physical risk factors of the building, isolated locations/job activities, lighting problems, high-risk activities or situations, problem workers, service/delivery personnel or customers, uncontrolled access, and areas of previous security problems.
- Identify low-risk positions for light or relief duty or restricted activity work positions when injuries do occur.

- Determine if risk factors have been reduced or eliminated to the extent feasible. Identify existing programs in place, and analyze effectiveness of those programs, including engineering control measures and their effectiveness.
- Apply analysis to all newly planned and modified facilities or any public services program to ensure that hazards are reduced or eliminated before involving the public, customers, or employees.
- Conduct periodic surveys at least annually (or whenever there are operation changes) to identify new or previously unnoticed risks and deficiencies and to assess the effects of changes in the building designs, work processes, patient services, and security practices. Evaluation and analysis of information gathered, and incorporation of all this information into a plan of correction and ongoing surveillance, should be the result of the work site analysis.

22.5.3   HAZARD PREVENTION AND CONTROL

Select work settings to apply methods of reducing hazards. You will need to make use of general engineering concepts, specific engineering and administrative controls, work practice controls, and personal protective equipment as appropriate to control hazards.

22.5.3.1   General Building, Work Station, and Area Designs

Workplace designs are appropriate when they provide secure, well-lighted and protected areas, which do not facilitate assaults or other uncontrolled activity.

- Design of facilities should ensure uncrowded conditions for workers and customers. Areas for privacy and protection are needed, although isolation should be avoided. For example, doors must be fitted with windows so that other workers can view any aberrant behavior.
- Work areas should be designed and furniture arranged to prevent entrapment of workers and others.
- Reception areas should be protected by enclosures that prevent molesting, throwing objects, reaching into the work area, or otherwise creating a hazard or nuisance to workers: such barriers should not restrict communication but should be protective (see Figure 22.2).
- Lockable and secure bathroom facilities and other amenities must be provided for workers separate from customer or public restrooms.
- Public or customer access to workers' workstations and other facility areas must be controlled; that is, doors from waiting rooms must be locked and all outside doors locked from the outside to prevent unauthorized entry, but permit exit in cases of emergency or fire.
- Metal bars or protective decorative grating should be installed on outside ground-level windows (in accordance with fire department codes) to prevent unauthorized entry.

FIGURE 22.2   Reasonable secure reception area for workers and visitors.

- Bright and effective lighting systems must be provided for all indoor building areas as well as grounds around the facility or workplace, especially in parking areas.
- Curved mirrors should be installed at intersections of halls or in areas where an individual may conceal his or her presence.
- All permanent and temporary employees who work in secured areas should be provided with keys or swipe cards to gain access to work areas whenever on duty (Figure 22.3).
- Metal detectors should be installed to screen visitors, customers, service personnel, and visitors in high-security areas. In other situations, implement handheld metal detectors to use in identifying weapons.

22.5.3.2 Maintenance

Maintenance must be an integral part of any safety and security system. Prompt repair and replacement programs are needed to ensure the safety of workers and customers. Replacement of burned out lights, broken windows, etc., is essential to maintain the system in safe operating condition.

If an alarm system is to be effective, it must be used, tested, and maintained according to strict policy. Any personal alarm devices should be carried and tested as required by the manufacturer and facility policy. Maintenance on personal and other alarm systems must take place monthly. Batteries and operation of alarm devices must be checked by a security office to ensure the function and safety of the system. Any mechanical device used for security and safety must be routinely tested for effectiveness and maintained on a scheduled basis.

FIGURE 22.3 Punch key entrance to work areas provides extra security.

22.5.3.3 Engineering Control

Alarm systems are imperative for use in psychiatric units, hospitals, mental health clinics, high-hazard areas, emergency rooms, or where drugs are stored. Though alarm systems are not necessarily preventive, they may reduce serious injury when a person acts in an abusive manner or threatens with or without a weapon. Many other engineering controls can be used such as the following:

- Alarm systems that rely on the use of telephones, whistles, or screams are ineffective and dangerous. A proper system consists of an electronic device that activates an alert to a dangerous situation in two ways, visually and audibly. Such a system identifies the location of the room or action of the worker by means of an alarm sound and a lighted indicator, which visually identifies the location. In addition, the alarm should be sounded in a security (or other response team) area in order to summon aid. This type of alarm system typically uses a pen-like device, which is carried by the employee and can be triggered easily in an emergency. Backup security personnel must be available to respond to the alarm.

An emergency personal alarm system is of the highest priority. An alarm system may be of two types: the personal alarm device or the type that is triggered at a desk or counter. This desk system may be silent at the desk or counter, but audible in a central assistance area. It must clearly identify the location in which the problem is occurring."

These alarm systems must be relayed to security police or locations where assistance is available 24 h per day. A telephone link to the local police department should be established in addition to other systems.

- "Panic buttons" are needed at times when someone is confronted with an abusive person. Any such alarm system may incorporate a telephone paging system to direct others to the location of the disturbance, but alarm systems must not depend on the use of a telephone to summon assistance.
- Video screening of high-risk areas or activities may be of value and permits one security guard to visualize a number of high-risk areas, both inside and outside the building. Closed-circuit TV monitors may be used to survey concealed areas or areas where problems may occur (see Figure 22.4).
- Metal detection systems, such as handheld devices or other systems to identify persons with hidden weapons, should be considered. These systems are in use in courts, boards of supervisors, some Departments of Public Social Service, schools and emergency rooms. Although controversial, the fact remains that many people including homeless and mentally ill persons, carry weapons for defense while living on the streets. Some system of identifying persons who are carrying guns, knives, ice picks, screw drivers, etc., may be useful and should be considered when situations merit them.

FIGURE 22.4 Split screen monitor can provide pictures from four cameras at one time.

Signs posted at the entrance must notify workers, customers, and visitors that screening will be performed.

- Reception areas should be designed so that receptionists and staff may be protected by safety glass and locked doors to their work areas.
- First-aid kits should be available.
- Materials and equipment to meet the requirements of the bloodborne pathogen standard should be available.
- Strictly enforced limited access to work areas is needed to eliminate entry by unwanted or dangerous persons. Doors may be locked or key-coded.
- In order to provide some measure of safety and to keep the employee in contact with headquarters or another source of assistance, cellular car phones should be installed/provided for official use when workers are assigned to duties that take them into private homes and the community. These workers may include (to name a few) parking enforcers, union business agents, psychiatric evaluators, public social service workers, children's service workers, visiting nurses, and home health aides.
- Handheld alarm or noise devices or other effective alarm devices are highly recommended to be provided for all field personnel.
- Beepers or alarm systems should be investigated and provided to alert a central office of problems.
- Other protective devices such as pepper spray should be investigated and provided.

22.5.3.4   Administrative Controls and Work Practices

A sound overall security program includes administrative controls that reduce hazards from inadequate staffing, insufficient security measures, and poor work practices.

Employees are to be instructed not to enter any location where they feel threatened or unsafe. This decision must be the judgment of the employee. Procedures should be developed to assist the employee in evaluating the relative hazards in a given situation. In hazardous cases, the managers must facilitate and establish a "buddy system". This "buddy system" should be required whenever an employee feels insecure regarding the time of activity, the location of work, the nature of the individuals in that location, and past history of aggressive or assaultive behavior by these individuals.

- Employers must provide a program or personal safety education for the field staff. This program should be at the minimum one provided by local police departments or other agencies. It should include training on awareness, avoidance, and action to take to prevent mugging, robbery, rapes, and other assaults.
- Procedures should be established to assist employees to reduce the likelihood of assaults and robbery from those seeking drugs or money, as well as procedures to follow in the case of threatening behavior and provision for a fail-safe backup in administration offices.

- A fail-safe backup system is provided in the administrative office at all times of operation for employees in the field who may need assistance.
- All incidents of threats or other aggression must be reported and logged. Records must be maintained and used to prevent future security and safety problems.
- Police assistance and escorts should be required in dangerous or hostile situations or at night. Procedures for evaluating and arranging for such police accompaniment must be developed and training provided.
- Security guards must be provided. These security guards should be assigned to areas where there may be problems such as emergency rooms or psychiatric services.
- In order to staff safely, a written guideline that evaluates the level of staff or worker coverage needed should be established. Provision of sufficient staff interaction and clinical activity is important because patients/clients need access to medical assistance from staff. Possibility of violence often threatens staff when the structure of the patient/nurse relationship is weak. Therefore, sufficient staff members are essential to allow formation of therapeutic relationships and a safe environment.
- It is necessary to establish on-call teams, reserve, or emergency teams of staff who may provide services, such as responding to emergencies, transportation or escort services, dining room assistance, or many of the other activities where potential hazards exist.
- Methods should be developed to communicate to workers who come to work about any potential security breaches or violence potentials.
- Workers should be instructed to limit physical intervention in altercations whenever possible, unless there are adequate numbers to assist them, or emergency response teams and security are called. In a case where serious injury could occur, emergency alarm systems should always be activated. Administrators need to give clear messages to everyone that violence is not permitted and that legal charges will be pressed when violence or the threat of it occurs. Management should provide information to workers who may be in danger. Policies must be provided with regard to safety and security of workers in confronting or querying unrecognized individuals in the workplace, key and door opening policy, open vs. locked seclusion policies, and evacuation policy in emergencies.
- Escort services by security should be arranged so that workers should not have to walk alone in parking lots or other parking areas in the evening or late hours.
- Visitors and maintenance persons or crews should be escorted and observed while in any locked or secure facility. Often they have tools or possessions, which could be inadvertently left unattended and thus become weapons.
- Management needs to work with the local police to establish liaison and response mechanisms for police assistance when calls are made for help. They should also make clear policies on how they wish the workers to respond.

- It is not wise to allow workers to confront an aggressive or threatening individual, nor is it appropriate to allow aggressive behavior to go unchecked. Workers should respond according to the company's policy and procedures.
- It is a wise policy to require badging of all workers and require them to visibly wear their picture badges at all times. Anyone who does not have a badge in restricted work areas should be confronted and reported to the supervisor, or security should be called.
- Security guards trained in principles of human behavior and aggression should be provided where there are large numbers of customers, clients, patients, or visitors. Guards should be provided where there may be psychologically stressed clients or persons who have taken hostile actions, such as in emergency facilities, hospitals where there are acute or dangerous patients, or areas where drug or other criminal activity is commonplace.
- No employee should be permitted to work or stay alone in an isolated area without protection from some source.
- Clothing and apparel that will not contribute to injury should be worn, such as low-heeled shoes, conservative earrings, or jewelry, and clothing that is not provocative.
- Keys should be kept covered and worn in such a manner as to avoid incidents, yet be available.
- All protective devices and procedures should be required to be used by all workers.
- After dark, all unnecessary doors must be locked, and access into the workplace limited and the facility patrolled by security.
- Emergency or hospital staff that has been assaulted should be permitted and helped to request police assistance or to file charges of assault against any customer, client, visitor, patient or relative who injures, just as a private citizen has the right to do so. Being at work does not reduce the right of pressing charges or damages.
- Visitors should sign in and out and have an issued pass that identifies them as visitor and specifies the locations they are permitted to access in the workplace.

22.5.4 TRAINING AND EDUCATION

A major program element in an effective safety and security program is training and education. The purpose of training and education is to ensure that employees are sufficiently informed about the safety and security hazards to which they may be exposed and thus are able to participate actively in their own and their coworkers' protection. All employees should be periodically trained in the employer's safety and security program.

Training and education are critical components of a safety and security program for employees who are potential victims of assaults. Training allows managers, supervisors, and employees to understand security and other hazards associated with a job or location within the facility, the prevention and control of these hazards, and the medical and psychological consequences of assault.

22.5.4.1 Training Program

A training program should include all affected employees who could encounter or be subject to abuse or assaults. This means all employees, such as engineers, security officers, maintenance personnel, supervisors, managers, and workers at all levels.

- The program should be designed and implemented by qualified persons. Appropriate special training should be provided for personnel responsible for administering the training program.
- Several types of programs are available and have been utilized, such as Management of Assaultive Behavior (MAB), Professional Assault Response Training (PART), Police Department Assault Avoidance Programs, or Personal Safety training. A combination of such training may be incorporated depending on the severity of the risk and assessed risk. These management programs must be provided and attendance required at least yearly. Updates may be provided monthly/quarterly.
- The program should be presented in the language and at a level of understanding appropriate for the individuals trained. It should provide an overview of the potential risk of illness and injuries from assault, the causes and early recognition of escalating behavior, or recognition of situations that may lead to assaults. The means of preventing or defusing volatile situations, safe methods of restraint or escape, or use of other corrective measures or safety devices that may be necessary to reduce injury and control behavior are critical areas of training. Methods of self-protection and protection of coworkers, the proper treatment of staff, patient procedures, recordkeeping, and employee rights need to be emphasized.
- The training program should also include a means for adequately evaluating its effectiveness. The adequacy of the frequency of training should be reviewed. The whole program evaluation may be achieved by using employee interviews, testing and observing, and reviewing reports of behavior of individuals in situations that are reported to be threatening in nature.
- Employees who are potentially exposed to safety and security hazards should be given formal instruction on the hazards associated with the unit of job and facility. This includes information on the types of injuries or problems identified in the facility, the policy and procedures contained in the overall safety program of the facility, hazards unique to the unit or program, and the methods used by the facility to control the specific hazards. The information should discuss the risk factors that cause or contribute to assaults, etiology of violence and general characteristics of violent people, methods of controlling aberrant behavior, methods of protection, reporting procedures, and methods to obtain corrective action.

Training for affected employees should consist of both general and specific job training. "Specific job training" is discussed in the following section and may also be found in the section on administrative controls in the specific work location section.

22.5.4.2   Job-Specific Training

New employees and reassigned workers should receive an initial orientation and hands-on training before they are placed in a treatment unit or job. Each new employee should receive a demonstration of alarm systems, protective devices, and the required maintenance schedules and procedures. The training should also contain the use of administrative or work practice controls to reduce injury.

22.5.4.3   Initial Training Program

The initial training program should include

- Care, use, and maintenance of alarm tools and other protection devices
- Location and operation of alarm systems
- MAB, PART, or other training
- Communication systems and treatment plans
- Policies and procedures for reporting incidents and obtaining medical care and counseling
- Hazard communication program
- Bloodborne pathogen program, if applicable
- Rights of employees, treatment of injury, and counseling programs

On-the-job training should emphasize employee development and use of safe and efficient techniques, methods of de-escalating aggressive behavior, self-protection techniques, methods of communicating information, which will help other staff to protect themselves, and discussions of rights of employees in the work setting.

Specific measures at each location, such as protective equipment, location and use of alarm systems, determination of when to use the buddy system, and so on, as needed for safety, must be part of the specific training. Training unit coworkers from the same unit and shift may facilitate teamwork in the work setting.

22.5.4.4   Training for Supervisors and Managers, Maintenance and Security Personnel

Supervisors and managers are responsible for ensuring that employees are not placed in assignments that compromise safety and that employees feel comfortable in reporting incidents. They must be trained in methods and procedures that will reduce the security hazards and train employees to behave compassionately with coworkers when an incident occurs. They need to ensure that employees have safe work practices and receive appropriate training to enable them to do this. Supervisors and managers therefore, should undergo training comparable to that of the employee and such additional training as will enable them to recognize a potentially hazardous situation, make changes in the physical plant, patient care treatment program, staffing policy and procedures, or other such situations that contribute to hazardous conditions. They should be able to reinforce the employer's program of safety and security, assist security guards when needed, and train employees as the need arises.

Training for engineers and maintenance personnel should consist of an explanation or a discussion of the general hazards of violence, the prevention and correction of security problems, and personal protection devices and techniques. They need to be acutely aware of how to avoid creating hazards in the process of their work.

Security personnel need to be recruited and trained whenever possible for the specific job and facility. Security companies usually provide general training on guard or security issues. However, specific training should include psychological components of handling aggressive and abusive individuals, types of disorders, and the psychology of handling aggression and defusing hostile situations. If weapons are used by security staff, special training and procedures need to be developed to prevent inappropriate use of weapons and the creation of additional hazards. See the sample security program in Appendix C.

22.5.5 MEDICAL MANAGEMENT

A medical program that provides knowledgeable medical and emotional treatment should be established. This program should ensure that victimized employees are provided with the same degree concern that is shown to the Right of perpetrators or other victims. Violence is a major safety hazard in psychiatric and acute care facilities, emergency rooms, homeless shelters, and other health care settings. Medical and emotional evaluation and treatment are frequently needed, but often difficult to obtain.

The consequences to employees who are abused by others may include death and severe, life threatening injuries, in addition to short- and long-term psychological trauma, posttraumatic stress, anger, anxiety, irritability, depression, shock, disbelief, self-blame, fear of returning to work, disturbed sleep patterns, headache, and change in relationships with coworkers and family. All have been reported by workers after assaults, particularly if the attack has come without warning. They may also fear criticism by managers, increase their use of alcohol and medication to cope with stress, suffer from feelings of professional incompetence, physical illness, powerlessness, increase in absenteeism, and may experience performance difficulties.

Managers and supervisors have often ignored the needs of the physically or psychologically abused or assaulted staff, requiring them to continue working, obtain medical care from private medical doctors, or blame the individual for irresponsible behavior. Injured staff must have immediate physical evaluations, be removed form the unit, and treated for acute injuries. Referral should be made for appropriate evaluation, treatment, counseling, and assistance at the time of the incident and for any required follow-up treatment. Medical services include

- Provision of prompt medical evaluation and treatment whenever an assault takes place, regardless of the severity. A system of immediate treatment is required regardless of the time of day or night. Injured employees should be removed from the unit until order has been restored. Transportation of the injured to medical care must be provided if it is not available onsite or in an employee health service. Follow-up treatment must also be provided at no cost to employees.

- A trauma-crisis counseling or critical incident-debriefing program must be established and provided on an ongoing basis to staff who are victims of assaults. This counseling program may be developed and provided by in-house staff as part of an employee health service, by a trained psychologist, psychiatrist, or other clinical staff members such as a clinical nurse specialist, or a social worker. A referral may also be made to an outside specialist. In addition, peer counseling or support groups may be provided. Any counseling provided should be by well-trained psychosocial counselors whether through an employee assistance program, in-house programs, or by other professionals away from the facility, who must understand the issues of assault and its consequences.
- Reassignment of staff should be considered when assaults have taken place. At times, it is very difficult for staff to return to the same unit to face the assailant. Assailants often repeat threats and aggressive behavior and actions need to be taken to prevent this from occurring. Staff development programs should be provided to teach staff and supervisors to be more sensitive to the feelings and trauma experienced by victims of assaults. Some professionals advocate joint counseling sessions including the assaultive client and staff member to attempt to identify the motive when it occurs in inpatient facilities and to defuse situations that may lead to continued problems.
- Other workers should also receive counseling to prevent "blaming the victim syndrome" and to assist them with any stress problems they may be experiencing as a result of the assault. Violence often leaves staff fearful and concerned. They need to have the opportunity to discuss these fears and to know that administration is concerned and will take measures to correct deficiencies. This may be called a defusing or debriefing session, and unit staff members may need this activity immediately after an incident to enable them to continue working. First-aid kits or materials must be provided in each unit or facility.
- The replacement and transportation of the injured workers must be provided for at the earliest possible time. Do not leave the workplace understaffed in the event of an assault. The development of an employee health service, staffed by a trained occupational health specialist, may be an important addition to the hospital team. Such employee health staff can provide treatment, arrange for counseling, and refer to a specialist. They should have procedures in place for all shifts. Employee health nurses should be trained in posttraumatic counseling and may be used for group counseling programs or other assistance programs.
- Legal advice regarding pressing charges should be available, and information regarding workers' compensation benefits and other employee rights must be provided regardless of apparent injury. If assignment to light duty is needed or disability is incurred, these services are to be provided without

hesitation. Assistance in reporting to the appropriate local law enforcement agency is to be provided. Employees may not be discouraged or coerced when making reports or workers' compensation claims.

- All assaults must be investigated, reports made, and needed corrective action determined. However, methods of investigation must be such that the individual does not perceive blame or criticism for assaultive actions taken by the attacker. The circumstances of the incident or other information that will help to prevent further problems needs to be identified, but not to blame the worker for incompetence and compound the psychological injury that is most commonly experienced.

22.5.6 Recordkeeping

Within the major program elements, recordkeeping is the heart of the program, providing information for analysis, evaluation of methods of control, severity determinations, identifying training needs, and overall program evaluations.

Records shall be kept of the following:

- OSHA 300 Log. OHSA regulations require entry on the injury and illness log of any injury that requires more than first aid, is a lost time injury, requires modified duty, or causes loss of consciousness. Assaults should be entered on the log. Doctors' reports of work injury and supervisors' reports shall be kept of each recorded assault.
- Incidents of abuse, verbal attacks, or aggressive behavior, which may be threatening to the worker but not result in injury, such as pushing, shouting, or an act of aggression toward other clients requiring action by staff, should be recorded. This record may be an assaultive incident report documented in manner, which can be evaluated on a monthly basis by the department safety committee.
- A system of recording and communicating should be developed so that all workers who may provide care for escalating or potentially aggressive, abusive or violent individuals will be aware of the status of those individuals and of any problems experienced in the past. This information regarding history of past violence should be noted on those individual's records, communicated in the shift change report, and noted in an incident log.
- An information gathering system should be in place, which will enable incorporation of past history of violent behavior, incarceration, probation reports, or any other information that will assist health care staff to assess violence status. Employees are to be encouraged to seek and obtain information regarding history of violence whenever possible.
- Emergency room staff should be encouraged to obtain and record, from police and relatives, information regarding drug abuse, criminal activity, or other information to assist in assessing a patient adequately. This would enable them to appropriately house, treat, and refer potentially violent cases. They should document the frequency of admission of violent clients or hostile encounters with relatives and friends.

- Records need to be kept concerning assaults, including the type of activity, i.e., unprovoked sudden attack, patient-to-patient altercation, and management of assaultive behavior actions. Information needed includes who was assaulted and circumstances of the incident without focusing on any alleged wrong doing of staff. These records also need to include a description of the environment, location, or any contributing factors, corrective measures identified, including building design, or other measures needed. Determination must be made of the nature of the injuries sustained, whether severe, minor, or the cause of long-term disability, and the potential or actual cost to the facility and employee. Records of any lost time or other factors that may result from the incident should be maintained.
- Minutes of the safety meetings and inspections shall be kept. Corrective actions recommended as a result of reviewing reports or investigating accidents or inspections need to be documented with the management's response and completion dates of those actions should be included in the minutes and records.
- Records of training program contents and sign-in sheets of all attendees should be kept. Attendance records of all training should be retained. Information on qualifications of trainers shall be maintained along with records of training.

22.5.7 EVALUATION OF THE PROGRAM

Procedures and mechanisms should be developed to evaluate the implementation of the safety and security programs and to monitor progress and accomplishments. Top management and supervisors should review the program regularly. Semiannual reviews are recommended to evaluate success in meeting goals and objectives. Evaluation techniques include some of the following:

- Establishment of a uniform reporting system and regular review of reports.
- Review of reports and minutes of the safety and security committee.
- Analyses of trends and rates in illness/injury or incident reports.
- Surveys of employees.
- Before and after surveys/evaluations of job or worksite changes or new systems.
- Up-to-date records of job improvements or programs implemented.
- Evaluation of employee experiences with hostile situations and results of medical treatment programs provided. Follow-up should be repeated several weeks and several months after an incident.
- Results of management's review of the program should be a written progress report and program update, which should be shared with all responsible parties and communicated to employees. New or revised goals arising from the review, identifying jobs, activities, procedures and departments, should be shared with all employees. Any deficiencies should be identified and corrective action taken. Safety of employees should not be given a lesser priority than client safety as they are often dependent on one another.

FIGURE 22.5 Security cameras have gone a long way toward providing documentation, protection, and records of problems or potential security issues.

> If it is unsafe for employees, the same problem will be the source of risk to other clients or patients.
> - Managers and supervisors should review the program frequently to reevaluate goals and objectives and discuss changes. Regular meetings with all involved including the safety committee, union representatives, and employee groups at risk should be held to discuss changes in the program.

If you are to provide a safe work environment, it must be evident from managers, supervisors, and peer groups that hazards from violence will be controlled. Employees in psychiatric facilities, drug treatment programs, social services, customer relations, human resource management, emergency rooms, law enforcement, service industries, convalescent homes, taxi cab services, community clinics, or community settings are to be provided with a safe and secure work environment, and injury from assault is not to be accepted or tolerated and is no longer part of the job.

Procedures and mechanisms should be developed to evaluate the implementation of the security program and to monitor progress. This evaluation and recordkeeping program should be reviewed regularly by top management and the medical management team. At least semiannual reviews are recommended to evaluate success in meeting goals and objectives (see Figure 22.5).

22.6 TYPES OF WORKPLACE VIOLENCE EVENTS

When one examines the circumstances associated with workplace violence, the events can be divided into three major types. However, it is important to keep in mind that a particular occupation or workplace may be subject to more than one type. In all three types of workplace violence events, a human being, or hazardous agent, commits the assault.

In Type I, the agent has no legitimate business relationship to the workplace and usually enters the affected workplace to commit a robbery or other criminal act.

In Type II, the agent is either the recipient, or the object, of a service provided by the affected workplace or the victim; e.g., the assailant is a current or former client, patient, customer, passenger, criminal suspect, inmate or prisoner.

In Type III, the agent has some employment-related involvement with the affected workplace. Usually this involves an assault by a current or former employee, supervisor, or manager; by a current/former spouse or lover; a relative or friend or some other person who has a dispute with an employee of the affected workplace.

The characteristics of the establishments affected, the profile and motive of the agent or assailant, and the preventive measures differ for each of the three major types of workplace violence events.

22.6.1 TYPE I EVENTS

The majority (60%) of workplace homicides involve a person entering a small late-night retail establishment, e.g., liquor store, gas station, or a convenience food store, to commit a robbery. During the commission of the robbery, an employee or, more likely, the proprietor is killed or injured.

Employees or proprietors who have face-to-face contact and exchange money with the public, work late at night and into the early morning hours, and work alone or in very small numbers are at the greatest risk of a Type I event. While the assailant may feign to be a customer as a pretext to enter the establishment, he or she has no legitimate business relationship to the workplace.

Retail robberies resulting in workplace assaults usually occur between the hours of eleven in the evening and six in the morning and are most often armed (gun or knife) robberies. In addition to employees who are classified as cashiers, many victims of late-night retail violence are supervisors or proprietors who are attacked while locking up their establishment for the night and janitors who are assaulted while cleaning the establishment after it is closed.

Other occupations/workplaces may be at risk of a Type I event. For instance, assaults on taxicab drivers also involve a pattern similar to retail robberies. The attack is likely to involve an assailant pretending to be a bona fide passenger during the late night or early morning hours who enters the taxicab to rob the driver of his or her fare receipts. Type I events also involve assaults on security guards. It has been known for some time that security guards are at risk of assault when protecting valuable property, which may be the object of an armed robbery.

22.6.1.1 Prevention Strategies for Type I Events

To many people, Type I workplace violence appears to be part of society's crime problem, and not a workplace safety and health problem at all. Under this view, the workplace is an innocent bystander, and the solution to the problem involves societal changes, not occupational safety and health principles. The ultimate solution to Type

I events may indeed involve societal changes, but until such changes occur it is still the employer's legal responsibility to provide a safe and healthful place of employment for their employees. Employers with employees who are known to be at risk for Type I events should be required to address workplace security hazards to satisfy the regulatory requirement of establishing, implementing, and maintaining an effective security program. The first step toward accomplishing this goal is strong management commitment to violence prevention. Employers at risk for Type I (as well as Types II and III) events should include the following as a part of their establishment's security program:

- A system for ensuring that employees comply with safe and healthy work practices, including ensuring that all employees, including supervisors and managers, comply with work practices designed to make the workplace more secure and do not engage in threats or physical actions that create a security hazard to other employees, supervisors, or managers in the workplace.
- A system for communicating with employees about workplace security hazards, including a means that employees can use to inform the employer of security hazards at the worksite without fear of reprisal.
- Procedures for identifying workplace security hazards including scheduled periodic inspections to identify unsafe conditions and work practices whenever the employer is made aware of a new or a previously unrecognized hazard.
- Procedures for investigating occupational injury or illness arising from a workplace assault or threat of assault.
- Procedures for correcting unsafe conditions, work practices, and work procedures, including workplace security hazards, with attention to procedures for protecting employees from physical retaliation for reporting threats.
- Training and instruction about how to recognize workplace security hazards, measures to prevent workplace assaults, and what to do when an assault occurs, including emergency action and post-emergency procedures.

The cornerstone of an effective workplace security plan is appropriate training of all employees, supervisors, and managers. Employers with employees at risk of workplace violence must educate them about the risk factors associated with the various types of workplace violence and provide appropriate training in crime awareness, assault and rape prevention, and defusing hostile situations. In addition, employers must instruct their employees about the steps to take during an emergency incident.

22.6.2 Type II Events

A Type II workplace violence event involves an assault by someone who is either the recipient or the object of a service provided by the affected workplace or the victim. Type I events represent the most common type of fatality. Type II events involving victims who provide services to the public are also increasing. Type II events accounted for approximately 30% of workplace homicides. Further, when more occupation-specific data about nonfatal workplace violence becomes available, nonfatal Type II events involving assaults to service providers, especially to health care providers, may represent the most prevalent category of workplace violence resulting in physical injury. Type II events involve fatal or nonfatal injuries to individuals who provide services to the public. These events involve assaults on public safety and correctional personnel, municipal bus or railway drivers, health care and social service providers, teachers, sales personnel, and other public or private service sector employees who provide professional, public safety, administrative, or business services to the public.

Law enforcement personnel are at risk of assault from the "object" of public safety services (suspicious persons, detainees, or arrestees) when making arrests, conducting drug raids, responding to calls involving robberies or domestic disputes, serving warrants and eviction notices, and investigating suspicious vehicles. Similarly, correctional personnel are at risk of assault while guarding or transporting jail or prison inmates. Of increasing concern, though, are Type II events involving assaults to the following types of service providers:

- Medical care providers in acute care hospitals, long-term care facilities, outpatient clinics, and home health agencies
- Mental health and psychiatric care providers in inpatient facilities, outpatient clinics, residential sites, and home health agencies
- Alcohol and drug treatment providers
- Social welfare service providers in unemployment offices, welfare eligibility offices, homeless shelters, probation offices, and child welfare agencies.
- Teaching and administrative and support staff in schools where students have a history of violent behavior
- Other types of service providers, e.g., justice system personnel, customer service representatives, and delivery personnel

Unlike Type I events, which often represent irregular occurrences in the life of any particular at-risk establishment, Type II events occur on a daily basis in many service establishments and therefore represent a more pervasive risk for many service providers.

22.6.2.1 Prevention Strategies for Type II Events

An increasing number of fatal, nonfatal assaults, and threats involve an employee who provides a service to a client, patient, customer, passenger, or other type of service recipient. Employers who provide service to recipients, or service objects, known or suspected to have a history of violence must also integrate an effective workplace security component into their security program. An important component of a workplace security program for employers at risk for Type II events is supervisor and employee training in how to effectively defuse hostile situations involving their clients, patients, customers, passengers, and members of the general public to whom they must provide services.

Employers concerned with Type II events need to be aware that the control of physical access through workplace design is also an important preventive measure.

This can include controlling access into and out of the workplace and freedom of movement within the workplace, in addition to placing barriers between clients and service providers. Escape routes can also be a critical component of workplace design. In certain situations, the installation of alarm systems or panic buttons may be an appropriate backup measure. Establishing a buddy system to be used in specified emergencies is often advisable as well. The presence of security personnel should also be considered where appropriate.

22.6.3   TYPE III EVENTS

A Type III workplace violence event consists of an assault by an individual who has some employment-related involvement with the workplace. Generally, a Type III event involves a threat of violence or a physical act of violence resulting in a fatal or nonfatal injury to an employee, supervisor, or manager of the affected workplace by the following types of individuals:

- A current or former employee, supervisor, or manager.
- Some other person who has a dispute with an employee of the affected workplace, e.g., current/former spouse or lover, relative, friend, or acquaintance.
- Type III events account for a much smaller proportion of fatal workplace injuries. Type III events accounted for only 10% of workplace homicides. Nevertheless, Type III fatalities often attract significant media attention and are incorrectly characterized by many as representing "the" workplace violence problem. In fact, their media visibility makes them appear much more common than they actually are.
- Most commonly, the primary target of a Type III event is a co-employee, a supervisor, or manager of the assailant. In committing a Type III assault, an individual may be seeking revenge for what he or she perceives as unfair treatment by a co-employee, a supervisor, or a manager. Increasingly, Type III events involve domestic or romantic disputes in which an employee is threatened in the workplace by an individual with whom he/she has a personal relationship outside of work.
- At first glance, a Type III assailant's actions may defy reasonable explanation. Often, his or her actions are motivated by perceived difficulties in his or her relationship with the victim, or with the affected workplace, and by psychosocial factors that are peculiar to the assailant.

Even though incomplete, existing data indicate that the number of Type III events resulting in nonfatal injury, or in no physical injury at all, greatly exceeds the number of fatal Type III events. Indeed, the most prevalent Type III event may involve threats and other types of verbal harassment.

22.6.3.1   Prevention Strategies for Type III Events

In a Type III event, the assailant has an employment-related involvement with the workplace. Usually, a Type III event involves a threat of violence, or a physical act of violence resulting in fatal or nonfatal injury, to an employee of the affected workplace by a current/former employee, supervisor or manager, or by some other person who has a dispute with an employee of the affected workplace, e.g., a current/former spouse or lover, relative, friend, or acquaintance.

Employers who have employees with a history of assaults or who have exhibited belligerent, intimidating, or threatening behavior in the workplace need to establish and implement procedures to respond to workplace security hazards when they are present and to provide training as necessary to their employees, supervisors, and managers to satisfy the regulatory requirement of establishing, implementing, and maintaining an effective IIP Program.

Since Type III events are more closely tied to employer–employee relations than are Type I or II events, an employer's considerate and respectful management of his or her employees represents an effective strategy for preventing Type III events. Some workplace violence researchers have pointed out that employer actions that are perceived by an employee to place his or her continuing employment status in jeopardy can be triggering events for a workplace violence event, e.g., layoffs or reduction-in-force actions and disciplinary actions such as suspensions and terminations. Thus, where actions such as these are contemplated, they should be carried out in a manner that is designed to minimize the potential for related Type III events.

Some mental health professionals believe that belligerent, intimidating, or threatening behavior by an employee or supervisor is an early warning sign of an individual's propensity to commit a physical assault in the future, and that monitoring and appropriately responding to such behavior is a necessary part of effective prevention.

Many management consultants who advise employers about workplace violence stress that to effectively prevent Type III events from occurring, employers need to establish a clear anti-violence management policy, apply the policy consistently and fairly to all employees, including supervisors and managers, and provide appropriate supervisory and employee training in workplace violence prevention.

Lastly, an important subset of Type III workplace violence events affects women disproportionately. Domestic violence is now spilling over into the workplace and employers need to take appropriate precautions to protect at-risk employees. For instance, when an employee reports threats from an individual with whom he or she has (or had) a personal relationship, employers should take appropriate precautions to ensure the safety of the threatened employee, as well as other employees who are in the zone of danger and who may be harmed if a violent incident occurs in the workplace. One option is to seek a temporary restraining order (TRO) and an injunction on behalf of the affected employee. Any employer may seek a TRO/injunction on behalf of an employee when he or she has suffered unlawful violence (assault, battery, or stalking).

22.6.4 TYPES I, II, AND III VIOLENCE EVENTS CHECKLIST

22.6.4.1 Pre-Event Measures

- Make your establishment unattractive to robbers by
  - Remove clutter, obstructions, and signs from the windows so that an unobstructed view of the establishment counter or cash register exists.

- Keep the store and parking lot as brightly lit as local law allows.
- Keep an eye on what is going on outside the establishment and report any suspicious persons or activities to the police.
- When there are no customers in the store, keep busy with other tasks away from the cash register.
- Post emergency police and fire department numbers and the establishment's address by the phone.
- Mount mirrors on the ceiling to help you keep an eye on hidden corners of the store. Consider surveillance cameras to record what goes on in the store and to act as a deterrent.
- Post signs that inform customers that you have a limited amount of cash on hand. Make sure they are placed so that they are easy to spot from the outside of the store.
- Limit accessible cash to a small amount and keep only small bills in the cash register.
- Use a time access safe for larger bills and deposit them as they are received.
- Use only one register after dark and leave unused registers open with empty cash drawers tilted up for all to see.
- Let your customers know that you keep only a small amount of cash on hand.
- Event measures
  - If you are robbed at gunpoint, stay calm and speak to the robber in a cooperative tone. Do not argue or fight with the robber and offer no resistance whatsoever. Hand over the money.
  - Never pull a weapon during the event—it will only increase your chances of getting hurt.
  - Always move slowly and explain each move to the robber before you do it.
- Post-event measures
  - Make no attempt to follow or chase the robber.
  - Stay where you are until you are certain the robber has left the immediate area, then lock the door of your establishment and call the police immediately.
  - Do not touch anything the robber has handled.
  - Write down everything you remember about the robber and the robbery while you wait for the police to arrive.
  - Do not open the door of the establishment until the police arrive.

Effective security management to prevent all three types of workplace violence events also includes post-event measures such as emergency medical care and debriefing employees about the incident. After a workplace assault occurs, employers should provide post-event trauma counseling to those who desire such intervention in order to reduce the short- and long-term physical and emotional effects of the incident.

Workplace safety and health hazards affecting employees have traditionally been viewed as arising from unsafe work practices, hazardous industrial conditions, or exposures to harmful chemical, biological, or physical agents, not from violent acts committed by other human beings. Recently, though, employees as well as supervisors and managers have become, all too frequently, victims of assaults or other violent acts in the workplace that entail a substantial risk of physical or emotional harm. Many of these assaults result in fatal injury, but an even greater number result in nonfatal injury, or in the threat of injury, which can lead to medical treatment, missed work, lost wages, and decreased productivity.

A single explanation for the increase in workplace violence is not readily available. Some episodes of workplace violence, like robberies of small retail establishments, seem related to the larger societal problems of crime and substance abuse. Other episodes seem to arise more specifically from employment-related problems.

What can be done to prevent workplace violence? Any preventive measure must be based on a thorough understanding of the risk factors associated with the various types of workplace violence. Moreover, even though our understanding of the factors that lead to workplace violence is not perfect, sufficient information is available that, if used effectively, can reduce the risk of workplace violence. However, strong management commitment and the day-to-day involvement of managers, supervisors, employees, and labor unions are required to reduce the risk of workplace violence.

Workplace violence has become a serious occupational health problem requiring the combined efforts of employers, employees, labor unions, government, academic researchers, and security professionals. The problem cannot be solved by government alone.

23 Other Hazards

Handrails and a ramped entrance make this doctor's office safer for clients.

23.1 AISLES AND PASSAGEWAYS (29 CFR 1910.17, .22, AND .176)

Aisles and passageways must be free from debris and kept clear for travel. Where mechanical material-handling equipment is used, sufficient safe clearance must be allowed for aisles, at loading docks, through doorways, and wherever turns or passage must be made. Aisles and passageways used by mechanical equipment are to be kept clear and in good repair with no obstructions across or in aisles that could create hazards. All permanent aisles and passageways should be appropriately marked. Areas where workers walk should be covered and guardrails are to be provided to protect workers from the hazards of open pits, tanks, vats, ditches, etc.

23.2 COMPRESSORS AND COMPRESSED AIR (29 CFR 1910.242)

The supplying of compressed air is accomplished by a compressor. Great care must be taken to ensure that such types of equipment operate in a safe manner. Safety devices for a compressed-air system should be checked frequently. Compressors should be equipped with pressure relief valves and pressure gauges.

Air intakes must be installed and equipped to ensure that only clean uncontaminated air enters the compressor. Air filters installed on the compressor intake facilitate this.

Before any repair work is done on the pressure system of a compressor, the pressure is to be bled off and the system locked out. All compressors must be operated and lubricated in accordance with the manufacturer's recommendations.

Signs are to be posted to warn of the automatic starting feature of compressors. The belt drive system is to be totally enclosed to provide protection from any contact.

No worker should direct compressed air toward a person, and employees are prohibited from using highly compressed air for cleaning purposes. If compressed air is used for cleaning of clothing, the pressure is to be reduced to less than 10 psi. When using compressed air for cleaning, employees are to wear protective chip guarding eyewear and personal protective equipment (PPE).

Safety chains or other suitable locking devices are to be used at couplings of high-pressure hose lines where a connection failure would create a hazard. Before compressed air is used to empty containers of liquid, the safe working pressure of the container is to be checked.

When compressed air is used with abrasive blast cleaning equipment, the operating valve type must be held open manually. When compressed air is used to inflate auto tires, a clip-on chuck and an inline regulator preset to 40 psi is required. Using compressed air to clean up or move combustible dust must be prohibited because such action could cause the dust to be suspended in the air and cause a fire or explosion hazard.

23.3 COMPRESSED-GAS CYLINDERS (29 CFR 1910.101 AND .253)

Incidents have occurred where compressed gas cylinders have exploded and have become airborne. There is a lot of stored energy in a compressed gas cylinder, which is why they should be handled with great care. Cylinders with a water weight capacity over 30 lb, must be equipped with means for connecting a valve protector device or with a collar or recess to protect the valve. Cylinders should be legibly marked to clearly identify the gas contained within the cylinder. Compressed gas cylinders are to be stored in areas that are to be protected from external heat sources such as flame impingement, intense radiant heat, electric arcs, or high-temperature lines. Inside of buildings, cylinders are to be stored in a well-protected, well-ventilated, dry location away from combustible materials by 20 ft. Also, the in-plant handling, storage, and utilization of all compressed gases in cylinders, portable tanks, rail tank cars, or motor vehicle cargo tanks should be in accordance with Compressed Gas Association pamphlet P-1-1965 (see Figure 23.1).

Cylinders are to be located or stored in areas where they will not be damaged by passing or falling objects or subject to tampering by unauthorized persons. Cylinders are to be stored or transported in a manner to prevent them from creating a hazard by tipping, falling, or rolling and stored 20 ft away from highly combustible materials. Where a cylinder is designed to accept a valve protection cap, caps are to be in place except when the cylinder is in use or is connected for use.

Cylinders containing liquefied fuel gas are to be stored or transported in a position such that the safety relief device is always in direct contact with the vapor space in the cylinder. All valves must be closed off before a cylinder is moved, when the cylinder is empty, and at the completion of each job. Low-pressure fuel-gas

FIGURE 23.1   Safe use and storage of compressed-gas cylinder around the public is good safety practices.

cylinders should be checked periodically for corrosion, general distortion, cracks, or any other defect that might indicate a weakness or render it unfit for service.

23.4   COMPRESSED GASES (29 CFR 1910.101, .102, .103, .104, .106, AND .253)

There are several hazards associated with compressed gases, including oxygen displacement, fires, explosions, toxic effects from certain gases, as well as the physical hazards associated with pressurized systems. Special storage, use, and handling precautions are necessary to control these hazards.

There are specific safety requirements for many of the compressed gases such as acetylene, hydrogen, nitrous oxide, and oxygen.

Acetylene cylinders are to be stored and used in a vertical, valve-end-up position only. Under no conditions should acetylene be generated, piped (except in approved cylinder manifolds), or used at a pressure in excess of 15 or 30 psi. The use of liquid acetylene is prohibited. The in-plant transfer, handling, and storage of acetylene in cylinders is to be in accordance with the Compressed Gas Association pamphlet C-1.3-1959.

Hydrogen containers must comply with one of the following: (1) designed, constructed, and tested in accordance with appropriate requirements of The American Society of Mechanical Engineers (ASME's) *Boiler and Pressure Vessel Code, Section VIII Unfired Pressure Vessels—1968*; or (2) designed, constructed, tested, and maintained in accordance with U.S. Department of Transportation specifications and regulations.

Hydrogen systems are to be located so that they are readily accessible to delivery equipment and to authorized personnel and must be located aboveground and not beneath electric power lines. Systems must not be located close to flammable liquid piping or piping of other flammable gases. Permanently installed containers are to be provided with substantial noncombustible supports on firm noncombustible foundations.

Nitrous oxide piping systems for the in-plant transfer and distribution of nitrous oxide are to be designed, installed, maintained, and operated in accordance with the Compressed Gas Association pamphlet G-8.1-1964.

Oxygen cylinders in storage must be separated from fuel-gas cylinders or combustible materials (especially oil or grease) a minimum distance of 20 ft. or by a noncombustible barrier at least 5 ft. high with a fire-resistance rating of 1/2 h.

23.5 CONTROL OF HAZARDOUS ENERGY SOURCES [LOCKOUT/TAGOUT] (29 CFR 1910.147)

Lockout/tagout deals with the preventing of the release of energy from machines, equipment, and electrical circuits that are perceived to be de-energized. OSHA estimates that compliance with the lockout/tagout standard will prevent about 120 fatalities and approximately 28,000 serious and 32,000 minor injuries each year. About 39 million general industry workers are to be protected from accidents during maintenance and servicing of equipment under this ruling.

The standard for the control of hazardous energy sources (lockout–tagout) covers servicing and maintenance of machines and equipment in which the unexpected energization or start-up of the machines or equipment or release of stored energy could cause injury to employees. The rule generally requires that energy sources for equipment be turned off or disconnected and that the switch be either locked or labeled with a warning tag. About 3 million workers servicing equipment face the greatest risk. These include craft workers, machine operators, and laborers. OSHA's data show that packaging and wrapping equipment, printing presses, and conveyors account for a high proportion of the accidents associated with lockout/tagout failures.

Typical injuries include fractures, lacerations, contusions, amputations, and puncture wounds with the average lost time for injuries running to 24 days. Agriculture, maritime, and construction employers are not covered under standard 29 CFR 1910.147. In addition, the generation, transmission, and distribution of electric power by utilities and work on electric conductors and equipment are excluded. The general requirements under the regulation require employers to

- Develop an energy control program
- Use locks when equipment can be locked out
- Ensure that new equipment or overhauled equipment can accommodate locks
- Employ additional means to ensure safety when tags rather than locks are used by using an effective tagout program

- Identify and implement specific procedures (generally in writing) for the control of hazardous energy including preparation for shutdown, equipment isolation, lockout/tagout application, release of stored energy, and verification of isolation
- Institute procedures for release of lockout/tagout including machine inspection, notification, safe positioning of employees, and removal of the lockout/tagout device
- Obtain standardized locks and tags, which indicate the identity of the employee using them and which are of sufficient quality and durability, to ensure their effectiveness
- Require that each lockout/tagout device be removed by the employee who applied the device
- Conduct inspections of energy control procedures at least annually
- Train employees in the specific energy control procedures with training reminders as part of the annual inspections of the control procedures
- Adopt procedures to ensure safety when equipment must be tested during servicing, when outside contractors are working at the site, when a multiple lockout is needed for a crew servicing equipment, and when shifts or personnel change

Excluded from coverage are normal production operations including

- Repetitive, routine minor adjustments, which would be covered under OSHA's machine guarding standards.
- Work on cord- and plug-connected electric equipment when it is unplugged, and the employee working on the equipment has complete control over the plug.
- Hot tap operations involving gas, steam, water, or petroleum products when the employer shows that continuity of service is essential, shutdown is impractical, and documented procedures are followed to provide proven effective protection for employees.

In summary, all machinery or equipment capable of movement is required to be de-energized or disengaged and locked out during cleaning, servicing, adjusting, or setting up operations, whenever required. Where the power disconnecting means for equipment does not also disconnect the electrical control circuit, the appropriate electrical enclosures must be identified. A means should be provided to ensure that the control circuit can also be disconnected and locked out. The locking out of control circuits in lieu of locking out main power disconnects must be prohibited. All equipment control valve handles are to be provided with a means for locking out. The lockout procedure requires that stored energy (mechanical, hydraulic, air, etc.) must be released or blocked before equipment is locked out for repairs. Appropriate employees must be provided with individually keyed personal safety locks and they must be expected to keep personal control of their key(s) while they have safety locks in use. It should be required that only the employee exposed to the hazard places or removes the safety lock. Employees must check the safety of the lockout by attempting a start-up after making sure that no one is exposed. Employees must be instructed to always push the control circuit stop button immediately after checking the safety of the lockout. A means is to be provided to identify any or all employees who work on locked out equipment by their locks or accompanying tags. A sufficient number of accident preventive signs or tags and safety padlocks need to be provided for any reasonably foreseeable repair emergency. When machine operations, configuration, or size requires the operator to leave his or her control station to install tools or perform other operations, and that part of the machine could move if accidentally activated, that element is required to be separated, locked, or blocked out. In the event that equipment or lines cannot be shut down, locked out, and tagged, a safe job procedure is to be established and rigidly followed.

23.6 ELEVATED SURFACES

Elevated surfaces present a real potential for falls and falling objects from above. Surfaces elevated more than 30 in. above the floor or ground must be provided with standard guardrails. Elevated surfaces (beneath which people or machinery could be exposed to falling objects) are to be provided with standard 4 in. toeboards. Material on elevated surfaces must be piled, stacked, or racked to prevent them from tipping, falling, collapsing, rolling, or spreading (see Figure 23.2).

23.7 FLAMMABLE AND COMBUSTIBLE LIQUIDS (29 CFR 1910.106)

Flammable liquids are to be kept in covered containers or tanks when not in use. The quantity of flammable or combustible liquid that may be located outside or inside a

FIGURE 23.2   Unsafe elevated work area due to lack of guardrails and toeboards.

storage room or storage cabinet in any one fire area of a building cannot exceed the following specifications:

- Class IA liquids in containers: 25 gal
- Class IB, IC, II, or III liquids in containers: 120 gal
- Class IB, IC, II, or III liquids in a single portable tank: 660 gal

Flammable and combustible liquids are to be drawn from or transferred into containers within buildings only through a closed piping system, from safety cans, by means of a device drawing through the top, or by gravity through an approved self-closing valve. Transfer by means of air pressure is prohibited. Not more than 60 gal of Class I or Class II liquids, nor more than 120 gal of Class III liquids may be stored in a storage cabinet. Inside storage rooms for flammable and combustible liquids are to be constructed to meet the required fire-resistive rating or wiring for their uses.

Outside storage areas must be graded to divert spills away from buildings or other exposures or be surrounded with curbs at least 6 in. high with appropriate drainage to a safe location for accumulated liquids The areas shall be protected against tampering or trespassing, where necessary, and shall be kept free of weeds, debris, and other combustible material not necessary to the storage.

Adequate precautions are to be taken to prevent the ignition of flammable vapors. Sources of ignition include, but are not limited to, open flames, lightning, smoking, cutting, and welding, hot surfaces, frictional heat, static, electrical, and mechanical sparks, spontaneous ignition, including heat-producing chemical reactions, and radiant heat.

Class I liquids are not to be dispensed into containers unless the nozzle and container are electrically interconnected. All bulk drums of flammable liquids are to be grounded and bonded to containers during dispensing.

23.8 FLAMMABLE AND COMBUSTIBLE MATERIALS

Combustible scrap, debris, and waste materials (oily rags, etc.) stored in covered metal receptacles are to be removed from the worksite promptly. Proper storage must be practiced to minimize the risk of fire including spontaneous combustion. Fire extinguishers are to be selected and provided for the types of materials in areas where they are to be used. NO SMOKING rules should be enforced in areas involving storage and use of hazardous materials.

23.9 FLOORS [GENERAL CONDITIONS] (29 CFR 1910.22 AND .23)

All floor surfaces are to be kept clean, dry, and free from protruding nails, splinters, loose boards, holes, or projections. Where wet processes are used, drainage is to be maintained, and false floors, platforms, mats, or other dry standing places are to be provided where practical.

In every building or other structures, or part thereof, used for mercantile, business, industrial, or storage purposes, the loads approved by the building official are to be marked on plates of approved design, which are to be supplied and securely affixed by the owner of the building, or their duly authorized agent, in a conspicuous place in each space to which they relate. Such plates must not be removed or defaced but, if lost, removed, or defaced, shall be replaced by the owner or his/her agent.

Every stairway and ladderway floor opening is to be guarded by standard railings, with standard toeboards on all exposed sides except at the entrance. For infrequently used stairways, the guard may consist of a hinged cover and removable standard railings. The entrance to ladderway openings must be guarded to prevent a person walking directly into the opening.

Every hatchway and chute floor opening is to be guarded by a hinged floor opening cover equipped with standard railings, to leave only one exposed side or a removable railing with toeboard on not more than two sides and a fixed standard railing with toeboards on all other exposed sides.

Every floor hole into which persons can accidentally walk shall be guarded by either a standard railing with standard toeboard on all exposed sides, or a floor hole cover that should be hinged in place. While the cover is not in place, the floor hole is to be attended or to be protected by a removable standard railing.

Every open-sided floor, platform, or runway 4 ft. or more above adjacent floor or ground level is to be guarded by a standard railing with toeboard on all open sides, except where there is an entrance to a ramp, stairway, or fixed ladder. Runways not less than 18 in. wide used exclusively for special purposes may have the railing on one side omitted where operating conditions necessitate. Regardless of height, open-sided floors, walkways, platforms, or runways above or adjacent to dangerous equipment are to be guarded with a standard railing and toeboard.

23.10 FORKLIFT TRUCKS (POWERED INDUSTRIAL TRUCKS) (29 CFR1910.178)

The American Society of Mechanical Engineers (ASME) defines a powered industrial truck as a mobile, power-propelled truck used to carry, push, pull, lift, stack, or tier materials. Powered industrial trucks are also commonly known as forklifts, pallet trucks, rider trucks, fork trucks, or lift trucks. Each year, tens of thousands of forklift-related injuries occur in US workplaces. Injuries usually involve employees being struck by lift trucks or falling while standing or working from elevated pallets and tines. Many employees are injured when lift trucks are inadvertently driven off loading docks or when the lift falls between a dock and an unchocked trailer. Most incidents also involve property damage, including damage to overhead sprinklers, racking, pipes, walls, machinery, and other equipment. Unfortunately, a majority of employee injuries and property damage can be attributed to lack of procedures, insufficient or inadequate training, and lack of safety-rule enforcement.

If at any time, a powered industrial truck is found to be in need of repair, defective, or in any way unsafe, the truck is to be taken out of service until it has been restored to a safe operating condition.

High-lift rider trucks shall be equipped with substantial overhead guards unless operating conditions do not permit. Fork trucks are to be equipped with vertical-load backrest extensions when the types of loads present a hazard to the operators. Each industrial truck is to have a warning alarm, whistle, gong, or other device that can be clearly heard above the normal noise in the areas where it is operated. The brakes of trucks are to be set and wheel chocks placed under the rear wheels to prevent the movement of trucks, trailers, or railroad cars while loading or unloading.

Only trained and authorized operators are permitted to operate a powered industrial truck. Methods are to be devised to train operators in the safe operation of powered industrial trucks.

23.11   HAND TOOLS (29 CFR 1910.242)

Hand and power tools are a common part of our everyday lives and are present in nearly every industry. These tools help us to easily perform tasks that otherwise would be difficult or impossible. However, these simple tools can be hazardous and have the potential for causing severe injuries when used or maintained improperly. Special attention toward hand and power tool safety is necessary to reduce or eliminate these hazards (see Figure 23.3).

Hand tools are nonpowered. They include anything from axes to wrenches. The greatest hazards posed by hand tools result from misuse and improper maintenance. For example, using a screwdriver as a chisel may cause the tip of the screwdriver to break and fly, hitting the user or other employees; if a wooden handle on a tool such as a hammer or an axe is loose, splintered, or cracked, the head of the tool may fly off and strike the user or another worker; A wrench must not be used if its jaws are sprung, because it might slip; or impact tools such as chisels, wedges, or drift pins are unsafe if they have mushroomed heads. The heads might shatter on impact, sending sharp fragments flying.

The employer is responsible for the safe condition of tools and equipment used by employees but the employees have the responsibility for properly using and

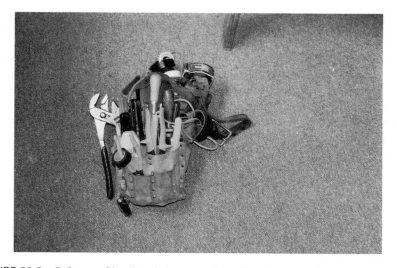

FIGURE 23.3   Safe use of hand tools is a part of people-oriented services.

maintaining tools. Employers should caution employees that saw blades, knives, or other tools must be directed away from aisle areas and other employees working in close proximity. Knives and scissors must be sharp. Dull tools can be more hazardous than sharp ones. Appropriate PPE, for example, safety goggles, gloves, etc., should be worn because of hazards that may be encountered while using portable power tools and hand tools.

Safety requires that floors be kept as clean and dry as possible to prevent accidental slips with or around dangerous hand tools. Around flammable substances, sparks produced by iron and steel hand tools can be a dangerous ignition source. Where this hazard exists, spark-resistant tools made from brass, plastic, aluminum, or wood will provide for safety.

Employees who use hand and power tools and who are exposed to the hazards of falling, flying, abrasive, and splashing objects, or exposed to harmful dusts, fumes, mists, vapors, or gases must be provided with the particular personal equipment necessary to protect them from the hazard.

23.12 HOUSEKEEPING (29 CFR 1910.22)

Housekeeping is by far one of the easiest preventive methods to implement and probably has the most immediate impact on the workplace. Poor housekeeping leads to a myriad of hazards from slips, trips, and falls to fire hazards. Housekeeping is an area of hazard removal in which everyone in the workplace can take an active role. It affects more than accident prevention and increases morale and productivity.

23.13 LADDERS, FIXED (29 CFR 1910.27)

A fixed ladder must be able to support at least two loads of 250 lb each, concentrated between any two consecutive attachments. Fixed ladders must also support added anticipated loads caused by ice buildup, winds, rigging, and impact loads resulting from using ladder safety devices. Fixed ladders must be used at a pitch no greater than 90° from the horizontal, measured from the backside of the ladder.

Individual rung/stepladders must extend at least 42 in. above an access level or landing platform either by the continuation of the rung spacing as horizontal grab bars or by providing vertical grab bars, which must have the same lateral spacing as the vertical legs of the ladder rails. Each step or rung of a fixed ladder must be able to support a load of at least 250 lb applied in the middle of the step or rung.

The minimum clear distance between the sides of individual rung/stepladders and between the side rails of other fixed ladders must be 16 in. The rungs of individual rung/stepladders must be shaped to prevent slipping off the end of the rungs. The rungs and steps of fixed metal ladders manufactured after March 15, 1991, must be corrugated, knurled, dimpled, coated with skid-resistant material, or treated to minimize slipping. The minimum perpendicular clearance between fixed ladder rungs, cleats, and steps and any obstruction behind the ladder must be 7 in., except for the clearance of an elevator pit ladder, which must be 4.5 in. The minimum perpendicular clearance between the centerline of fixed ladder rungs, cleats, steps, and any obstruction on the climbing side of the ladder must be 30 in.

If obstructions are unavoidable, clearance may be reduced to 24 in., provided a deflection device is installed to guide workers around the obstruction. The step-across distance between the center of the steps or rungs of fixed ladders and the nearest edge of a landing area must be no less than 7 in. and no more than 12 in. A landing platform must be provided if the step-across distance exceeds 12 in. Fixed ladders without cages or wells must have at least a 15 in. clear width to the nearest permanent object on each side of the centerline of the ladder.

Fixed ladders must be provided with cages, wells, ladder safety devices, or self-retracting lifelines where the length of climb is less than 24 ft. but the top of the ladder is at a distance greater than 24 ft. above lower levels. If the total length of the climb on a fixed ladder equals or exceeds 24 ft., the following requirements must be met: fixed ladders must be equipped with either (a) ladder safety devices; (b) self-retracting lifelines and rest platforms at intervals not to exceed 150 ft.; or (c) a cage or well and multiple ladder sections, each ladder section not to exceed 50 ft. in length. These ladder sections must be offset from adjacent sections, and landing platforms must be provided at maximum intervals of 50 ft.

The side rails of through or side-step fixed ladders must extend 42 in. above the top level or landing platform served by the ladder. Parapet ladders must have an access level at the roof if the parapet is cut to permit passage through it; if the parapet is continuous, the access level is the top of the parapet. Steps or rungs for through-fixed-ladder extensions must be omitted from the extension; and the extension of side rails must be flared to provide between 24 and 30 in. clearance between side rails. When safety devices are provided, the maximum clearance distance between side rail extensions must not exceed 36 in.

Cages must not extend less than 27 in., or more than 30 in. from the centerline of the step or rung, and must not be less than 27 in. wide. The inside of the cage must be clear of projections.

Horizontal bands must be fastened to the side rails of rail ladders or directly to the structure, building, or equipment for individual-rung ladders. Horizontal bands must be spaced at intervals not more than 4 ft. apart, measured from centerline to centerline. Vertical bars must be on the inside of the horizontal bands and must be fastened to them. Vertical bars must be spaced at intervals not more than 9.5 in., measured from centerline to centerline.

The bottom of the cage must be between 7 and 8 ft. above the point of access to the bottom of the ladder. The bottom of the cage must be flared not less than 4 in. between the bottom horizontal band and the next higher band. The top of the cage must be a minimum of 42 in. above the top of the platform or the point of access at the top of the ladder. Provisions must be made for access to the platform or other point of access.

Wells must completely encircle the ladder. Wells must be free of projections. The inside face of the well on the climbing side of the ladder must extend between 27 and 30 in. from the centerline of the step or rung. The inside width of the well must be at least 30 in. The bottom of the well above the point of access to the bottom of the ladder must be between 7 and 8 ft.

All safety devices must be able to withstand, without failure, a drop test consisting of a 500 lb weight dropping 18 in. All safety devices must permit the worker to ascend or descend without continually having to hold, push, or pull any part of the device, leaving both hands free for climbing. All safety devices must be activated within 2 ft. after a fall occurs and limit the descending velocity of an employee to 7 ft./s or less. The connection between the carrier or lifeline and the point of attachment to the body harness must not exceed 9 in. in length.

Mountings for rigid carriers must be attached at each end of the carrier, with intermediate mountings, spaced along the entire length of the carrier, to provide the necessary strength to stop workers' falls. Mountings for flexible carriers must be attached at each end of the carrier. Cable guides for flexible carriers must be installed with a spacing between 25 and 40 ft. along the entire length of the carrier, to prevent wind damage to the system. The design and installation of mountings and cable guides must not reduce the strength of the ladder. Side rails and steps or rungs for side-step fixed ladders must be continuous in extension.

Fixed ladders with structural defects—such as broken or missing rungs, cleats, or steps, broken or split rails, or corroded components—must be withdrawn from service until repaired. Defective fixed ladders are considered withdrawn from use when they are (1) immediately tagged with Do Not Use or similar language; (2) marked in a manner that identifies them as defective; or (3) blocked such as with a plywood attachment that spans several rungs.

23.14   LADDERS, PORTABLE (29 CFR 1910.25 AND .26)

Nonself-supporting and self-supporting portable ladders must support at least 4 times the maximum intended load; extra heavy-duty type 1A metal or plastic ladders must sustain 3.3 times the maximum intended load. The ability of a self-supporting ladder to sustain loads must be determined by applying the load to the ladder in a downward vertical direction. The ability of a nonself-supporting ladder to sustain loads must be determined by applying the load in a downward vertical direction when the ladder is placed at a horizontal angle of 75.5°.

When portable ladders are used for access to an upper landing surface, the side rails must extend at least 3 ft. above the upper landing surface. When such an extension is not possible, the ladder must be secured, and a grasping device such as a grab rail must be provided to assist workers in mounting and dismounting the ladder. A ladder extension must not deflect under a load that would cause the ladder to slip off its supports.

Ladders must be maintained free of oil, grease, and other slipping hazards. Ladders must not be loaded beyond the maximum intended load for which they were built or beyond the manufacturer's rated capacity. Ladders must be used only for the purpose for which they were designed. Nonself-supporting ladders must be used at an angle where the horizontal distance from the top support to the foot of the ladder is approximately one quarter of the working length of the ladder. Wood job-made ladders with spliced side rails must be used at an angle where the horizontal distance is one eighth the working length of the ladder.

Ladders must be used only on stable and level surfaces unless secured to prevent accidental movement. Ladders must not be used on slippery surfaces unless secured or provided with slip-resistant feet to prevent accidental movement. Slip-resistant feet must not be used as a substitute for the care in placing, lashing, or holding a ladder upon slippery surfaces. Ladders placed in areas such as passageways, doorways, or driveways, or where they can be displaced by workplace activities or traffic must be secured to prevent accidental movement or a barricade must be used to keep traffic or activities away from the ladder. The area around the top and bottom of the ladders must be kept clear.

The top of a nonself-supporting ladder must be placed with two rails supported equally unless it is equipped with a single support attachment. Ladders must not be moved, shifted, or extended while in use. Ladders must have nonconductive side rails if they are used where the worker or the ladder could contact exposed energized electrical equipment.

The top or top step of a stepladder must not be used as a step. Crossbracing on the rear section of stepladders must not be used for climbing unless the ladders are designed and provided with steps for climbing on both front and rear sections.

Ladders must be inspected by a competent person for visible defects on a periodic basis and after any incident that could affect their safe use. Single-rail ladders must not be used.

When ascending or descending a ladder, the worker must face the ladder. Each worker must use at least one hand to grasp the ladder when climbing. A worker on a ladder must not carry any object or load that could cause him/her to lose balance and fall.

A double-cleated ladder or two or more ladders must be provided when ladders are the only way to enter or exit a work area with 25 or more employees, or when a ladder serves simultaneous two-way traffic. Ladder rungs, cleats, and steps must be parallel, level, and uniformly spaced when the ladder is in position for use. Rungs, cleats, and steps of portable and fixed ladders (except as provided below) must not be spaced less than 10 in. apart, nor more than 14 in. apart, along the ladder's side rails. Rungs, cleats, and steps of step stools must not be less than 8 in. apart, nor more than 12 in. apart, between centerlines of the rungs, cleats, and steps.

Ladders must not be tied or fastened together to create longer sections unless they are specifically designed for such use. A metal spreader or locking device must be provided on each stepladder to hold the front and back sections in an open position when the ladder is used. Two or more separate ladders used to reach an elevated work area must be offset with a platform or landing between the ladders, except when portable ladders are used to gain access to fixed ladders.

Ladder components must be surfaced to prevent injury from punctures or lacerations and prevent snagging of clothing. Wood ladders must not be coated with any opaque covering, except for identification or warning labels, which may be placed only on one face of a side rail.

Portable ladders with structural defects—such as broken or missing rungs, cleats, or steps, broken or split rails, corroded components, or other faulty or defective components—must immediately be marked defective or tagged with Do Not Use or similar language and withdrawn from service until repaired. Ladder repairs must restore the ladder to a condition meeting its original design criteria before the ladder is returned to use.

Under the provisions of the OSHA standard, employers must provide a training program for each employee using ladders and stairways. The program must enable each employee to recognize hazards related to ladders and stairways and to use proper procedures to minimize these hazards. For example, employers must ensure that each employee is trained by a competent person in the following areas, as applicable:

- Nature of fall hazards in the work area
- Correct procedures for erecting, maintaining, and disassembling the fall protection systems to be used
- Proper construction, use, placement, and care in handling of all stairways and ladders
- Maximum intended load-carrying capacities of ladders used

23.15 MATERIAL HANDLING (29 CFR 1910.176)

Some form of material handling transpires in all workplaces. Material handling is one of the leading causes of occupational injuries whether it be manual lifting or the use of lifting equipment. Great care must be taken to assure safe clearance for equipment through aisles and doorways. Aisleways are to be designated, permanently marked, and kept clear to allow unhindered passage.

Any motorized vehicles and mechanized equipment used for material handling should be inspected daily or before use. Vehicles must be shut off and brakes set prior to loading or unloading. Trucks and trailers are to be secured from movement during loading and unloading operations. Dockboards (bridge plates) are to be used when loading or unloading operations take place between vehicles and docks. Dockplates and loading ramps are constructed and maintained with sufficient strength to support imposed loading.

Chutes are to be equipped with sideboards of sufficient height to prevent the materials handled from falling off. Chutes and gravity roller sections must be firmly placed or secured to prevent displacement. At the delivery end of the rollers or chutes, provisions must be made to break the movement of the handled materials.

Hooks with safety latches or other arrangements used when hoisting materials must be designed such that slings or load attachments would not accidentally slip off the hoist hooks. Securing chains, ropes, chokers, or slings must be adequate for the job to be performed. When hoisting material or equipment, provisions shall be made to ensure that no one passes under the suspended loads.

Containers of combustibles or flammables, when stacked while they are moved, are always separated by dunnage sufficient to provide stability. Material safety data sheets (MSDSs) need to be available to employees handling hazardous substances.

Material handling is one of the hazards in the workplace that affect everyone. Seldom is anyone not exposed to the potential hazards at some period during their work (see Figure 23.4).

23.16 MOTOR VEHICLE SAFETY

According to the Bureau of Labor Statistics (BLS), over 2000 deaths a year result from occupational motor vehicle incidents, more than 30% of the total annual

FIGURE 23.4 Pushing is always better than pulling a load.

number of fatalities from occupational injuries. These deaths include driver and passenger deaths in highway crashes, farm equipment accidents, and industrial vehicle incidents as well as pedestrian fatalities. There are no specific OSHA standards concerning workplace motor vehicle safety; however, most of the occupational fatalities occur on public highways where there are seat belt requirements and traffic laws. OSHA issued a notice of proposed rulemaking in July 1990 for a standard that would have required seat belt use and driver awareness programs. Nothing has transpired regarding this proposal.

23.17   PORTABLE (POWER-OPERATED) TOOLS AND EQUIPMENT (29 CFR 1910.243)

Tools are such a common part of our lives that it is difficult to remember that they may pose hazards. All tools are manufactured with safety in mind but, tragically, a serious accident often occurs before steps are taken to search out and avoid or eliminate tool-related hazards. In the process of removing or avoiding the hazards, workers must learn to recognize the hazards associated with the different types of tools and the safety precautions necessary to prevent those hazards. All hazards involved in the use of power tools can be prevented by following five basic safety rules:

- Keep all tools in good condition with regular maintenance
- Use the right tool for the job
- Examine each tool for damage before use
- Operate according to the manufacturer's instructions
- Provide and use the proper protective equipment

Employees and employers have a responsibility to work together to establish safe working procedures. If a hazardous situation is encountered, it should be brought to the attention of the proper individual immediately.

Power tools can be hazardous when improperly used. There are several types of power tools, based on the power source they use: electric, pneumatic, liquid fuel, hydraulic, and powder-actuated. Employees should be trained in the use of all tools—not just power tools. They should understand the potential hazards as well as the safety precautions to prevent those hazards from occurring. The following general precautions should be observed by power tool users:

- Never carry a tool by the cord or hose.
- Never yank the cord or the hose to disconnect it from the receptacle.
- Keep cords and hoses away from heat, oil, and sharp edges.
- Disconnect tools when not in use, before servicing, and when changing accessories such as blades, bits, and cutters.
- Keep all observers at a safe distance away from the work area.
- Secure work with clamps or a vise, freeing both hands to operate the tool.
- Avoid accidental starting. The worker should not hold a finger on the switch button while carrying a plugged-in tool.
- Maintain tools with care. They should be kept sharp and clean for the best performance. Follow instructions in the user's manual for lubricating and changing accessories.
- Be sure to keep good footing and maintain good balance.
- Wear proper apparel. Loose clothing, ties, or jewelry can be caught in moving parts.
- Remove from use all portable electric tools that are damaged and tag them "Do Not Use."

Hazardous moving parts of a power tool need to be safeguarded. For example, belts, gears, shafts, pulleys, sprockets, spindles, drums, flywheels, chains, or other reciprocating, rotating, or moving parts of equipment must be guarded if such parts are exposed to contact by employees. Guards, as necessary, should be provided to protect the operator and others from the following:

- Point of operation
- In-running nip points
- Rotating parts
- Flying chips and sparks

Safety guards must never be removed when a tool is used. For example, portable circular saws must be equipped with guards. An upper guard must cover the entire blade of the saw. A retractable lower guard must cover the teeth of the saw, except when it makes contact with the work material. The lower guard must automatically return to the covering position when the tool is withdrawn from the work.

The following handheld powered tools must be equipped with a momentary contact on–off control switch: drills, tappers, fastener drivers, horizontal, vertical and angle grinders with wheels larger than 2 in. in diameter, disc and belt sanders, reciprocating saws, saber saws, and other similar tools.

These tools also may be equipped with a lock-on control provided that turnoff can be accomplished by a single motion of the same finger or fingers that turn it on.

The following handheld powered tools may be equipped with only a positive on–off control switch: platen sanders, disc sanders with discs 2 in. or less in diameter, grinders with wheels 2 in. or less in diameter, routers, planers, laminate trimmers, nibblers, shears, scroll saws and jigsaws with blade shanks 1/4 in. wide or less.

Other handheld powered tools such as circular saws with a blade diameter greater than 2 in., chain saws, and percussion tools without positive accessory holding means must be equipped with a constant pressure switch that will shut off the power when the pressure is released.

Employees using electric tools must be aware of several dangers; the most serious is the possibility of electrocution. Among the chief hazards of electric-powered tools are burns and slight shocks, which can lead to injuries or even heart failure. Under certain conditions, even a small amount of current can result in fibrillation of the heart and eventual death. A shock can also cause the user to fall off a ladder or other elevated work surface.

To protect the user from shock, tools must either have a three-wire cord with ground and be grounded, double insulated, or powered by a low-voltage isolation transformer. Three-wire cords contain two current-carrying conductors and a grounding conductor. One end of the grounding conductor connects to the tool's metal housing. The other end is grounded through a prong on the plug. Anytime an adapter is used to accommodate a two-hole receptacle, the adapter wire must be attached to a known ground. The third prong should never be removed from the plug. Double insulation is more convenient. The user and the tools are protected in two ways: by normal insulation on the wires inside and by a housing that cannot conduct electricity to the operator in the event of a malfunction. These general practices should be followed when using electric tools:

- Electric tools should be operated within their design limitations.
- Gloves and safety footwear are recommended during use of electric tools.
- When not in use, tools should be stored in a dry place.
- Electric tools should not be used in damp or wet locations.
- Work areas should be well lighted.

Powered abrasive grinding, cutting, polishing, and wire buffing wheels create special safety problems because they may throw off flying fragments. Before an abrasive wheel is mounted, it should be inspected closely and sound- or ring-tested to be sure that it is free from cracks or defects. To test, wheels should be tapped gently with a light non-metallic instrument. If the wheel sounds cracked or dead, it could fly apart in operation and so it must not be used. A sound and undamaged wheel gives a clear metallic tone or ring. To prevent the wheel from cracking, the user should be sure it fits freely on the spindle. The spindle nut must be tightened enough to hold the wheel in place, without distorting the flange. Follow the manufacturer's recommendations. Care must be taken to assure that the spindle wheel does not exceed the abrasive wheel specifications. Due to the possibility of a wheel disintegrating (exploding) during start-up, the employee should never stand directly in front of the wheel as it accelerates to full operating speed.

Portable grinding tools need to be equipped with safety guards to protect workers not only from the moving wheel surface, but also from flying fragments in case of breakage. In addition, when using a powered grinder

- Always use eye protection
- Turn off the power when not in use
- Never clamp a handheld grinder in a vise

23.18   PRESSURE VESSELS (29 CFR 1910.106, .216, AND .217)

Generally, a pressure vessel is a storage tank or vessel that has been designed to operate at pressures above 15 psig. Recent inspections of pressure vessels have shown that there are a considerable number of cracked and damaged vessels in workplaces. Cracked and damaged vessels can result in leakage or rupture failures. Potential health and safety hazards of leaking vessels include poisonings, suffocations, fires, and explosion hazards. Rupture failures can be much more catastrophic and can cause considerable damage to life and property. The safe design, installation, operation, and maintenance of pressure vessels in accordance with the appropriate codes and standards are essential to worker safety and health.

Pressure vessel design, construction, and inspection are referenced in the ASME *Boiler and Pressure Vessel Code*, 1968 and current 1910.106(b), 1910.217(b)(12), and 1910.261(a)(3) and *OSHA Technical Manual* CPL 2–2.208, Chapter 10 (Pressure Vessel Guidelines). These set the guidelines for pressure vessel safety. Two consequences result from a complete rupture:

- Blast effects due to sudden expansion of the pressurized fluid
- Fragmentation damage and injury, if vessel rupture occurs

For a leakage failure, the hazard consequences can range from no effect to very serious effects:

- Suffocation or poisoning, depending on the nature of the contained fluid, if the leakage occurs into a closed space
- Fire and explosion (physical hazards for a flammable fluid)
- Chemical and thermal burns from contact with process liquids

Most of the pressure or storage vessels in service in the United States will have been designed and constructed in accordance with one of the following pressure vessel design codes:

- The ASME Code, or Section VIII of the ASME *Boiler and Pressure Vessel Code*.
- The API Standard 620 or the *American Petroleum Institute Code*, which provides rules for lower-pressure vessels not covered by the ASME Code.

Information that identifies the specific vessel that is assessed and provides general information about it includes the following items:

- Current owner of the vessel
- Vessel location
  - Original location and current location if it has been moved
- Vessel identification
  - Manufacturer's serial number
  - National Board number if registered with NB
- Manufacturer identification
  - Manufacturer's serial number
  - Name and address of manufacturer
  - Authorization or identification number of the manufacturer
- Date of manufacture of the vessel
- Data report for the vessel
  - ASME U-1 or U-2, API 620 form, or other applicable report
- Date vessel was placed in service
- Interruption dates if not in continuous service

Information on the conditions of operating history of the vessel or tank, which will be helpful in safety assessment, includes the following items:

- Fluids handled
  - Type and composition, temperature, and pressures
- Type of service
  - Continuous, intermittent, or irregular
- Significant changes in service conditions
  - Changes in pressures, temperatures, and fluid compositions and the dates of the changes
- Vessel history
  - Alterations, reratings, and repairs performed
  - Date(s) of changes or repairs

Information about inspections performed on the vessel or tank and the results obtained, which will assist in the safety assessment, includes the following items:

- Inspection(s) performed
  - Type, extent, and dates
- Examination methods
  - Preparation of surfaces and welds
  - Techniques used (visual, magnetic particle, penetrant test, radiography, ultrasonic)
- Qualifications of personnel
  - American Society for Nondestructive Testing (ASNT) levels or equivalent of examining and supervisory personnel

- Inspection results and report
  - Report form used (NBIC NB-7, API 510, or other)
  - Summary of type and extent of damage or cracking
  - Disposition (no action, delayed action, or repaired)

The information acquired for the above items is not adaptable to any kind of numerical ranking for quantitative safety assessment purposes. However, the information can reveal the owner or user's apparent attention to good practice, careful operation, regular maintenance, and adherence to the recommendations and guidelines developed for susceptible applications. If the assessment indicated cracking and other serious damage problems, it is important that the inspector obtain qualified technical advice and opinion. After all, safety and health is of primary concern with reference to pressure vessels.

23.19   RAILINGS (29 CFR 1910.23)

The general requirements apply to all stairrails and handrails, and stairways with four or more risers, or rising more than 30 in. in height—whichever is less—must have at least one handrail. A stairrail must also be installed along each unprotected side or edge. When the top edge of a stairrail system also serves as a handrail, the height of the top edge must be no more than 37 in. nor less than 36 in. from the upper surface of the stairrail to the surface of the tread. Winding or spiral stairways must have a handrail to prevent using areas where the tread width is less than 6 in. Stairrails installed after March 15, 1991, must be not less than 36 in. in height.

Midrails, screens, mesh, intermediate vertical members, or equivalent intermediate structural members must be provided between the top rail and stairway steps to the stairrail system. Midrails, when used, must be located midway between the top of the stairrail system and the stairway steps. Screens or mesh, when used, must extend from the top rail to the stairway step and along the opening between top rail supports.

Intermediate vertical members, such as balusters, when used, must not be more than 19 in. apart. Other intermediate structural members, when used, must be installed so that there are no openings more than 19 in. wide.

Handrails and the top rails of the stairrail systems must be able to withstand, without failure, at least 200 lb of weight applied within 2 in. of the top edge in any downward or outward direction, at any point along the top edge. The height of handrails must not be more than 37 in. nor less than 30 in. from the upper surface of the handrail to the surface of the tread. The height of the top edge of a stairrail system used as a handrail must not be more than 37 in. nor less than 36 in. from the upper surface of the stairrail system to the surface of the tread.

Stairrail systems and handrails must be surfaced to prevent injuries such as punctures or lacerations and to keep clothing from snagging. Handrails must provide an adequate handhold for employees to grasp to prevent falls. The ends of stairrail systems and handrails must be built to prevent dangerous projections, such as rails protruding beyond the end posts of the system. Temporary handrails must have a minimum clearance of 3 in. between the handrail and walls, stairrail systems, and other objects. Unprotected sides and edges of stairway landings must be provided with standard 42 in. guardrail systems.

23.20   SCAFFOLDS (29 CFR 1910.28)

Analysis of 1986 BLS data to support OSHA's scaffolding standard estimates that of the 500,000 injuries and illnesses that occur in the construction industry annually, 10,000 are related to scaffolds. In addition, of the estimated 900 occupational fatalities occurring annually, at least 80 are associated with work on scaffolds. Seventy-two percent of the workers injured in scaffold accidents covered by the BLS study attributed the accident either to the planking or support giving way, or to the employee slipping or being struck by a falling object. Plank slippage was the most commonly cited cause.

All scaffolds and their supports must be capable of supporting the load they are designed to carry with a safety factor of at least 4. All planking is to be of scaffold grade, as recognized by grading rules for the species of wood used.

The maximum permissible span for a 1¼ in. by 9 in. or wider plank for full thickness is 4 ft., with a medium loading of 50 lb/sq ft. Scaffold planks are to extend over their supports not less than 6 in. nor more than 18 in. Scaffold planking is to overlap a minimum of 12 in. or secured from movement.

23.21   SKYLIGHTS (29 CFR 1910.23)

Over a number of years, many workers have fallen through skylights when they assumed that they were designed to support the human weight. This misconception has resulted in many death and injuries. Every skylight floor opening and hole is to be guarded by a standard skylight screen or a fixed standard railing on all exposed sides.

23.22   SPRAY-FINISHING OPERATIONS (29 CFR 1910.107)

In conventional dry-type spray booths, overspray dry filters or rolls, if installed, are to conform to the following: The spraying operations, except electrostatic spraying, must ensure an average air velocity over the open face of the booth of not less than 100 ft/min. Electrostatic spraying operations may be conducted with an air velocity of not less than 60 ft/min, depending on the volume of the finishing material applied and its flammability and explosion characteristics. Visible gauges, or audible alarm or pressure-activated devices, are to be installed to indicate or ensure that the required air velocity is maintained. Filter pads must be inspected after each period of use and clogged filter pads must be discarded and replaced. Filter pads are to be inspected to ensure proper replacement of filter media.

Spray booths are to be so installed that all portions are readily accessible for cleaning. A clear space of not less than 3 ft on all sides is to be kept free from storage or combustible construction.

Space within the spray booth on the downstream and upstream sides of filters is to be protected with approved automatic sprinklers. There shall be no open flame or spark-producing equipment in any spraying area nor within 20 ft thereof, unless separated by a partition. Electrical wiring and equipment not subject to deposits of combustible residues but located in a spraying area are to be explosion proof.

The quantity of flammable or combustible liquids kept in the vicinity of spraying operations is to be the minimum required for operations and should ordinarily not exceed a supply for 1 day or 1 shift. Bulk storage of portable containers of flammable or combustible liquids is to be in a separated, constructed buildings detached from other important buildings or cut off in a standard manner. Whenever flammable or combustible liquids are transferred from one container to another, both containers must be effectively bonded and grounded to prevent discharge sparks of static electricity.

All spraying areas are to be kept as free from the accumulation of deposits of combustible residues as practical, with cleaning conducted daily if necessary. Scraper, spuds, or other tools used for cleaning purposes are to be of nonsparking materials. Residue scrapings and debris contaminated with residue must be immediately removed from the premises. NO SMOKING signs in large letters on contrasting color background shall be conspicuously posted in all spraying areas and paint storage rooms.

Adequate ventilation is to be ensured before spray operations are started. Mechanical ventilation is to be provided when spraying operations are carried out in enclosed areas. When mechanical ventilation is provided during spraying operations, it is to be so arranged that it will not circulate the contaminated air.

23.23   STAIRS, FIXED INDUSTRIAL (29 CFR 1910.23 AND .24)

Fixed stairways are to be provided for access from one structure to another where operations necessitate regular travel between levels and for access to operating platforms at any equipment that requires attention routinely during operations. Fixed stairs must also be provided where access to elevations is required daily or at each shift, where such work may expose employees to harmful substances, or where the carrying of tools or equipment by hand is normally required. Spiral stairways are not permitted except for special limited usage and secondary access situations where it is not practical to provide a conventional stairway (see Figure 23.5).

Every flight of stairs with four or more risers is to be provided with a standard railing on all open sides. Handrails are to be provided on at least one side of closed stairways, preferably on the right side descending. Fixed stairways are to have a minimum width of 22 in. Stairs shall be constructed such that the riser height and tread width are uniform throughout and do not vary more than 1/4 in. Other general requirements include

- A stairway or ladder must be provided at all worker points of access where there is a break in elevation of 19 in. or more and no ramp, runway, embankment, or personnel hoist is provided.
- When there is only one point of access between levels, it must be kept clear to permit free passage by workers. If free passage becomes restricted, a second point of access must be provided and used.
- Where there are more than two points of access between levels, at least one point of access must be kept clear.

FIGURE 23.5 Example of a fixed set of stairs.

- Stairways must be installed at least 30°, and no more than 50°, from the horizontal.
- Where doors or gates open directly onto a stairway, a platform that extends at least 20 in. beyond the swing of the door must be provided.
- All stairway parts must be free of dangerous projections such as protruding nails.
- Slippery conditions on stairways must be corrected.

Extreme care must be taken by workers when ascending and descending a set of stairs. Many serious injuries and even fatalities occur when workers slip and fall on stairways.

23.24 STORAGE (29 CFR 1910.176)

Stored materials stacked in tiers are to be stacked, blocked, interlocked, and limited in height so that they are secure against sliding or collapse. Storage areas must be kept free from accumulation of materials that constitute hazards from tripping, fire explosion, or pest harborage. Vegetation control is to be exercised when necessary. Where mechanical handling equipment is used, sufficient safe clearance is to be allowed for aisles, at loading docks, through doorways, and whenever turns or passage must be made.

23.25 TIRE INFLATION

Because tires have the potential to release a large amount of energy if they explode, and they can cause injury due to poor inflation procedure, care must be taken to follow safety procedures. Tires that are mounted or inflated on drop center wheels must have a safe practice procedure posted and enforced. When tires are mounted or inflated on wheels with split rims or retainer rings, a safe practice procedure is to be posted and enforced. Each tire inflation hose should have a clip-on chuck with at least 24 in. of hose between the chuck and an inline hand valve and gauge. The tire inflation control valve must be of automatic shutoff type so that the airflow stops when the valve is released. Employees should be strictly forbidden from taking a position directly over or in front of a tire while it is inflated. A tire-restraining device such as a cage, rack, or other effective means must be used while inflating tires mounted on split rims or rims using retainer rings.

23.26 TOEBOARDS (29 CFR 1910.23)

Toeboards are used to protect workers from being struck by objects falling from elevated areas. Railings protecting floor openings, platforms, and scaffolds are to be equipped with toeboards whenever persons can pass beneath the open side, wherever there is moving machinery, or wherever there is equipment with which falling material could cause a hazard. A standard toeboard is to be at least 4 in. in height and may be of any substantial material, either solid or open, with openings not exceeding 1 in. in the greatest dimension.

23.27 TRANSPORTING EMPLOYEES AND MATERIALS

All employees who operate vehicles on public thoroughfares must have valid operator's licenses. When seven or more employees are regularly transported in a van, bus, or truck, the operator's license must be appropriate for the class of vehicle driven. Each van, bus, or truck used regularly to transport employees must be equipped with an adequate number of seats and seat belts. When employees are transported by truck, provisions are to be made to prevent their falling from the vehicle. Vehicles used to transport employees must be equipped with lamps, brakes, horns, mirrors, windshields, turn signals, and be in good repair. Transport vehicles are to be provided with handrails, steps, stirrups, or similar devices, so placed and arranged that employees can safely mount or dismount.

Employee transport vehicles need to be equipped at all times with at least two reflective type as well as a full-charged fire extinguisher, in good condition, with at least 4 B:C rating. Cutting tools or tools with sharp edges are not to be carried in passenger compartments of employee transport vehicles and are to be placed in closed boxes or containers that are secured in place. Employees must be prohibited from riding on top of any load that can shift, topple, or otherwise become unstable.

23.28 WALKING/WORKING SURFACES (29 CFR 1910.21 AND .22)

Slips, trips, and falls constitute the majority of general industry accidents. They cause 15% of all accidental deaths and are second only to motor vehicles as a cause of fatalities. The OSHA standards for walking and working surfaces apply to all permanent places of employment, except where only domestic, mining, or agricultural work is performed.

Working/walking surfaces that are wet need to be covered with nonslip materials. All spilled materials are to be cleaned up immediately. Any holes in the floor, sidewalk, or other walking surfaces must be repaired properly, covered, or otherwise made safe. All aisles and passageways are to be kept clear and marked as appropriate. There is to be safe clearance for walking in aisles where motorized or mechanical handling equipment are operated.

Materials or equipment should be stored in such a way that sharp projections will not interfere with the walkway. Changes of direction or elevations should be readily identifiable. There should be adequate headroom provided for the entire length of any aisle or walkway.

23.29 WELDING, CUTTING, AND BRAZING (29 CFR 1910.251, .252, .253, .254, AND .255)

Welding, cutting, and brazing are hazardous activities, which pose a unique combination of both safety and health risks to more than 500,000 workers in a wide variety of industries. The risk from fatal injuries alone is more than four deaths per 1000 workers over a working lifetime. An estimated 562,000 employees are at risk for exposure to chemical and physical hazards of welding, cutting, and brazing.

Moreover, 58 deaths from welding and cutting incidents, including explosions, electrocutions, asphyxiation, falls, and crushing injuries, were reported by the BLS in 1993.

There are numerous health hazards associated with exposure to fumes, gases, and ionizing radiation formed or released during welding, cutting, and brazing, including heavy metal poisoning, lung cancer, metal fume fever, flash burns, and others. These risks vary, depending on the type of welding materials and welding surfaces. Only authorized and trained personnel are permitted to use welding, cutting, or brazing equipment.

Each operator has a copy of the appropriate operating instructions and they are directed to follow them. Only approved apparatuses (torches, regulators, pressure-reducing valves, acetylene generators, and manifolds) are to be used.

Compressed-gas cylinders are to be regularly examined for obvious signs of defects, deep rusting, or leakage, and care should be used in handling and storing of cylinders, safety valves, and relief valves to prevent damage. Precautions are to be taken to prevent the mixing of air or oxygen with flammable gases, except at a burner or in a standard torch. Cylinders must be kept away from sources of heat. Cylinders are to be kept away from elevators, stairs, or gangways. Cylinders are prohibited from being used as rollers or supports. Empty cylinders are to be appropriately marked and their valves closed. Signs reading DANGER—NO SMOKING,

MATCHES, OR OPEN LIGHTS, or the equivalent, are to be posted. Cylinders, cylinder valves, couplings, regulators, hoses, and apparatus must be kept free of oily or greasy substances. Do not drop or strike cylinders. Unless secured on special trucks, regulators are to be removed and valve-protection caps put in place before moving cylinders. Liquefied gases are to be stored and shipped valve-end up with valve covers in place.

Provisions must be made to never crack a fuel-gas cylinder valve near sources of ignition. Before a regulator is removed, the valve is to be closed and gas is to be released from the regulator. Red is used to identify the acetylene (and other fuel-gas) hose, green for oxygen hose, and black for inert gas and air hose. Pressure-reducing regulators should be used only for the gas and pressures for which they are intended.

When the object to be welded cannot be moved and fire hazards cannot be removed, shields are to be used to confine heat, sparks, and slag. A fire watch is to be assigned when welding or cutting is performed in locations where a serious fire might develop.

When welding is done on metal walls, precautions must be taken to protect combustibles on the other side. Before hot work is begun, used drums, barrels, tanks, and other containers are to be so thoroughly cleaned that no substances remain that could explode, ignite, or produce toxic vapors.

Arc welding and cutting operations must be shielded by noncombustible or flame-retardant screens, which will protect employees and other persons working in the vicinity from the direct rays of the arc. Arc welding and cutting cables are to be of the completely insulated, flexible type, capable of handling the maximum current requirement of the work in progress. Cables in need of repair should not be used. When the welder or cutter has occasion to leave work or to stop work for any appreciable length of time, or when the welding or cutting machine is to be moved, the power supply switch to the equipment is to be opened. All ground return cables and all arc welding and cutting machine grounds must be in accordance with regulatory requirements. Ground connections are to be made directly to the material that is welded. Open circuit (no load) voltage of arc welding and cutting machines must be as low as possible and not in excess of the recommended limits.

Under wet conditions, automatic controls for reducing no load voltage should be used. Electrodes are to be removed from the holders when not in use. It is required that electric power to the welder be shut off when no one is in attendance. Suitable fire extinguishing equipment is to be available for immediate use.

Welders shall be forbidden to coil or loop welding electrode cable around their body. Wet machines must be thoroughly dried and tested before they are used, decreasing the risk of electrical shock or electrocution.

It is required that eye protection helmets, hand shields, and goggles meet appropriate standards. Employees exposed to the hazards created by welding, cutting, or brazing operations must be protected with PPE and clothing.

24 Summary

This sign says it all that this part of the service sector provides services, safety, and health for people.

In summarizing the people-oriented service sector, it seems appropriate to provide some tools that will help in the assessment of the hazards that exist or could exist in these sectors of the service industry. The following is a hazard identification and action item checklist for determining the degree of Occupational Safety and Health Administration (OSHA) compliance or the effective amount of safeguards in place to protect the workforce.

24.1 BIOLOGICAL HAZARDS

Yes ☐   No ☐   Are there biological organisms present?
Yes ☐   No ☐   Are any biologicals potential pathogens?
Yes ☐   No ☐   Have all biologicals been accounted for?
Yes ☐   No ☐   Have the biological risks been evaluated?
Yes ☐   No ☐   Are there hygiene facilities present?
Yes ☐   No ☐   Is hygiene practiced?
Yes ☐   No ☐   Do specific methods for handling of organisms exist?
Yes ☐   No ☐   Are biohazard procedures in use?
Yes ☐   No ☐   Have workers been trained regarding biological hazards?
Yes ☐   No ☐   Are biological warning signs posted?
Yes ☐   No ☐   Do biohazard-warning containers exist?

Yes ☐   No ☐   Are there procedures for disposal of biologicals?
Yes ☐   No ☐   Are laboratory facilities where biologicals exist secure?

24.2   BLOODBORNE PATHOGENS

Yes ☐   No ☐   Is there a written exposure control plan consisting of the following:
Yes ☐   No ☐   A list of employees whose job exposes them to bloodborne diseases?
Yes ☐   No ☐   A list of all tasks that present exposure potential?
Yes ☐   No ☐   A procedure for evaluating exposure potential?
Yes ☐   No ☐   Do all personnel take precautions to prevent blood and body fluid contact among employees?
Yes ☐   No ☐   Are hand washing facilities that are easily accessible to all employees provided?
Yes ☐   No ☐   Are contaminated needles bent or recapped by hand?
Yes ☐   No ☐   Are contaminated needles disposed of in approved medical waste containers?
Yes ☐   No ☐   When there is an exposure hazard, is the following allowed: eating, drinking, smoking, applying cosmetics or lip balm, or handling contact lenses?
Yes ☐   No ☐   Is food and drink stored in areas where blood or any other body fluid is stored?
Yes ☐   No ☐   Is suction pipetting of any body fluid prohibited?
Yes ☐   No ☐   Are containers specifically designed for the fluid stored in it?
Yes ☐   No ☐   Are containers labeled properly before storage or transfer?
Yes ☐   No ☐   Are labels placed on any container that contains or may contain contaminated items or waste products?
Yes ☐   No ☐   If any contact with blood or infectious material is possible, does the employer provide personal protective equipment (PPE) at no charge?
Yes ☐   No ☐   Is the employer responsible for the maintenance and repair of all PPE?
Yes ☐   No ☐   Is any item that contacts blood or any other body fluid disinfected or discarded?
Yes ☐   No ☐   Do employers ensure that the work site is in a clean and sanitary condition?
Yes ☐   No ☐   Are proper disinfectants that kill pathogens used to clean working surfaces, bins, or other areas where contamination may occur?
Yes ☐   No ☐   Is broken glassware picked up with a device; do you ever use unprotected hands?
Yes ☐   No ☐   Are Hepatitis B vaccinations made available to employees who may be exposed to the virus during employment?
Yes ☐   No ☐   Have the employers provided an occupational exposure training program for all employees during working hours?
Yes ☐   No ☐   Are any exposures noted and kept in the employee's medical record?

24.3 CHEMICALS

Yes ☐ No ☐ If hazardous substances are used in your processes, do you have a medical or biological monitoring system in operation?

Yes ☐ No ☐ Are you familiar with threshold limit values or permissible exposure?

Yes ☐ No ☐ Are limits of airborne contaminants and physical agents used in your workplace?

Yes ☐ No ☐ Have control procedures been instituted for hazardous materials, where appropriate, such as respirators, ventilation systems, and handling practices?

Yes ☐ No ☐ Whenever possible, are hazardous substances handled in properly designed and exhausted booths or similar locations?

Yes ☐ No ☐ Are there written standard operating procedures for the selection and use of respirators where needed?

Yes ☐ No ☐ If you have a respirator protection program, are your employees instructed on the correct usage and limitations of the respirators?

Yes ☐ No ☐ Are the respirators NIOSH-approved for this particular application?

Yes ☐ No ☐ Are they regularly inspected and cleaned, sanitized, and maintained?

Yes ☐ No ☐ If hazardous substances are used in your processes, do you have a medical or biological monitoring system in operation?

Yes ☐ No ☐ Do you use general dilution or local exhaust ventilation systems to control dusts, vapors, gases, fumes, smoke, solvents, or mists that may be generated in your workplace?

Yes ☐ No ☐ Is ventilation equipment provided for removal of contaminants from such operations as production grinding, buffing, spray painting, and vapor degreasing, and is it operating properly?

Yes ☐ No ☐ Do employees complain about dizziness, headaches, nausea, irritation, or other factors of discomfort when they use solvents or other chemicals?

Yes ☐ No ☐ Is there a dermatitis problem?

Yes ☐ No ☐ Do employees complain about dryness, irritation, or sensitization of the skin?

Yes ☐ No ☐ Are combustion engines used, and is carbon monoxide kept within acceptable levels?

Yes ☐ No ☐ Is vacuuming used, rather than blowing or sweeping dusts, whenever possible for cleanup?

Yes ☐ No ☐ Have you considered the use of an industrial hygienist or environmental scientist?

24.4 COMPRESSED-GAS CYLINDERS (CGCs)

Yes ☐ No ☐ Are CGCs kept away from radiators and other sources of heat?

Yes ☐ No ☐ Are CGCs stored in well-ventilated, dry locations at least 20 ft. away from materials such as oil, grease, excelsior, reserve stocks of carbide, acetylene, or other fuels as they are likely to cause acceleration of fires?

Yes ☐   No ☐   Are CGCs stored only in assigned areas?

Yes ☐   No ☐   Are CGCs stored away from elevators, stairs, and gangways?

Yes ☐   No ☐   Are CGCs stored in areas where they will not be dropped, knocked over, or tampered with?

Yes ☐   No ☐   Are CGCs stored in areas with poor ventilation?

Yes ☐   No ☐   Are storage areas marked with signs such as: "Oxygen, no smoking, or no open flames"?

Yes ☐   No ☐   Are CGCs not stored outside generator houses?

Yes ☐   No ☐   Do storage areas have wood and grass cut back within 15 ft?

Yes ☐   No ☐   Are CGCs secured to prevent falling?

Yes ☐   No ☐   Are CGCs stored in a vertical position?

Yes ☐   No ☐   Are protective caps in place at all times except when in use?

Yes ☐   No ☐   Are threads on caps or cylinders lubricated?

Yes ☐   No ☐   Are all CGCs legibly marked for identifying the gas content with the chemical or trade name of the gas?

Yes ☐   No ☐   Are the markings on CGCs by stenciling, stamping, or labeling?

Yes ☐   No ☐   Are markings located on the slanted area directly below the cap?

Yes ☐   No ☐   Does each employee determine that CGCs are in a safe condition by means of a visual inspection?

Yes ☐   No ☐   Is each portable tank and all piping, valves, and accessories visually inspected at intervals not exceeding 2.5 years?

Yes ☐   No ☐   Are inspections conducted by the owner, agent, or approved agency?

Yes ☐   No ☐   On insulated tanks, is the insulation not to be removed if, in the opinion of the person performing the visual inspection, external corrosion is likely to be negligible?

Yes ☐   No ☐   If evidence of any unsafe condition is discovered, is the portable tank not to be returned to service until it meets all corrective standards?

24.5   ELECTRICAL

Yes ☐   No ☐   Do you specify compliance with OSHA for all contract electrical work?

Yes ☐   No ☐   Are all employees required to report as soon as practicable any obvious hazard to life or property observed in connection with electrical equipment or lines?

Yes ☐   No ☐   Are employees instructed to make preliminary inspections and appropriate tests to determine what conditions exist before starting work on electrical equipment or lines?

Yes ☐   No ☐   When electrical equipment or lines are to be serviced, maintained, or adjusted, are necessary switches opened, locked out, and tagged whenever possible?

Yes ☐ No ☐ Are portable electrical tools and equipment grounded or of the double-insulated type?

Yes ☐ No ☐ Are electrical appliances such as vacuum cleaners, polishers, and vending machines grounded?

Yes ☐ No ☐ Do the extension cords used have a grounding conductor?

Yes ☐ No ☐ Are multiple plug adaptors prohibited?

Yes ☐ No ☐ Are ground-fault circuit interrupters installed on each temporary 15 or 20 A, 120 V AC circuit at locations where construction, demolition, modifications, alterations, or excavations are performed?

Yes ☐ No ☐ Are all temporary circuits protected by suitable disconnecting switches or plug connectors at the junction with permanent wiring?

Yes ☐ No ☐ Do you have electrical installations in hazardous dust or vapor areas? If so, do they meet the National Electrical Code (NEC) for hazardous locations?

Yes ☐ No ☐ Are exposed wiring and cords with frayed or deteriorated insulation repaired or replaced promptly?

Yes ☐ No ☐ Are flexible cords and cables free of splices or taps?

Yes ☐ No ☐ Are clamps or other securing means provided on flexible cords or cables at plugs, receptacles, tools, equipment, etc., and is the cord jacket securely held in place? Are all cord, cable, and raceway connections intact and secure?

Yes ☐ No ☐ In wet or damp locations, are electrical tools and equipment appropriate for the use or location or otherwise protected?

Yes ☐ No ☐ Is the location of electrical power lines and cables (overhead, underground, underfloor, other side of walls) determined before digging, drilling, or similar work is begun?

Yes ☐ No ☐ Are metal measuring tapes, ropes, handlines, or similar devices with metallic thread woven into the fabric prohibited where they could come in contact with energized parts of equipment or circuit conductors?

Yes ☐ No ☐ Is the use of metal ladders prohibited in areas where the ladder or the person using the ladder could come in contact with energized parts of equipment, fixtures, or circuit conductors?

Yes ☐ No ☐ Are all disconnecting switches and circuit breakers labeled to indicate their use or equipment served?

Yes ☐ No ☐ Are disconnecting means always opened before fuses are replaced?

Yes ☐ No ☐ Do all interior wiring systems include provisions for grounding metal parts of electrical raceways, equipment, and enclosures?

Yes ☐ No ☐ Are all electrical raceways and enclosures securely fastened in place?

Yes ☐ No ☐ Are all energized parts of electrical circuits and equipment guarded against accidental contact by approved cabinets or enclosures?

Yes ☐ No ☐ Are sufficient access and working space provided and maintained about all electrical equipment to permit ready and safe operations and maintenance?

Yes ☐   No ☐   Are all unused openings (including conduit knockouts) in electrical enclosures and fittings closed with appropriate covers, plugs, or plates?

Yes ☐   No ☐   Are electrical enclosures such as switches, receptacles, and junction boxes provided with tight fitting covers or plates?

Yes ☐   No ☐   Are disconnecting switches for electrical motors in excess of two horsepower capable of opening the circuit when the motor is in a stalled condition, without exploding? (Switches must be horsepower rated equal to or in excess of the motor hp rating.) Is low-voltage protection provided in the control device of motors driving machines or equipment that could cause probable injury from inadvertent starting?

Yes ☐   No ☐   Is each motor disconnecting switch or circuit breaker located within sight of the motor control device?

Yes ☐   No ☐   Is each motor located within sight of its controller, is the controller disconnecting means capable of being locked in the open position, or is a separate disconnecting means installed in the circuit within sight of the motor?

Yes ☐   No ☐   Is the controller for each motor in excess of two horsepower, rated in horsepower equal to or in excess of the rating of the motor it serves?

Yes ☐   No ☐   Are employees who regularly work on or around energized electrical equipment or lines instructed in cardiopulmonary resuscitation (CPR) methods?

Yes ☐   No ☐   Are employees prohibited from working alone on energized lines or equipment over 600 V?

24.6   EMERGENCY RESPONSE AND PLANNING

Yes ☐   No ☐   Is there a written emergency response planning that is available to all employees?

Yes ☐   No ☐   Is there an established procedure specifically outlining the steps to be taken by all employees including route of evacuation, place to meet outside building, and designation of person responsible for verifying that employees are all accounted for?

Yes ☐   No ☐   Have proper evacuation procedures been communicated to everyone before the need for an actual evacuation, and have those procedures been actively practiced in a mock evacuation situation?

Yes ☐   No ☐   Is there an established protocol for determining the need for evacuation?

Yes ☐   No ☐   Is there a designated person responsible for making an evacuation decision?

Yes ☐   No ☐   Is the need for evacuation communicated to employees in such a way that everyone (other than those designated as the initial contacts) receives the same information at the same time?

Yes ☐   No ☐   In the event of electrical failure, is there a backup system for both broadcasting of messages and lighting of escape routes?

Yes ☐   No ☐   Are established escape routes clearly marked, and are maps posted outlining the entire route?

Yes ☐   No ☐   Are escape routes determined to be the shortest safe route possible, allowing adequate room and number of routes for the number of employees?

Yes ☐   No ☐   Are all emergency exits clearly marked and functioning properly?

Yes ☐   No ☐   Are all escape routes free of clutter and tripping hazards?

Yes ☐   No ☐   Is there adequate emergency lighting along the routes?

Yes ☐   No ☐   Is emergency equipment such as fire extinguishers and flashlights located at predetermined sites along escape routes and is this equipment routinely tested for proper operation?

Yes ☐   No ☐   In the event that employees are required to remain within hallways/stairways of the escape route for longer than expected, is there adequate ventilation, temperature control, and some type of communication equipment?

Yes ☐   No ☐   Are all established meeting places outside of the building a reasonably safe distance away?

Yes ☐   No ☐   Is there an established method for verification that all employees have left the building and a way to communicate to emergency personnel the identities and possible locations of those who have not?

24.7 ERGONOMICS

24.7.1 MANUAL MATERIAL HANDLING

Yes ☐   No ☐   Is there lifting of loads, tools, or pans?
Yes ☐   No ☐   Is there lowering of tools, loads, or parts?
Yes ☐   No ☐   Is there overhead reaching for tools, loads, or parts?
Yes ☐   No ☐   Is there bending at the waist to handle tools, loads, or parts?
Yes ☐   No ☐   Is there twisting at the waist to handle tools, loads, or parts?

24.7.2 PHYSICAL ENERGY DEMANDS

Yes ☐   No ☐   Do tools and parts weigh more than 10 lb?
Yes ☐   No ☐   Is reaching greater than 20 in.?
Yes ☐   No ☐   Is bending, stooping, or squatting a primary task activity?
Yes ☐   No ☐   Is lifting or lowering loads a primary task activity?
Yes ☐   No ☐   Is walking or carrying loads a primary task activity?
Yes ☐   No ☐   Is stair or ladder climbing with loads a primary task activity?
Yes ☐   No ☐   Is pushing or pulling loads a primary task activity?

Yes ☐   No ☐   Is reaching overhead a primary task activity?
Yes ☐   No ☐   Do any of the above tasks require five or more complete work cycles to be completed within a minute?
Yes ☐   No ☐   Do workers complain that rest breaks and fatigue allowances are insufficient?

24.7.3   OTHER MUSCULOSKELETAL DEMANDS

Yes ☐   No ☐   Do manual jobs require frequent, repetitive motions?
Yes ☐   No ☐   Do work postures require frequent bending of the neck, shoulder, elbow, wrist, or finger joints?
Yes ☐   No ☐   For seated work, do reaches for tools and materials exceed 15 in. from the worker's position?
Yes ☐   No ☐   Is the worker unable to change his or her position often?
Yes ☐   No ☐   Does the work involve forceful, quick, or sudden motions?
Yes ☐   No ☐   Does the work involve shock or rapid buildup of forces?
Yes ☐   No ☐   Is finger-pinch gripping used?
Yes ☐   No ☐   Do job postures involve sustained muscle contraction of any limb?

24.7.4   COMPUTER WORKSTATION

Yes ☐   No ☐   Do operators use computer workstations for more than 4 h a day?
Yes ☐   No ☐   Are there complaints of discomfort from those working at these stations?
Yes ☐   No ☐   Is the chair or desk nonadjustable?
Yes ☐   No ☐   Is the display monitor, keyboard, or document holder nonadjustable?
Yes ☐   No ☐   Does lighting cause glare or make the monitor screen hard to read?
Yes ☐   No ☐   Is the room temperature too hot or too cold?
Yes ☐   No ☐   Is there irritating vibration or noise?

24.7.5   ENVIRONMENT

Yes ☐   No ☐   Is the temperature too hot or too cold?
Yes ☐   No ☐   Are workers' hands exposed to temperatures less than 70°F?
Yes ☐   No ☐   Is the workplace poorly lit?
Yes ☐   No ☐   Is there glare?
Yes ☐   No ☐   Is there excessive noise that is annoying, distracting, or producing hearing loss?
Yes ☐   No ☐   Is there upper extremity or whole body-vibration?
Yes ☐   No ☐   Is air circulation too high or too low?

24.7.6   GENERAL WORKPLACE

Yes ☐   No ☐   Are walkways uneven, slippery, or obstructed?
Yes ☐   No ☐   Is housekeeping poor?

Yes ☐   No ☐   Is there inadequate clearance or accessibility for performing tasks?
Yes ☐   No ☐   Are stairs cluttered or lacking railings?
Yes ☐   No ☐   Is proper footwear worn?

24.7.7 TOOLS

Yes ☐   No ☐   Is the handle too small or too large?
Yes ☐   No ☐   Does the handle shape cause the operator to bend the wrist in order to use the tool?
Yes ☐   No ☐   Is the tool hard to access?
Yes ☐   No ☐   Does the tool weigh more than 9 lb?
Yes ☐   No ☐   Does the tool vibrate excessively?
Yes ☐   No ☐   Does the tool cause excessive kickback to the operator?
Yes ☐   No ☐   Does the tool become too hot or too cold?

24.7.8 GLOVES

Yes ☐   No ☐   Do the gloves require the worker to use more force when performing job tasks?
Yes ☐   No ☐   Do the gloves provide inadequate protection?
Yes ☐   No ☐   Do the gloves present a hazard of catch points on the tool or in the workplace?

24.7.9 ADMINISTRATION

Yes ☐   No ☐   Is there little worker control over the work process?
Yes ☐   No ☐   Is the task highly repetitive and monotonous?
Yes ☐   No ☐   Does the job involve critical tasks with high accountability and little or no tolerance for error?
Yes ☐   No ☐   Are work hours and breaks poorly organized?

24.8 FIRE PROTECTION AND PREVENTION

Yes ☐   No ☐   Do the employers provide portable fire extinguishers for small fires?
Yes ☐   No ☐   Are all fire extinguishers clearly marked with symbols that distinctly reflect the type of fire hazard for which they are intended?
Yes ☐   No ☐   Are portable fire extinguishers located where they are readily accessible to employees without subjecting them to possible injury?
Yes ☐   No ☐   Are fire extinguishers fully charged and operable at all times?
Yes ☐   No ☐   Are Class A and D fire extinguishers no more than 75 ft apart?
Yes ☐   No ☐   Are Class B fire extinguishers no more than 50 ft apart?

Yes ☐   No ☐   Are Class C fire extinguishers patterned among Class A and B extinguishers where a Class C fire hazard exists?

Yes ☐   No ☐   Are protective clothing worn to protect the entire body including respiratory, head, hand, foot, leg, eye, and face?

Yes ☐   No ☐   Are fixed extinguishing systems used on specific fire hazards?

Yes ☐   No ☐   Is an alarm with a delay in place to warn employees before a fixed extinguisher is to be discharged?

Yes ☐   No ☐   Are hazard warning or caution signs posted at the entrance to, and inside, areas protected by systems that use agents known to be hazardous to employee safety and health?

Yes ☐   No ☐   Are fire detection systems installed and maintained to ensure best detection of a fire?

Yes ☐   No ☐   Is an employee alarm system that is capable of warning every employee of an emergency installed?

Yes ☐   No ☐   Is the alarm system such that it can be heard above the sound level of the work area?

Yes ☐   No ☐   Are warning lights installed, if there are hearing impaired employees?

Yes ☐   No ☐   Is all fire fighting equipment inspected at least annually and records kept?

Yes ☐   No ☐   Are portable fire extinguishers inspected at least monthly and records kept?

Yes ☐   No ☐   Is any damaged equipment removed immediately from service and replaced?

Yes ☐   No ☐   Is hydrostatic testing done on each extinguisher at least every 5 years?

Yes ☐   No ☐   Are fixed extinguishing systems inspected annually by a qualified person?

Yes ☐   No ☐   Are fire detection systems tested monthly if they are battery operated?

Yes ☐   No ☐   Is training provided on the use of portable fire extinguishers and records of attending employees kept?

Yes ☐   No ☐   Is training provided to employees designated to inspect, maintain, operate, or repair fixed extinguishing systems?

Yes ☐   No ☐   Is an annual review training required to keep them up to date?

Yes ☐   No ☐   Are all employees trained to recognize the alarm signals for each emergency (fire, tornado, chemical release, etc.)?

Yes ☐   No ☐   Are employees trained in how to report an emergency, where the alarms are, and how to sound them?

Yes ☐   No ☐   Is training provided on evacuation procedures?

Yes ☐   No ☐   Are drills performed periodically to ensure employees are aware of their duties?

Yes ☐   No ☐   Is all training conducted by a qualified/competent person?

Yes ☐   No ☐   Has the employer established and maintained a written policy that establishes the existence of a fire brigade?

Yes ☐   No ☐   Does the employer use employees who are physically capable of performing the duties as a member of a fire brigade that may be assigned to them during an emergency?

Yes ☐   No ☐   Is training of the duties provided by the employer before the
                 employee is asked to do any emergency response duties?
Yes ☐   No ☐   Are all fire brigade members trained at least annually and interior
                 structural fire fighters provided with an education session or
                 training at least quarterly?
Yes ☐   No ☐   Has the employer informed the fire brigade members of special
                 hazards, such as storage and use of flammable liquids and gases,
                 toxic chemicals, radioactive sources, and water-reactive
                 substances that they may encounter during an emergency?

24.9   HAZARD COMMUNICATION

Yes ☐   No ☐   Is there a list of hazardous substances used in your workplace?
Yes ☐   No ☐   Is there a written hazard communication program dealing with
                 material safety data sheets (MSDS), labeling, and employee
                 training?
Yes ☐   No ☐   Is each container for a hazardous substance (i.e., vats, bottles,
                 storage tanks, etc.) labeled with product identity and a hazard
                 warning (communication of the specific health hazards and
                 physical hazards)?
Yes ☐   No ☐   Is there a MSDS readily available for each hazardous substance
                 used?
Yes ☐   No ☐   Is there an employee training program for hazardous substances?
                 Does this program include
Yes ☐   No ☐   An explanation of what an MSDS is and how to use and
                 obtain one
Yes ☐   No ☐   MSDS contents for each hazardous substance or class of
                 substances
Yes ☐   No ☐   Explanation of "Right to Know?"
Yes ☐   No ☐   Identification of where an employee can see the employers' written
                 hazard communication program and where hazardous substances
                 are present in their work areas
Yes ☐   No ☐   The physical and health hazards of substances in the work area,
                 and specific protective measures to be used
Yes ☐   No ☐   Details of the hazard communication program, including how to
                 use the labeling system and MSDS's
                 Are employees trained in the following:
Yes ☐   No ☐   How to recognize tasks that might result in occupational exposure?
Yes ☐   No ☐   How to use work practice, engineering controls, and PPE and how
                 to know their limitations?
Yes ☐   No ☐   How to obtain information on the types selection, proper use,
                 location, removal, handling, decontamination, and disposal
                 of PPE?
Yes ☐   No ☐   Who to contact and what to do in an emergency?

24.10   HEAT HAZARDS

Yes ☐   No ☐   Are there thermal processes in use?
Yes ☐   No ☐   Are heat sources shielded or guarded?
Yes ☐   No ☐   Do burn risks exist?
Yes ☐   No ☐   Have workers been burned?
Yes ☐   No ☐   Are there extreme heat hazards?
Yes ☐   No ☐   Is heat stress a possible problem?
Yes ☐   No ☐   Does appropriate PPE exist?
Yes ☐   No ☐   Are workers wearing the required PPE?
Yes ☐   No ☐   Have the heat processes caused eye injuries?
Yes ☐   No ☐   Have workers been trained on avoidance of thermal hazards?

24.11   IONIZING RADIATION

Yes ☐   No ☐   Are there sources of ionizing radiation?
Yes ☐   No ☐   Are radioisotopes in use?
Yes ☐   No ☐   Are medical sources of radiation in use?
Yes ☐   No ☐   Are x-rays in use?
Yes ☐   No ☐   Are radiation surveys conducted?
Yes ☐   No ☐   Are signs warning of radiation posted?
Yes ☐   No ☐   Are workers protected by shielding, time, or distance?
Yes ☐   No ☐   Have workers been trained on the hazards and safeguards?
Yes ☐   No ☐   Is personal PPE needed?
Yes ☐   No ☐   Have workers been trained in the use of PPE?
Yes ☐   No ☐   Are workers wearing the appropriate PPE?
Yes ☐   No ☐   Are nuclear reactors in use?
Yes ☐   No ☐   Have or do workers exhibit any signs and symptoms of radiation sickness or exposures?

24.12   MACHINE GUARDING AND SAFETY

Yes ☐   No ☐   Do the safeguards provided meet the minimum OSHA requirements?
Yes ☐   No ☐   Do the safeguards prevent workers' hands, arms and other body parts from making contact with dangerous moving parts?
Yes ☐   No ☐   Are the safeguards firmly secured and not easily removable?
Yes ☐   No ☐   Do the safeguards ensure that no objects will fall into the moving parts?
Yes ☐   No ☐   Do the safeguards permit safe, comfortable, and relatively easy operation of the machine?
Yes ☐   No ☐   Can the machine be oiled without removing the safeguard?
Yes ☐   No ☐   Is there a system for shutting down the machinery before safeguards are removed?

Yes ☐ No ☐   Can the existing safeguards be improved?

Yes ☐ No ☐   Is there a point-of-operation safeguard provided for the machine?

Yes ☐ No ☐   Does it keep the operator's hands, fingers, and body out of the danger area?

Yes ☐ No ☐   Is there evidence that the safeguards have been tampered with or removed?

Yes ☐ No ☐   Could you suggest a more practical, effective safeguard?

Yes ☐ No ☐   Could changes be made on the machine to eliminate the point-of-operation hazard entirely?

Yes ☐ No ☐   Are there any unguarded gears, sprockets, pulleys, or flywheels on the apparatus?

Yes ☐ No ☐   Are there any exposed belts or chain drives?

Yes ☐ No ☐   Are there any exposed setscrews, key ways, collars, etc.?

Yes ☐ No ☐   Are starting and stopping controls within easy reach of the operator?

Yes ☐ No ☐   If there is more than one operator, are separate controls provided?

Yes ☐ No ☐   Are safeguards provided for all hazardous moving parts of the machine including auxiliary parts?

Yes ☐ No ☐   Have appropriate measures been taken to safeguard workers against noise hazards?

Yes ☐ No ☐   Have special guards, enclosures, or PPE been provided, where necessary, to protect workers from exposure to harmful substances used in machine operation?

Yes ☐ No ☐   Is the machine installed in accordance with National Fire Protection Association and NEC requirements?

Yes ☐ No ☐   Are there loose conduit fittings?

Yes ☐ No ☐   Is the machine properly grounded?

Yes ☐ No ☐   Is the power supply correctly fused and protected?

Yes ☐ No ☐   Do workers occasionally receive minor shocks while operating any of the machines?

Yes ☐ No ☐   Do operators and maintenance workers have the necessary training in how to use the safeguards and why?

Yes ☐ No ☐   Have operators and maintenance workers been trained in where the safeguards are located, how they provide protection, and what hazards they protect against?

Yes ☐ No ☐   Have operators and maintenance workers been trained in how and under what circumstances guards can be removed?

Yes ☐ No ☐   Have workers been trained in the procedures to follow if they notice guards that are damaged, missing, or inadequate?

Yes ☐ No ☐   Is protective equipment required?

Yes ☐ No ☐   It protective equipment is required, is it appropriate for the job, in good condition, kept clean and sanitary, and stored carefully when not in use?

Yes ☐ No ☐   Is the operator dressed safely for the job (i.e., no loose-fitting clothing or jewelry)?

Yes ☐   No ☐   Have maintenance workers received up-to-date instruction on the machines they service?

Yes ☐   No ☐   Do maintenance workers lock out the machine from its power sources before beginning repairs?

Yes ☐   No ☐   Where several maintenance persons work on the same machine, are multiple lockout devices used?

Yes ☐   No ☐   Do maintenance parsons use appropriate and safe equipment in their repair work?

Yes ☐   No ☐   Is the maintenance equipment itself property guarded?

Yes ☐   No ☐   Are maintenance and servicing workers trained in the requirements of 29 CFR 1910.147, lockout/tagout hazard, and do the procedures for lockout/tagout exist before they attempt their tasks?

24.13   MATERIAL HANDLING

24.13.1   MATERIAL-HANDLING EQUIPMENT

Yes ☐   No ☐   Are all operators of material-handling equipment trained (includes: hand trucks, cranes, hoists, fork trucks, or any motorized equipment)?

Yes ☐   No ☐   Are all operators of forklifts trained by a certified instructor?

Yes ☐   No ☐   Is all material-handling equipment kept in good repair and maintained by trained personnel?

Yes ☐   No ☐   Is all material-handling equipment inspected before use, daily, monthly, and annually as required?

Yes ☐   No ☐   Is all material-handling equipment properly marked with load ratings?

Yes ☐   No ☐   Are forklifts marked "Flammable," if they use propane or any other compressed gas source?

Yes ☐   No ☐   Are railroad cars, heavy equipment, and rolling hoists or cranes chocked or blocked to prevent rolling?

Yes ☐   No ☐   Are grading or ramps installed between two working levels for safe vehicle movement?

Yes ☐   No ☐   Is material-handling equipment that poses a danger to equipment or personnel guarded to prevent access within a safe distance?

24.13.2   STORAGE AREAS

Yes ☐   No ☐   Are maximum safe load limits observed?

Yes ☐   No ☐   Are load limits posted for platforms and floors?

Yes ☐   No ☐   Are storage racks stable and secure?

Yes ☐   No ☐   Is stored material neatly stacked, racked, blocked, or interlocked?

Yes ☐   No ☐   Are height limits set and posted to ensure stability of stacked material?

Yes ☐   No ☐   Do all aisles, loading docks, doorways, turns, and passages have safe clearances for equipment and material?

Yes ☐   No ☐   Are clearance signs posted in a visible place to warn employees of clearance limits?

Yes ☐   No ☐   Are all ramps, open pits, tanks, vats, ditches, and elevated surfaces 4 ft. or more guarded?

24.13.3   HOUSEKEEPING

Yes ☐   No ☐   Are storage areas kept clean, dry, and in good condition?

Yes ☐   No ☐   Are storage areas kept free of tripping and slipping hazards?

Yes ☐   No ☐   Are storage areas kept free of fire hazards (trash, paper, oily rags, or empty flammable liquid containers)?

Yes ☐   No ☐   Are storage areas kept free of explosion hazards (unsecured compressed-gas cylinders, flammable vapors, or dusts)?

Yes ☐   No ☐   Are storage areas kept free of pests such as rats, mice, roaches, other vermin?

24.14   MEANS OF EXIT

Yes ☐   No ☐   Do all exits have an illuminated sign above them that states, "Exit"?

Yes ☐   No ☐   Are there signs that state "Not an exit," placed over doors if there is the possibility that they could be mistaken for exits, e.g., closets, stairways, and doors?

Yes ☐   No ☐   Are exits locked while the building is occupied?

Yes ☐   No ☐   Are all emergency exit doors equipped with panic bars?

Yes ☐   No ☐   Do all emergency exit doors, designated for fire escape, lead to a safe area of refuge?

Yes ☐   No ☐   Do all emergency exit doors or passageways have emergency illumination, in case of power loss?

Yes ☐   No ☐   Is there access to exits that are unobstructed at all times?

Yes ☐   No ☐   Are all floor areas around exits clean and dry at all times?

Yes ☐   No ☐   Is an inspection by a fire marshal done at least once a year?

Yes ☐   No ☐   Is a general inspection of exit signs, exit doors, exit accesses, and alarm systems conducted by a trained person who has the authority to rectify any problems?

Yes ☐   No ☐   Is training done on the identification of all exits and their locations?

24.15   MEDICAL SERVICES AND FIRST AID

Yes ☐   No ☐   Are medical facilities and medically trained personnel onsite if possible?

Yes ☐   No ☐   In the absence of a medical facility that is close and available, are adequately trained personnel readily available to render first aid?

Yes ☐ No ☐ Are physician-approved first-aid supplies readily available?
Yes ☐ No ☐ Are quick drenching or flushing facilities available in work areas where the eyes or body may be exposed to injurious corrosive materials or chemicals?
Yes ☐ No ☐ Is a first-aid log kept on employees?
Yes ☐ No ☐ Is an inventory checklist kept of all first-aid supplies?
Yes ☐ No ☐ Are all employees trained on basic first-aid techniques and procedures?
Yes ☐ No ☐ Are all employees trained on usage of PPE while first aid is performed?

24.16 NONIONIZING RADIATION

Yes ☐ No ☐ Are there radio frequency sources?
Yes ☐ No ☐ Are there ultraviolet sources?
Yes ☐ No ☐ Are there infrared sources?
Yes ☐ No ☐ Are there lasers in use?
Yes ☐ No ☐ Are there microwaves in use?
Yes ☐ No ☐ Is there known risk of nonionizing radiation sources?
Yes ☐ No ☐ Are safeguards in place?
Yes ☐ No ☐ Are interlocks in place?
Yes ☐ No ☐ Are caution, warning, and information signs in place?
Yes ☐ No ☐ If PPE is needed, is it provided?
Yes ☐ No ☐ Are workers wearing the appropriate PPE?
Yes ☐ No ☐ Have workers complained of medically related signs or symptoms from exposure to nonionizing sources?
Yes ☐ No ☐ Does welding take place?
Yes ☐ No ☐ Have flash-related injuries occurred?

24.17 PPE

Yes ☐ No ☐ Are employers assessing the workplace to determine if hazards that require the use of PPE (for example, head, eye, face, hand, or foot protection) are present or are likely to be present?
Yes ☐ No ☐ If hazards or the likelihood of hazards is found, are employers selecting and having affected employees use properly fitted PPE suitable for protection from these hazards?
Yes ☐ No ☐ Has the employee been trained on PPE procedures; that is, what PPE is necessary for a job task, when they need it, and how to properly adjust it?
Yes ☐ No ☐ Are protective goggles or face shields provided and worn where there is any danger of flying particles or corrosive materials?

Yes ☐ No ☐ Are approved safety glasses required to be worn at all times in areas where there is a risk of eye injuries such as punctures, abrasions, contusions, or burns?

Yes ☐ No ☐ Are employees who need corrective lenses (glasses or contacts) in working environments with harmful exposures required to wear only approved safety glasses and protective goggles or use other medically approved precautionary procedures?

Yes ☐ No ☐ Are protective gloves, aprons, shields, or other means provided and required where employees could be cut or where there is reasonably anticipated exposure to corrosive liquids, chemicals, blood, or other potentially infectious materials?

Yes ☐ No ☐ Are hard hats provided and worn where danger of falling objects exists?

Yes ☐ No ☐ Are hard hats inspected periodically for damage to the shell and suspension system?

Yes ☐ No ☐ Is appropriate foot protection required where there is a risk of foot injuries from hot, corrosive, or poisonous substances, falling objects, or crushing or penetrating actions?

Yes ☐ No ☐ Are approved respirators provided for regular or emergency use where needed?

Yes ☐ No ☐ Is all protective equipment maintained in a sanitary condition and ready for use?

Yes ☐ No ☐ Do you have eye wash facilities and a quick drench shower within the work area where employees are exposed to injurious corrosive materials? Where special equipment is needed for electrical workers, is it available?

Yes ☐ No ☐ Where food or beverages are consumed on the premises, are they consumed in areas where there is no exposure to toxic material, blood, or other potentially infectious materials?

Yes ☐ No ☐ Is protection against the effects of occupational noise exposure provided when sound levels exceed those of the OSHA noise standard?

Yes ☐ No ☐ Are adequate work procedures, protective clothing, and equipment provided and used when cleaning up spilled toxic or otherwise hazardous materials or liquids?

Yes ☐ No ☐ Are there appropriate procedures in place for disposing of or decontaminating PPE contaminated with, or reasonably anticipated to be contaminated with, blood or other potentially infectious materials?

24.18 SECURITY

Yes ☐ No ☐ Is there a plan to provide for security of workers?

Yes ☐ No ☐ Is there a trained security force on duty?

Yes ☐   No ☐   Are workers isolated from the public and outside as much as possible?
Yes ☐   No ☐   Is there a separated sign-in/-out screening performed for all visitors?
Yes ☐   No ☐   Do surveillance cameras exist?
Yes ☐   No ☐   Are alarms available for emergencies?
Yes ☐   No ☐   Are emergency telephone numbers posted?
Yes ☐   No ☐   Have workers been trained regarding personal security?
Yes ☐   No ☐   Is there escort service for employees?
Yes ☐   No ☐   Are all areas well lighted?
Yes ☐   No ☐   Are parking areas well lighted?
Yes ☐   No ☐   Does landscaping permit high visibility?

24.19   SLIPS, TRIPS, AND FALLS

Yes ☐   No ☐   Are all walking surfaces clear of debris, tools, etc.?
Yes ☐   No ☐   Are travel surfaces even?
Yes ☐   No ☐   Are spills cleaned immediately?
Yes ☐   No ☐   Do workers wear nonslip footwear?
Yes ☐   No ☐   Are materials stored in passageways and aisles?
Yes ☐   No ☐   Are stairs clear of items such as tools?
Yes ☐   No ☐   Are floors and walking surfaces free of spills, grease, or oil?
Yes ☐   No ☐   Are handrails in place?
Yes ☐   No ☐   Are signs or mirrors used for blind corners?

24.20   VIOLENCE

Yes ☐   No ☐   Is there zero tolerance for workplace violence?
Yes ☐   No ☐   Are all weapons precluded from the workplace?
Yes ☐   No ☐   Have workers and supervisors been trained regarding acts of violence?
Yes ☐   No ☐   Is there a protocol for violence incidents?
Yes ☐   No ☐   Are protective steps taken to prevent violent acts?
Yes ☐   No ☐   Are potentials for violence reported?
Yes ☐   No ☐   Is action taken immediately when potential for violent acts are possible?
Yes ☐   No ☐   Are there emergency procedures to handle cases of workplace violence?
Yes ☐   No ☐   Are emergency telephone numbers posted?

24.21   WALKING–WORKING SURFACES

Yes ☐   No ☐   Is a documented, functioning housekeeping program in place?
Yes ☐   No ☐   Are all worksites clean, sanitary, and orderly?

Yes ☐ No ☐ Are work surfaces kept dry or are appropriate steps taken to ensure that surfaces are slip-resistant?

Yes ☐ No ☐ Are all spilled hazardous materials or liquids, including blood and other potentially infectious materials, cleaned up immediately and according to proper procedures?

Yes ☐ No ☐ Is combustible scrap, debris, and waste stored safely and removed from the worksite properly?

Yes ☐ No ☐ Is all regulated waste, as defined in the OSHA bloodborne pathogens standard (1910.1030), discarded according to federal, state, and local regulations?

Yes ☐ No ☐ Are accumulations of combustible dust routinely removed from elevated surfaces including the overhead structure of buildings, etc.?

Yes ☐ No ☐ Is combustible dust cleaned up with a vacuum system to prevent the dust from going into suspension?

Yes ☐ No ☐ Is metallic or conductive dust prevented from entering or accumulating on or around electrical enclosures or equipment?

Yes ☐ No ☐ Are covered metal waste cans used for oily and paint-soaked waste?

24.21.1 WALKWAYS

Yes ☐ No ☐ Are aisles and passageways kept clear?

Yes ☐ No ☐ Are aisles and walkways marked as appropriate?

Yes ☐ No ☐ Are wet surfaces covered with nonslip materials?

Yes ☐ No ☐ Are holes in the floor, sidewalk, or other walking surfaces repaired properly, covered, or otherwise made safe?

Yes ☐ No ☐ Is there safe clearance for walking in aisles where motorized or mechanical handling equipment is operating?

Yes ☐ No ☐ Are materials or equipment stored in such a way that sharp projectives do not interfere with the walkway?

Yes ☐ No ☐ Are spilled materials cleaned up immediately?

Yes ☐ No ☐ Are changes of direction or elevation readily identifiable?

Yes ☐ No ☐ Are aisles or walkways that pass near moving or operating machinery, welding operations, or similar operations arranged such that employees will not be subjected to potential hazards?

Yes ☐ No ☐ Is adequate headroom provided for the entire length of any aisle or walkway?

Yes ☐ No ☐ Are standard guardrails provided wherever aisle or walkway surfaces are elevated more than 30 in. above any adjacent floor or the ground?

Yes ☐ No ☐ Are bridges provided over conveyors and similar hazards?

24.21.2 FLOOR AND WALL OPENINGS

Yes ☐ No ☐ Are floor openings guarded by a cover, a guardrail, or equivalent on all sides (except at the entrance to stairways or ladders)?

Yes ☐ No ☐ Are toeboards installed around the edges of permanent floor openings (where persons may pass below the opening)?

Yes ☐   No ☐   Are skylight screens of such construction and mounting that they will withstand a load of at least 200 lb?

Yes ☐   No ☐   Is the glass in the windows, doors, glass walls, etc., which are subject to human impact, of sufficient thickness and type for the condition of use?

Yes ☐   No ☐   Are grates or similar type covers over floor openings such as floor drains of such design that foot traffic or rolling equipment will not be affected by the grate spacing?

Yes ☐   No ☐   Are unused portions of service pits and pits not actually in use either covered or protected by guardrails or equivalent?

Yes ☐   No ☐   Are manhole covers, trench covers, and similar covers, and their supports designed to carry a truck rear axle load of at least 20,000 lb when located in roadways and subject to vehicle traffic?

Yes ☐   No ☐   Are floor or wall openings in fire-resistive construction provided with doors or covers compatible with the fire rating of the structure and provided with a self-closing feature when appropriate?

24.21.3 STAIRS AND STAIRWAYS

Yes ☐   No ☐   Do standard stair rails or handrails on all stairways have four or more risers?

Yes ☐   No ☐   Are all stairways at least 22 in. wide?

Yes ☐   No ☐   Do stairs have landing platforms not less than 30 in. in the direction of travel and extend 22 in. in width at every 12 ft or less of vertical rise?

Yes ☐   No ☐   Do stairs angle no more than 50° and no less than 30°?

Yes ☐   No ☐   Are step risers on stairs uniform from top to bottom?

Yes ☐   No ☐   Are steps on stairs and stairways designed or provided with a surface that renders them slip resistant?

Yes ☐   No ☐   Are stairway handrails located between 30 and 34 in. above the leading edge of stair treads?

Yes ☐   No ☐   Do stairway handrails have at least 3 in. of clearance between the handrails and the wall or surface they are mounted on?

Yes ☐   No ☐   Where doors or gates open directly on a stairway, is there a platform provided so that the swing of the door does not reduce the width of the platform to less than 21 in.?

Yes ☐   No ☐   Where stairs or stairways exit directly into any area where vehicles may be operated, are adequate barriers and warnings provided to prevent employees stepping into the path of traffic?

Yes ☐   No ☐   Do stairway landings have a dimension (measured in the direction of travel) at least equal to the width of the stairway?

24.21.4 ELEVATED SURFACES

Yes ☐ No ☐ Are signs posted, when appropriate, showing the elevated surface load capacity?

Yes ☐ No ☐ Are surfaces elevated more than 30 in. above the floor or ground provided with standard guardrails?

Yes ☐ No ☐ Are all elevated surfaces (beneath which people or machinery could be exposed to falling objects) provided with standard 4 in. toeboards?

Yes ☐ No ☐ Is a permanent means of access and egress provided to elevated storage and work surfaces?

Yes ☐ No ☐ Is required headroom provided where necessary?

Yes ☐ No ☐ Is material on elevated surfaces piled, stacked, or racked in a manner to prevent it from tipping, falling, collapsing, rolling, or spreading?

Yes ☐ No ☐ Are dockboards or bridge plates used when transferring materials between docks and trucks or rail cars?

Appendix A   Common Exposures or Accident Types

The common exposures or accident types help standardize the review of hazards. There are 11 basic types of accidents:

- Struck-against
- Struck-by
- Contact-with
- Contacted-by
- Caught-in
- Caught-on
- Caught-between
- Fall-to-same-level
- Fall-to-below
- Overexertion
- Exposure

Hazards should be looked with these common accident types in mind to identify procedures, processes, occupations, and tasks, which present a hazard that could cause one of the accident types in the following section.

A.1   ACCIDENT TYPES

A.1.1   STRUCK-AGAINST TYPES OF ACCIDENTS

Look at these first four basic accident types—struck-against, struck-by, contact-with, and contacted-by—in more detail, with the job step walk-round inspection in mind. Can the worker strike against anything while doing the job step? Think of the worker moving and contacting something forcefully and unexpectedly—an object capable of causing injury. Can he/she forcefully contact anything that will cause injury? This forceful contact may be with machinery, timber or bolts, protruding objects or sharp, jagged edges. Identify not only what the worker can strike against, but also how the contact can come about. This does not mean that every object around the worker must be listed.

A.1.2   STRUCK-BY TYPES OF ACCIDENTS

Can the worker be struck by anything while doing the job step? The phrase "struck by" means that something moves and strikes the worker abruptly with force. Study the work environment for what is moving in the vicinity of the worker, what is about to move, or what will move as a result of what the worker does. Is unexpected movement possible from normally stationary objects? Examples are ladders, tools, containers, supplies, and so on.

A.1.3 CONTACT-BY AND CONTACT-WITH TYPES OF ACCIDENTS

The subtle difference between contact with and contact-by injuries is that in the first, the agent moves to the victim, while in the second, the victim moves to the agent. Can the worker be contacted by anything while doing the job step? The contacted by accident is one in which the worker could be contacted by some object or agent. This object or agent is capable of injuring by nonforceful contact. Examples of items capable of causing injury are chemicals, hot solutions, fire, electrical flashes, and steam.

Can the worker come in contact with some agent that will injure without forceful contact? Any type of work that involves materials or equipment, which may be harmful without forceful contact, is a source of contact with accidents. There are two kinds of work situations, which account for most of the contact with accidents. One situation is working on or near electrically charged equipment, and the other is working with chemicals or handling chemical containers.

A.1.4 CAUGHT-IN AND CAUGHT-ON TYPES OF ACCIDENTS

The next three accident types involve "caught" accidents. Can the person be caught in, caught on, or caught between objects? A caught in-accident is one in which the person, or some part of his/her body, is caught-in an enclosure or opening of some kind. Can the worker be caught on anything while doing the job step? Most caught-on accidents involve worker's clothing being caught on some projection of a moving object. This moving object pulls the worker into an injury contact. Or, the worker may be caught on a stationary protruding object, causing a fall.

A.1.5 CAUGHT-BETWEEN TYPES OF ACCIDENTS

Can the worker be caught between any objects while doing the job step? Caught-between accidents involve having a part of the body caught between something moving and something stationary, or between two moving objects. Always look for pinch points.

A.1.6 FALL-TO-SAME-LEVEL AND FALL-TO-BELOW TYPES OF ACCIDENTS

Slip, trip, and fall accident types are one of the most common accidents occurring in the workplace. Can the worker fall while doing a job step? Falls are such frequent accidents that we need to look thoroughly for slip, trip, and fall hazards. Consider whether the worker can fall from something above ground level, or whether the worker can fall to the same level. Two hazards account for most fall-to-same-level accidents: slipping hazards and tripping hazards. The fall-to-below accidents occur in situations where employees work above ground or above floor level, and the results are usually more severe.

A.1.7 OVEREXERTION AND EXPOSURE TYPES OF ACCIDENTS

The next two accident types are overexertion and exposure. Can the worker be injured by overexertion; that is, can he/she be injured while lifting, pulling, or pushing? Can awkward body positioning while doing a job step cause a sprain or strain? Can the repetitive nature of a task cause injury to the body? An example of this is excessive flexing of the wrist, which can cause carpal tunnel syndrome (which is abnormal pressure on the tendons and nerves in the wrist).

Finally, can exposure to the work environment cause injury to the worker? Environmental conditions such as noise, extreme temperatures, poor air, toxic gases and chemicals, or harmful fumes from work operations should also be listed as hazards.

Appendix B Glove Selection Chart

TABLE B.1
Glove Chart

Type	Advantages	Disadvantages	Use Against
Natural rubber	Low cost, good physical properties, dexterity	Poor vs. oils, greases, organics. Frequently imported; may be poor quality	Bases, alcohols, dilute water solutions; fair vs. aldehydes, ketones
Natural rubber blends	Low cost, dexterity, better chemical resistance than natural rubber vs. some chemicals	Physical properties frequently inferior to natural rubber	Same as natural rubber
Polyvinylchloride (PVC)	Low cost, very good physical properties, medium cost, medium chemical resistance	Plasticizers can be stripped; frequently imported may be poor quality	Strong acids and bases, salts, other water solutions, alcohols
Neoprene	Medium cost, medium chemical resistance, medium physical properties	NA	Oxidizing acids, anilines, phenol, glycol ethers
Nitrile	Low cost, excellent physical properties, dexterity	Poor vs. benzene, methylene chloride, trichloroethylene, many ketones	Oils, greases, aliphatic chemicals, xylene, perchloroethylene, trichloroethane; fair vs. toluene
Butyl	Specialty glove, polar organics	Expensive, poor vs. hydrocarbons, chlorinated solvents	Glycol ethers, ketones, esters
Polyvinyl alcohol (PVA)	Specialty glove, resists a very broad range of organics, good physical properties	Very expensive, water sensitive, poor vs. light alcohols	Aliphatics, aromatics, chlorinated solvents, ketones (except acetone), esters, ethers
Fluoroelastomer (Viton)	Specialty glove, organic solvents	Extremely expensive, poor physical properties, poor vs. some ketones, esters, amines	Aromatics, chlorinated solvents, also aliphatics and alcohols
Norfoil (silver shield)	Excellent chemical resistance	Poor fit, easily punctures, poor grip, stiff	Use for Hazmat work

Source: Courtesy of the Department of Energy.

TABLE B.2
Glove Type and Chemical Use

Chemical	Neoprene	Natural Latex or Rubber	Butyl	Nitrile Latex
Acetaldehyde[a]	VG	G	VG	G
Acetic acid	VG	VG	VG	VG
Acetone	G	VG	VG	P
Ammonium hydroxide	VG	VG	VG	VG
Amyl acetate	F	P	F	P
Aniline	G	F	F	P
Benzaldehyde	F	F	G	G
Benzene	F	F	F	P
Butyl acetate	G	F	F	P
Butyl alcohol	VG	VG	VG	VG
Carbon disulfide	F	F	F	F
Carbon tetrachloride	F	P	P	G
Castor oil	F	P	F	VG
Chlorobenzene	F	P	F	P
Chloroform	G	P	P	E
Chloronaphthalene	F	P	F	F
Chromic acid (50%)	F	P	F	F
Citric acid (10%)	VG	VG	VG	VG
Cyclohexanol	G	F	G	VG
Dibutyl phthalate	G	P	G	G
Diesel fuel	G	P	P	VG
Diisobutyl ketone	P	F	G	P
Dimethylformamide	F	F	G	G
Dioctyl phthalate	G	P	F	VG
Dioxane	VG	G	G	G
Epoxy resins, dry	VG	VG	VG	VG
Ethyl acetate	G	F	G	F
Ethyl alcohol	VG	VG	VG	VG
Ethyl ether	VG	G	VG	G
Ethylene dichloride	F	P	F	P
Ethylene glycol	VG	VG	VG	VG
Formaldehyde	VG	VG	VG	VG
Formic acid	VG	VG	VG	VG
Freon 11	G	P	F	G
Freon 12	G	P	F	G
Freon 21	G	P	F	G
Freon 22	G	P	F	G
Furfural	G	G	G	G
Gasoline, leaded	G	P	F	VG
Gasoline, unleaded	G	P	F	VG
Glycerine	VG	VG	VG	VG
Hexane	F	P	P	G
Hydrochloric acid	VG	G	G	G
Hydrofluoric acid (48%)	VG	G	G	G

TABLE B.2 (continued)
Glove Type and Chemical Use

Chemical	Neoprene	Natural Latex or Rubber	Butyl	Nitrile Latex
Hydrogen peroxide (30%)	G	G	G	G
Hydroquinone	G	G	G	F
Isooctane	F	P	P	VG
Isopropyl alcohol	VG	VG	VG	VG
Kerosene	VG	F	F	VG
Ketones	G	VG	VG	P
Lacquer thinners	G	F	F	P
Lactic acid (85%)	VG	VG	VG	VG
Lauric acid (36%)	VG	F	VG	VG
Lineoleic acid	VG	P	F	G
Linseed oil	VG	P	F	VG
Maleic acid	VG	VG	VG	VG
Methyl alcohol	VG	VG	VG	VG
Methylamine	F	F	G	G
Methyl bromide	G	F	G	F
Methyl chloride	P	P	P	P
Methyl ethyl ketone	G	G	VG	P
Methyl isobutyl ketone	F	F	VG	P
Methyl methacrylate	G	G	VG	F
Monoethanolamine	VG	G	VG	VG
Morpholine	VG	VG	VG	G
Naphthalene	G	F	F	G
Naphthas, aromatic	G	P	P	G
Nitric acid	G	F	F	F
Nitromethane (95.5%)	F	P	F	F
Nitropropane (95.5%)	F	P	F	F
Octyl alcohol	VG	VG	VG	VG
Oleic acid	VG	F	G	VG
Oxalic acid	VG	VG	VG	VG
Palmitic acid	VG	VG	VG	VG
Perchloric acid (60%)	VG	F	G	G
Perchloroethylene	F	P	P	G
Petroleum distillates (naphtha)	G	P	P	VG
Phenol	VG	F	G	F
Phosphoric acid	VG	G	VG	VG
Potassium hydroxide	VG	VG	VG	VG
Propyl acetate	G	F	G	F
Propyl alcohol	VG	VG	VG	VG
Propyl alcohol (iso)	VG	VG	VG	VG
Sodium hydroxide	VG	VG	VG	VG
Styrene	P	P	P	F
Stryene (100%)	P	P	P	F
Sulfuric acid	G	G	G	G

(*continued*)

TABLE B.2 (continued)
Glove Type and Chemical Use

Chemical	Neoprene	Natural Latex or Rubber	Butyl	Nitrile Latex
Tannic acid (65%)	VG	VG	VG	VG
Tetrahydrofuran	P	F	F	F
Toluene	F	P	P	F
Toluene diisocyanate	F	G	G	F
Trichloroethylene	F	F	P	G
Triethanolamine	VG	G	G	VG
Tung oil	VG	P	F	VG
Turpentine	G	F	F	VG
Xylene	P	P	P	F

Source: Courtesy of the Department of Energy.

VG, very good; G, good; F, Fair; P, poor (not recommended).

[a] Limited service.

Appendix C   Workplace Security Program

Our establishment's program for workplace security addresses the hazards known to be associated with the three major types of workplace violence. Type I workplace violence involves a violent act by an assailant with no legitimate relationship to the workplace, who enters the workplace to commit a robbery or other criminal acts. Type II involves a violent act or threat of violence by a recipient of a service provided by our establishment, such as a client, patient, customer, passenger, or a criminal suspect or prisoner. Type III involves a violent act or threat of violence by a current or former worker, supervisor or manager, or another person who has some employment-related involvement with our establishment, such as a worker's spouse or lover, a worker's relative or friend, or another person who has a dispute with one of our workers.

C.1   RESPONSIBILITY

We have decided to assign responsibility for security in our workplace. The security program administrator for workplace security is_____and he/she has the authority and responsibility for implementing the provisions of this program for_____.

All managers and supervisors are responsible for implementing and maintaining this security program in their work areas and for answering worker questions about it. A copy of this security program is available from each manager and supervisor.

C.2   COMPLIANCE

We have established the following policy to ensure compliance with our rules on workplace security.

Management of our establishment is committed to ensuring that all safety and health policies and procedures involving workplace security are clearly communicated and understood by all workers.

All workers are responsible for using safe work practices, for following all directives, policies and procedures, and for assisting in maintaining a safe and secure work environment. Our system of ensuring that all workers, including supervisors and managers, comply with work practices that are designed to make the workplace more secure, and do not engage in threats or physical actions that create a security hazard for others in the workplace, includes

1. Informing workers, supervisors, and managers of the provisions of our security program
2. Evaluating the performance of all workers in complying with our establishment's workplace security measures
3. Recognizing workers who perform work practices that promote security in the workplace
4. Providing training and counseling to workers whose performance is deficient in complying with work practices designed to ensure workplace security
5. Disciplining workers who fail to comply with workplace security practices
6. The following practices that ensure worker compliance with workplace security directives, policies, and procedures:_____.

C.3 COMMUNICATION

At our establishment, we recognize that to maintain a safe, healthy, and secure workplace we must have open, two-way communication between all workers, including managers and supervisors, on all workplace safety, health, and security issues. Our establishment has a communication system designed to encourage a continuous flow of safety, health, and security information between management and our workers without fear of reprisal and in a form that is readily understandable. Our communication system consists of the following checked items:

- New worker orientation on our establishment's workplace security policies, procedures, and work practices
- Periodic review of our security program with all personnel
- Training programs designed to address specific aspects of workplace security unique to our establishment
- Regularly scheduled safety meetings with all personnel, which include workplace security discussions
- A system to ensure that all workers, including managers and supervisors, understand the workplace security policies
- Posted or distributed workplace security information
- A system for workers to inform management about workplace security hazards or threats of violence
- Procedures for protecting workers who report threats from retaliation by the person making the threats
- Addressing security issues at our workplace security team meetings
- Our establishment has fewer than ten workers and communicates with and instructs workers orally about general safe work practices with respect to workplace security
- Other:_____
  _____
  _____
  _____

C.4 HAZARD ASSESSMENT

We perform workplace hazard assessment for workplace security in the form of periodic inspections. Periodic inspections to identify and evaluate workplace security hazards and threats of workplace violence are performed by the following observer (s) in the following areas of our workplace:

Observer	Area

Periodic inspections are performed according to the following schedule:

1. _____
2. Frequency (daily, weekly, monthly, etc.)
3. When we initially established our security program
4. When new, previously unidentified security hazards are recognized
5. When occupational injuries or threats of injury occur
6. Whenever workplace security conditions warrant an inspection

Periodic inspections for security hazards consist of identification and evaluation of workplace security hazards and changes in worker work practices and may require assessing for more than one type of workplace violence. Our establishment performs inspections for each type of workplace violence by using the methods specified below to identify and evaluate workplace security hazards.

Inspections for Type I workplace security hazards include assessing:

1. The exterior and interior of the workplace for its attractiveness to robbers
2. The need for security surveillance measures, such as mirrors or cameras
3. Posting of signs notifying the public that limited cash is kept on the premises
4. Procedures for worker response during a robbery or other criminal act
5. Procedures for reporting suspicious persons or activities
6. Posting of emergency telephone numbers for law enforcement, fire, and medical services where workers have access to a telephone with an outside line
7. Limiting the amount of cash on hand and using time access safes for large bills
8. Other:_____
   _____
   _____
   _____

Inspections for Type II workplace security hazards include assessing:

1. Access to, and freedom of movement within, the workplace
2. Adequacy of workplace security systems, such as door locks, security windows, physical barriers, and restraint systems

3. Frequency and severity of threatening or hostile situations that may lead to violent acts by persons who are service recipients of our establishment
4. Workers' skill in safely handling threatening or hostile service recipients
5. Effectiveness of systems and procedures to warn others of a security danger or to summon assistance, e.g., alarms or panic buttons
6. The use of work practices such as "buddy" systems for specified emergency events
7. The availability of worker escape routes
8. Other: _____
   _____
   _____
   _____

Inspections for Type III workplace security hazards include assessing:

1. How well our establishment's antiviolence policy has been communicated to workers, supervisors, or managers
2. How well our establishment's management and workers communicate with each other
3. Our workers', supervisors', and managers' knowledge of the warning signs of potential workplace violence
4. Access to, and freedom of movement within, the workplace by nonworkers, including recently discharged workers or persons with whom one of our workers has a dispute
5. Frequency and severity of worker reports of threats of physical or verbal abuse by managers, supervisors, or other workers
6. Any prior violent acts, threats of physical violence, verbal abuse, property damage, or other signs of strain or pressure in the workplace
7. Worker disciplinary and discharge procedures
8. Other: _____
   _____
   _____
   _____

C.5   INCIDENT INVESTIGATIONS

We have established the following policy for investigating incidents of workplace violence. Our procedures for investigating incidents of workplace violence, which include threats and physical injury, include

1. Reviewing all previous incidents
2. Visiting the scene of an incident as soon as possible
3. Interviewing threatened or injured workers and witnesses
4. Examining the workplace for security risk factors associated with the incident, including any previous reports of inappropriate behavior by the perpetrator

5. Determining the cause of the incident
6. Taking corrective action to prevent the incident from recurring
7. Recording the findings and corrective actions taken
8. Other:_____

_____

_____

_____

C.6   HAZARD CORRECTION

Hazards that threaten the security of workers shall be corrected in a timely manner based on severity when they are first observed or discovered. Corrective measures for Type I workplace security hazards can include

1. Making the workplace unattractive to robbers
2. Using surveillance measures, such as cameras or mirrors, to provide information as to what is going on outside and inside the workplace
3. Procedures for reporting suspicious persons or activities
4. Posting of emergency telephone numbers for law enforcement, fire, and medical services where workers have access to a telephone with an outside line
5. Posting of signs notifying the public that limited cash is kept on the premises
6. Limiting the amount of cash on hand and using time access safes for large bills
7. Worker, supervisor, and management training on emergency action procedures
8. Other:_____

_____

_____

_____

Corrective measures for Type II workplace security hazards can include:

1. Controlling access to the workplace and freedom of movement within it, consistent with business necessity
2. Ensuring the adequacy of workplace security systems, such as door locks, security windows, physical barriers, and restraint systems
3. Providing worker training in recognizing and handling threatening or hostile situations that may lead to violent acts by persons who are service recipients of our establishment
4. Placing effective systems to warn others of a security danger or to summon assistance, e.g., alarms or panic buttons
5. Providing procedures for a "buddy" system for specified emergency events
6. Ensuring adequate worker escape routes
7. Other:_____

_____

_____

_____

Corrective measures for Type III workplace security hazards include:

1. Effectively communicating our establishment's antiviolence policy to all workers, supervisors, or managers
2. Improving the communication between our establishment's management and workers
3. Increasing awareness of workers, supervisors, and managers of the warning signs of potential workplace violence
4. Controlling access to, and freedom of movement within, the workplace by nonworkers, including recently discharged workers or persons with whom one of our workers has a dispute
5. Providing counseling to workers, supervisors, or managers who exhibit behavior that represents strain or pressure, which may lead to physical or verbal abuse of coworkers
6. Ensuring that all reports of violent acts, threats of physical violence, verbal abuse, property damage, or other signs of strain or pressure in the workplace are handled effectively by management and that the person making the report is not subject to retaliation by the person making the threat
7. Ensuring that worker disciplinary and discharge procedures address the potential for workplace violence
8. Other:_____
   _____
9. _____
   _____
10. _____
    _____

C.7   TRAINING AND INSTRUCTION

We have established the following policy on training all workers with respect to workplace security.

All workers, including managers and supervisors, shall have training and instruction on general and job-specific workplace security practices. Training and instruction shall be provided when the security program is first established and periodically thereafter. Training shall also be provided to all new workers and to other workers for whom training has not previously been provided and to all workers, supervisors, and managers given new job assignments for which specific workplace security training has not been provided previously. Additional training and instruction will be provided to all personnel whenever the employer is made aware of new or previously unrecognized security hazards.

General workplace security training and instruction includes, but is not limited to, the following:

1. Explanation of the security program including measures for reporting any violent acts or threats of violence

2. Recognition of workplace security hazards including the risk factors associated with the three types of workplace violence
3. Measures to prevent workplace violence, including procedures for reporting workplace security hazards or threats to managers and supervisors
4. Ways to defuse hostile or threatening situations
5. Measures to summon others for assistance
6. Worker routes of escape
7. Notification of law enforcement authorities when a criminal act may have occurred
8. Emergency medical care provided in the event of any violent act upon an worker
9. Post-event trauma counseling for workers desiring such assistance

In addition, we provide specific instructions to all workers regarding workplace security hazards unique to their job assignment, to the extent that such information was not already covered in other training.

We have chosen the following checked items for Type I training and instruction for managers, supervisors, and workers:

- Crime awareness
- Location and operation of alarm systems
- Communication procedures
- Proper work practices for specific workplace activities, occupations, or assignments, such as late night retail sales, taxicab drivers, or security guards
- Other:_____
  _____
  _____
  _____

We have chosen the following checked items for Type II training and instruction for managers, supervisors, and workers:

- Self-protection
- Dealing with angry, hostile, or threatening individuals
- Location, operation, care, and maintenance of alarm systems and other protective devices
- Communication procedures
- Determination of when to use the "buddy" system or other assistance from coworkers
- Awareness of indicators that lead to violent acts by service recipients
- Other:_____
  _____
  _____
  _____

We have chosen the following checked items for Type III training and instruction for managers, supervisors, and workers:

- Preemployment screening practices
- Worker assistance programs
- Awareness of situational indicators that lead to violent acts
- Managing with respect and consideration for worker well-being
- Review of antiviolence policy and procedures
- Other:_____
  _____
  _____
  _____

Bibliography

American Conference of Governmental Industrial Hygienists. *TLVs. Threshold Limit Values and Biological Exposure Indices for 1985–86*. Cincinnati, OH: ACGIH, 1991, pp. 91–98.

Bureau of Labor Statistics 1999, U.S. Department of Labor. *National Census of Fatal Occupational Injuries*. Washington, DC, 1998.

Bureau of Labor Statistics, U.S. Department of Labor, *Workplace Injuries and Illnesses in 2004*. Available at http://bls.gov.

Bureau of Labor Statistics, U.S. Department of Labor. *National Census of fatal Occupational Injuries in 2005*. Available at http://bls.gov.

Bureau of Labor Statistics, U.S. Department of Labor, *Career Guide to Industries, 2006–07 Edition*, Utilities, available at http://www.bls.gov/oco/cg/cgs018.htm (visited September 20, 2006).

California Department of Labor. *Guidelines for Security and Safety of Health Care and Community Service Workers*. Available at http://www.ca.gov. Sacramento, CA, March 1998.

California Department of Industrial Relations (Cal/OSHA), *Easy Ergonomics: A Practical Approach for Improving the Workplace*, Sacramento, CA, 1999.

Centers for Disease Control and Prevention, U.S. Department of Health and Human Services. *Biosafety in Microbiological and Biomedical Laboratories*, 4th ed. Washington, DC, 1999.

Ducatman, A.M. and Haes, D.L., Jr. Nonionizing radiation. In: *Textbook of Clinical Occupational and Environmental Medicine*, eds. L. Rosenstock and M.R. Cullen. Philadelphia: W.B. Saunders, 1994.

Eastman Kodak Company, 1983. *Ergonomic Design for People at Work*, Vols. 1 and 2. Belmont, CA: Lifetime Learning Publications.

The Hartford Loss Control Department, Loss Control Technical Information Paper Series. *Handling General Liability Incidents (TIP S 520.604)*. Hartford, 2002.

The Hartford Loss Control Department, Loss Control Technical Information Paper Series, *Safety for Company Visitors: Protecting Guests and Minimizing Liability (TIPS S 520.606)*. Hartford, 2002.

Michaelson, S.M. *Fundamentals and Applied Aspects of Nonionizing Radiation*. New York: Plenum, 1975.

Mine Safety and Health Administration, U.S. Department of Labor. *MSHA's Guide to Equipment Guarding for Metal and Nonmetal Mining*. Washington, DC, 1992.

Mine Health and Safety Administration, U.S. Department of Labor. *Hazard Recognition and Avoidance: Training Manual*. MSHA 0105, Revised May 1996.

National Institute for Occupational Safety and Health (NIOSH), DHHS, *Work Practices Guide for Manual Lifting 1981 and 1991*.

National Institute for Occupational Safety and Health, U.S. Department of Health and Human Services. *Criteria for a Recommended Standard; Welding, Brazing and Thermal Cutting (Pub. No. 88–110)*. Washington, DC, 1988.

National Institute for Occupational Safety and Health, U.S. Department of Health and Human Services. *NIOSH Current Intelligence Bulletin 57. Violence in the Workplace: Risk Factors and Prevention Strategies*, Atlanta, GA, 1996.

National Institute for Occupational Safety and Health, U.S. Department of Health and Human Services. *Violence in the Workplace: Risk Factors and Prevention Strategies (CIB 57)*. Washington, DC, June 1996.

National Safety Council. *Accident Prevention Manual for Business and Industry*, 11th edition. Itasco, IL, 1997.

National Institute for Occupational Safety and Health, U.S. Department of Health and Human Services. *Elements of Ergonomics Programs (DHHS 97–117)*, Atlanta, GA, 1997.

National Institute for Occupational Safety and Health, U.S. Department of Health and Human Services. *Worker Death by Electrocutions (Pub. No. 98–131)*. Morgantown, WV, 1998.

National Institute for Occupational Safety and Health, U.S. Department of Health and Human Services. *Electrical Safety: Safety and Health for Electrical Trades—Student Manual (Pub. No. 2002–123)*. Cincinnati, OH, 2002.

NIOSH 1976. *Standards for Occupational Exposures to Hot Environments—Proceedings of Symposium*. Cincinnati, OH: U.S. Department of Health, Education and Welfare, Public Health Service, Center for Disease Control, National Institute for Occupational Safety and Health, HEW(NIOSH) Publication No. 76–100.

NIOSH 1986. *Criteria for a Recommended Standard. Occupational Exposure to Hot Environments—Revised Criteria*. Cincinnati, OH: U.S. Department of Health and Human Services, Public Health Service, Centers for Disease Control, National Institute for Occupational Safety and Health, DHHS(NIOSH) Publication No. 86–113.

NIOSH 1986. *Working in Hot Environments*. Cincinnati, OH: U.S. Department of Health, Education and Welfare, Public Health Service, Center for Disease Control, National Institute for Occupational Safety and Health, HEW(NIOSH) Publication No. 86–112.

Northwestern University, ORS Laboratory Safety, *Chemical Hazards*. Evanston, IL, Available at http://www.nwu.edu/research-safety/chem/ethid2.htm 2007.

Nuclear Regulatory Commission. *Ionizing Radiation*. Available at http://www.nrc.gov, Washington, DC, 2007.

Occupational Safety and Health Administration, U.S. Department of Labor. *Essentials of Machine Guarding (OSHA 2227)*. Washington, DC, 1975.

Occupational Safety and Health Administration, U.S. Department of Labor. *An Illustrated Guide to Electrical Safety (OSHA 3073)*. Washington, DC, 1983.

Occupational Safety and Health Administration, U.S. Department of Labor. *Guidelines for Pressure Vessel Safety Assessment, Instruction Pub 8–1.5*. Washington, DC, 1989.

Occupational Safety and Health Administration, U.S. Department of Labor. *Draft Report by OSHA's Crane Safety Task Group*. Washington, DC, September 1990.

Occupational Safety and Health Administration, U.S. Department of Labor. *Concepts and Techniques of Machine Guarding (OSHA 3067)*. Washington, DC, 1992.

Occupational Safety and Health Administration, U.S. Department of Labor. Office of Training and Education. *OSHA Voluntary Compliance Outreach Program: Instructors Reference Manual*. Des Plaines, IL, 1993.

Occupational Safety and Health Administration, U.S. Department of Labor. *General Industry Digest (OSHA 2201)*. Washington, DC, 1995.

Occupational Safety and Health Administration, U.S. Department of Labor. Subject Index. "Internet." April, 1999. Available at http://www.osha.gov.

Occupational Safety and Health Administration, U.S. Department of Labor. *29 Code of Federal Regulations 1910*. Washington, DC, 1999.

Occupational Safety and Health Administration, U.S. Department of Labor. *29 Code of Federal Regulations 1926*. Washington, DC, 1999.

Occupational Safety and Health Administration, U.S. Department of Labor. *How to Plan for Workplace Emergencies and Evacuations (OSHA 3088)*. Washington, DC, 2001.

Occupational Safety and Health Administration, U.S. Department of Labor. *Safeguarding Equipment and Protecting Workers from Amputations (OSHA 3170)*. Washington, DC, 2001.

Occupational Safety and Health Administration, U.S. Department of Labor. *Trainer Manual OSHA # 501 for the General Industry*. OSHA Training Institute, Des Plaines, IL, 2001

Occupational Safety and Health Administration, U.S. Department of Labor. *Controlling Electrical Hazards (OSHA 3075)*. Washington, DC, 2002.

Occupational Safety and Health Administration, U.S. Department of Labor. *Assessing the Need for Personal Protective Equipment: A Guide for Small Business Employers (OSHA 3151)*. Washington, DC, 2003.

Occupational Safety and Health Administration, U.S. Department of Labor. *General Industry Standards (29 CFR 1910)*. Washington, DC, 2006.

Occupational Safety and Health Administration, U.S. Department of Labor. *Construction Standards (29 CFR 1926)*. Washington, DC, 2006.

Occupational Safety and Health Administration, U.S. Department of Labor. *OSHA Handbook for Small Businesses (OSHA 2209)*. Washington, DC, 2006.

Occupational Safety and Health Administration, U.S. Department of Labor. *Ionizing Radiation*. Available at http://www.osha.gov, Washington, DC, 2007.

Occupational Health and Safety Administration, U.S. Department of Labor. Available at http://www.osha.gov.

Petersen, D. *Techniques of Safety Management: A Systems Approach*, 3rd ed. Goshen, NY: Aloray Inc., 1989.

Reese, C.D. and Eidson, J.V. *Handbook of OSHA Construction Safety and Health*, 2nd ed. Boca Raton, FL: CRC/Lewis Publishers, 1999.

Reese, C.D. *Material Handling Systems: Designing for Safety and Health*. New York: Taylor & Francis, 2000.

Reese, C.D. *Accident/Incident Prevention Techniques*. New York: Taylor & Francis, 2001.

Reese, C.D. *Occupational Health and Safety Management: A Practical Approach*. Boca Raton, FL: CRC/Lewis Publishers, 2003.

Reese, C.D. *Office Building Safety and Health*. Boca Raton, FL: CRC/Lewis Publisher, 2004.

Reese, C.D. and Eidson, J.V. *Handbook of OSHA Construction Safety and Health,* 2nd ed. Boca Raton, FL: CRC/Lewis Publishers, 2006.

Saskatchewan Labour. *Identifying and Assessing Safety Hazards*. Available at http://www. labour.gov.sk.ca/safety, 2007.

U.S. Department of Energy. *OSH Technical Reference Manual*. Washington, DC, 1993.

U.S. Department of Labor. *Training Requirements in OSHA Standards and Training Guidelines (OSHA 2254)*. Washington, DC, 1998.

U.S. Office of Personnel Management. *Dealing with Workplace Violence: A Guide for Agency Planners*. Washington, DC, February 1998.

U.S. Department of Health and Human Services, Agency for Toxic Substances and Disease Registry. *Ionizing Radiation*. Available at http://www.atsdr.cdc.gov, Washington, DC, 2007.

U.S. Environmental Protection Agency. *Ionizing Radiation*. Available at http:// www.epa.gov, Washington, DC, 2007.

Westinghouse Electric Corporation 1986. *Heat Stress Management Program for Nuclear Power Plants*. University Park, PA: Pennsylvania State University, GPU Nuclear Corporation.

Wilkening, G.M. Nonionizing Radiation. In: *Patty's Industrial Hygiene and Toxicology,* eds. G.D. Clayton and F.E. Clayton. New York: Wiley, 1991.

Index

A

Accidents
  prevention
    hazards protection, 244–246
    protective clothing, 243–244
    safety practices and procedures, 246–248
    special considerations, 244
    thermal condition control, 243
    vs. incidents, 5–6
  types
    caught-in, caught-on and caught
      between, 422
    contact-by and contact-with, 422
    overexertion and exposure, 423
    slip, trip, and fall accident, 422–423
    struck-against and struck-by, 421–422
Accommodation and food services sectors, 37
  food services and drinking places
    applicable OSHA regulations, 64–66
    dining areas and work hazards, 57
    full-time and part-time workers, 56
    limited-service eating places, 55
    occupations, 62–64
    well-designed kitchens, 56–57
    worker profiles, 57–60
  hotels and other accommodations
    applicable OSHA regulations, 64–66
    casino, residential and extended-stay
      hotels, 53
    commercial hotels and motels, 52
    occupations, 58, 60–62
    part-time and full-time workers, 54
    resort hotels and motels, 52–53
    worker profiles, 57–60
    work hazards, 54–55
  NAICS's breakdown, 51–52
Acute back injuries
  causes, 190–191
  contributing factors, 192
  factors associated with, 191–192
  improper lifting injuries, 192–193
  mitigation issues, 194–195
  prevention and control
    engineering controls, 193–194
    recommendations, 195–196
  principle variables, 193
Acute toxicity, 163
Aisles and passageways, 373
Alpha particles
  air-purifying respirators, 261

characteristics, 260
properties, 253–254
vs. beta particles, 254
American National Standards Institute (ANSI)
  eye protection, 329
  protective headgear, 333
  safety footwear, 335
  safety standards, 327
Applicable OSHA regulations
  accommodation and food services sector,
    64–66
  arts, entertainment, and recreation, 49–50
  educational services, 21–22
  health care and social assistance, 33–35
Arts, entertainment, and recreation sectors
  applicable OSHA regulations, 49–50
  hazards faced by workers, 44–45
  historical/cultural/educational
    exhibits, 39
  live performances/events, 38–39
  NAICS categories, 37–38
  occupations
    management, business, and financial
      occupations, 48
    office and administrative support
      occupations, 47–48
    professional occupations, 46–47
    wage-and-salary workers, 45–46
  recreation/leisure activities, 39–40
  work conditions, 40–41
  worker profiles
    injuries, 42–44
    occupationally related deaths, 41–42
    occupational illnesses, 42, 44
Automobile repair services, 67, 77–79

B

Back and back injuries, *see* Acute back injuries
Beta particles
  beta-emitting radioisotopes, 266
  characteristics, 260
  low and high-energy beta emitters, 261
  properties, 254
Biological hazards, 136–137
Biological safety levels (BSL)
  biosafety level 1
    description, 143–144
    facilities, 145–146
    safety equipments and requirements, 145

441